COAL BED METHANE

COAL BED METHANE

FROM PROSPECT TO PIPELINE

Edited by

PRAMOD THAKUR
*Coal Degas Group, Murray American Energy Inc.,
Morgantown, West Virginia, USA*

STEVE SCHATZEL
NIOSH, US Department of Health & Human Service

KASHY AMINIAN
West Virginia University, Morgantown, West Virginia, USA

ELSEVIER

AMSTERDAM • BOSTON • HEIDELBERG • LONDON
NEW YORK • OXFORD • PARIS • SAN DIEGO
SAN FRANCISCO • SINGAPORE • SYDNEY • TOKYO

Elsevier
525 B Street, Suite 1900, San Diego, CA 92101-4495, USA
225 Wyman Street, Waltham, MA 02451, USA

Notice
No responsibility is assumed by the publisher for any injury and/or damage to persons or property as a matter of products liability, negligence or otherwise, or from any use or operation of any methods, products, instructions or ideas contained in the material herein. Because of rapid advances in the medical sciences, in particular, independent verification of diagnoses and drug dosages should be made.

Library of Congress Cataloging-in-Publication Data
Coal bed methane : from prospect to pipeline/edited by Pramod Thakur, Steve Schatzel, Kashy Aminian. – First edition.
        pages cm
    Includes bibliographical references.
    ISBN 978-0-12-800880-5
1. Coalbed methane. 2. Gas drilling (Petroleum engineering) I. Thakur, Pramod, editor of compilation. II. Schatzel, Steven J., editor of compilation. III. Aminian, Kashy, editor of compilation. IV. Title: Coalbed methane.
    TN844.5.C628 2014
    622'.3385--dc23

                            2014006132

British Library Cataloguing in Publication Data
A catalogue record for this book is available from the British Library

ISBN: 978-0-12-800880-5

For information on all Elsevier publications
visit our web site at store.elsevier.com

Transferred to Digital Printing, 2014

Working together
to grow libraries in
developing countries

www.elsevier.com • www.bookaid.org

Contents

10. Production Engineering Design

MARK V. LEIDECKER

11. Coalbed and Coal Mine Methane Gas Purification

MICHAEL MITARITEN

12. Current and Emerging Practices for Managing Coalbed Methane Produced Water in the United States

RICHARD HAMMACK

13. Plugging In-Mine Boreholes and CBM Wells Drilled from Surface

GARY DUBOIS, STEPHEN KRAVITS, JOE KIRLEY, DOUG CONKLIN, JOANNE REILLY

14. Economic Analysis of Coalbed Methane Projects

MICHAEL J. MILLER, DANNY A. WATSON II

15. Legal Issues Associated with Coalbed Methane Development

SHARON O. FLANERY, RYAN J. MORGAN

16. Permitting Coalbed Methane Wells

JOANNE REILLY

17. United States Lower 48 Coalbed Methane—Benchmark (2010)

STEPHEN W. LAMBERT

18. Worldwide Coal Mine Methane and Coalbed Methane Activities

CHARLEE BOGER, JAMES S. MARSHALL, RAYMOND C. PILCHER

Contributors

Kashy Aminian West Virginia University, Morgantown, West Virginia, USA

Charlee Boger Raven Ridge Resources Inc., Grand Junction, Colorado, USA

Charles Byrer U.S. Department of Energy (retired); Now managing a mineral title/leasing company: ArthurHenry, LLC

Doug Conklin Coal Gas Recovery, LP, Affiliate of Alpha NR, Bristol, VA, USA

Gary DuBois Target Drilling Inc., Smithton, PA, USA

Sharon O. Flanery Steptoe & Johnson PLLC, Charleston WV, USA

Ling Gao Department of Geological Sciences, Indiana University, Bloomington, IN, USA

Richard Hammack Office of Research and Development, National Energy Technology Laboratory, Pittsburgh, PA, USA

Ihor Havryluk CBM consultant

Anthony T. Iannacchione Mining Engineering Program, University of Pittsburgh, PA, USA

Joe Kirley Concrete Construction Materials, USA

Fred N. Kissell Mine Safety Consulting, PA, USA

Stephen Kravits Target Drilling Inc., Smithton, PA, USA

Stephen W. Lambert Schlumberger Data and Consulting Services, USA, Retired

Mark V. Leidecker Jesmar Energy Inc., Holbrook Pennsylvania, USA

James S. Marshall Raven Ridge Resources Inc., Grand Junction, Colorado, USA

Maria Mastalerz Indiana Geological Survey, Indiana University, Bloomington, IN, USA

Michael J. Miller Cardno MM&A (Cardno Ltd.), Kingsport, Tennessee, USA

Michael Mitariten Guild Associates Inc., USA

Ryan J. Morgan Steptoe & Johnson PLLC, Charleston WV, USA

Jack C. Pashin Devon Energy Chair of Basin Research, Boone Pickens School of Geology, Oklahoma State University, Stillwater, Oklahoma, USA

Raymond C. Pilcher Raven Ridge Resources Inc., Grand Junction, Colorado, USA

Joanne Reilly Coal Gas Recovery, LP, Affiliate of Alpha NR, Bristol, VA, USA

Gary Rodvelt Global Technical Services, Halliburton Energy Services, Canonsburg, Pennsylvania, USA

Arndt Schimmelmann Department of Geological Sciences, Indiana University, Bloomington, IN, USA

Todd Sutton Schlumberger, Pittsburgh, PA, USA

Pramod Thakur Coal Degas Group, Murray American Energy Inc., Morgantown, West Virginia, USA

David Uhrin Coalbed Methane Consulting, Pittsburgh, PA, USA

Danny A. Watson II Formerly of Cardno MM&A (Cardno, Ltd.) Kingsport, Tennessee, USA, currently of Gulfport Energy Corporation, Oklahoma City, Oklahoma, USA

Preface

Coal is the most abundant and economical fossil fuel in the world today. Over the past 250 years, it has played a vital role in the growth and stability of the world economy. Total minable reserves of coal are estimated at 1 trillion tons (to a depth of 3000 ft) while the estimated reserve to a depth of 10,000 ft ranges from 17 to 30 trillion tons. Coal bed methane (CBM) and coal are syngenetic in origin. The coal seams (minable and nonminable) are a reserve for vast quantities of methane (10,000 to 30,000 TCF). Seventy countries around the world mine almost 8000 million tons of coal and produce 3 TCF of CBM per year. At this rate, coal and CBM are likely to remain a dominant source of energy in this and the next century.

CBM was a bane for mining from the very beginning. As coal is mined, methane is released in mine air. Methane becomes explosive when mixed with air in the range 4.5–15% by volume. The history of coal mining around the world is replete with mine disasters when the methane–air mixture exploded. It is estimated that about 8000 people died in the US alone. Number of fatalities is much higher for many other countries. Efforts to mitigate this disaster started in Europe via gob gas drainage with cross-measure boreholes but serious efforts to degas the coal seam prior to mining and postmining began in the US only in the 1970s. The coal industry and the erstwhile US Bureau of Mines often pooled their resources to make the mines a safer place to work and boost the productivity as well. Major achievements are as follows:

1. In-mine horizontal drilling (1974–1980). The drilling rig and the instrument system can drill up to 3000 ft long boreholes in 5–6 ft thick coal seams and degas the coal prior to mining.
2. Vertical gob wells over mined out areas (1975–1983). The European cross-measure boreholes could not cope with the highly productive US longwall faces with their fast rate of mining and consequent high volumes of methane emissions. Vertical gob wells drilled over the longwall gobs with blowers could capture 70–80% of the total emissions allowing very high rate of mining and productivities of 70–80 tons/man-day.
3. Massive hydraulic fracturing of coal seams (1984–1994). This technique allowed degasification of deeper (2000–3300 ft) coal seams and gave rise to commercial exploitation of CBM.

4. Horizontal boreholes drilled from surface (2001–2010). With the development of new instrumentations that monitored the drill bit while drilling, it became possible to drill 5000 ft long horizontal boreholes in coal seams from surface. While this technique is applicable to coal seams at all depth, it is particularly suited to very deep, thick, and gassy coal seams. Deep coal seams of the world can be successfully exploited by combining this technique with hydrofracking of horizontal legs. This is the same technique that revolutionized the gas production from Marcellus shale in the USA.

Thus it would be no exaggeration to say that CBM that was a bane to mining industry has now become a boon—a viable source of additional energy.

It is clear now that these new technologies can open up the vast coal reserve for gas production. Currently 10% of US gas production is realized from coal seams but it can easily go to 20% if the deep and thick coal seams are put on production.

The North American Coal Bed Methane Forum was created in 1985 to promote mine safety and increase energy supply by producing methane from coal seams. It is a nonprofit organization based in West Virginia, USA. It has continuously offered a seminar once or twice a year on timely issues related to CBM production and methane control in mines. The book, "Coal Bed Methane: Prospects to Pipeline" is the proceedings of the 25th Anniversary of the Forum. It covers this vast subject in 18 chapters. The book is designed to be a textbook for undergraduate and graduate level courses taught in many US universities.

The editors would like to thank all authors who wrote and presented the papers at the 25th Anniversary Forum. The publication was inordinately delayed for lack of a viable publisher. We would like to thank Elsevier and particularly Louisa Hutchins for their tremendous support in getting the document ready for publication. In spite of our efforts to avoid, there may still be some minor errors in the text. We will be indebted to careful readers if they can point them out. It would be our sincere hope and commitment that we will remove them in the second edition of this very useful book.

Hoping to keep the coal mines safe and increasing the supply of natural gas to US homes, we remain grateful for the opportunity to work together and publish this book.

Pramod Thakur
Steve Schatzel
Kashy Aminian

1

Coalbed Methane: A Miner's Curse and a Valuable Resource

Charles Byrer[1], Ihor Havryluk[2], David Uhrin[3]

[1]U.S. Department of Energy (retired); Now managing a mineral title/leasing company: ArthurHenry, LLC, [2]CBM consultant, [3]Coalbed Methane Consulting, Pittsburgh, PA, USA

The US coal industry has been challenged over the years by the explosive coalbed methane (CBM) gas. It permeates most coalbeds, rendering mining process dangerous. Indeed, CBM was nicknamed the "miner's curse", because it escaped from mineable coal seams and exploded when mixed with air if a source of ignition was present.

The "miner's curse" is also an energy source—and possibly an environmental problem. Since the 1830s gas explosions in US coal mines have killed thousands of miners. Among the worst US mine disasters of the twentieth century was one that occurred in Northern Appalachia—362 miners perished in a gas and dust explosion at Monongah, West Virginia in 1907 (Figure 1.1). Over 600 mine explosions in US coal mines have been documented in the various coal fields (Table 1.1). Mine safety has come a long way since then. Methane in coal mines will always be a hazard, but the risk of explosion has been greatly minimized by increased safety regulations, sensitive gas detectors, improved ventilation, and methane drainage.

And therein lies an opportunity to utilize an energy source and further improve mine safety and possibly improve regional environments. Although venting gas into the atmosphere has helped to reduce underground explosions to infrequent events, it also discards potentially valuable fuel and adds "methane" (a classified greenhouse gas) to our regional atmosphere. Thus, the large volumes of CBM vented by mines represent both an economic loss and an environmental challenge.

In the 1990s, up to 300 US billion cubic feet of methane were vented from US coal mines (mostly underground operations). This was 15% of all

362 Miners killed

FIGURE 1.1   Monongah, WV mine explosion—1907.

TABLE 1.1   Coal Mine Disasters in US Since 1839

Time Period	Coal Mine Disasters
Through 1875	19
1876–1900	101
1901–1925	305
1926–1947	147
1951–1975	35
1976–2003	15
2004–2013	2
Total	624
Undocumented	>300
Estimated total US Coal mine deaths	>8000

global methane emissions from coal mining, and less than 1% of all methane released into the atmosphere by mankind.

Methane's "greenhouse gas" potential has been stated to be many times greater than CO_2, so its release during coal mining and processing is a concern. Currently, the atmospheric methane concentration is a lesser problem than CO_2, simply because methane is much scarcer in the atmosphere, with only 1/200th the concentration of CO_2. But this may be changing: the methane percentage is slowly increasing worldwide, at a faster rate than the CO_2 percentage. The US Geological Survey has been forecasting methane to surpass CO_2 as the dominant greenhouse in the second half of the twenty-first century—if its concentration continues to grow at the present rate.

1.1 ABUNDANCE OF METHANE IN MOST COALBEDS

As coal forms slowly from decaying plants over eons of years, methane forms along with it. Thus, most coalbeds are permeated with methane, so much so that a cubic foot of coal can contain six or seven times the methane that exists in a cubic foot of a conventional sandstone gas reservoir. However, the methane content in coal is highly variable, varying widely over short distances (a few hundred feet, for example). The higher grades of coal are richer in gas, and deeper coalbeds are for the most part "gassier" because they can store more gas on its internal surface.

The gas often occurs in concentrated pockets as well, creating a major mining hazard. When the mining of coal breaks open these pockets, or when coal is pulverized during mining and processing, methane is released into the mine and the atmosphere. In addition to ventilating the operating areas in mines, methane often is removed from the virgin coal in advance of mining by drilling extraction wells into the coal seam and capturing the gas or venting the gas into the atmosphere.

Recent estimates are that 400–700 trillion cubic feet (Tcf) of methane exist in US coalbeds, although only 90–100 Tcf may be economically recoverable with the current technology. This equates to a 4-year supply of natural gas for USA More realistically, it can be viewed as a 25-year supply for one sixth of US population. The world CBM resource spans all continents and is estimated at 4000–7500 Tcf. So CBM is an extremely large potential energy resource.

The energy (Btu) in CBM amounts to a very small percent of the energy in the coal itself. The amount varies widely, from little gas in a ton of coal, up to 1000 cubic feet per ton. As a general example, burning a ton of bituminous coal can release 21–30 million Btu of heat energy, depending on the coal's rank. The methane within that ton of coal—typically 250–500 cubic feet of the methane gas—can provide 250,000–500,000 Btu when burned. In many cases, this amount of energy potential can make the gas worth recovering as a fuel.

1.2 CBM: ALREADY A COMMERCIAL SUCCESS

CBM has strong economic potential. It can be used to generate electricity, either at mine sites or can be piped to commercial utilities. It can be "cofired" with coal at power plants to reduce SO_X and NO_X emissions. It can fuel gas turbines or fuel cells to generate power. At mines, it can fire drying units that remove moisture from washed coal. And it can be pipelined for utility and industrial use. Some of this potential is already being realized.

During the 1930s, the Big Run gas field in northern West Virginia began producing coalbed gas from the thick Pittsburgh coal seam and continues producing at a reduced rate to this day, demonstrating a common characteristic of coalbed wells: they tend to produce much longer than conventional reservoirs. By the late 1970s, some CBM was being produced commercially from coalbeds in Alabama. In northern New Mexico, over 40 US billion cubic feet of CBM has been produced from nearly 1700 wells. Self-supporting CBM projects also exist in Pennsylvania, West Virginia, Colorado, and Virginia. Currently, pipeline-quality coal mine methane (CMM) is being sold to distribution systems in the Appalachian coal basin. Today, the annual US demand for natural gas is over 21 Tcf, with 1.8 Tcf being produced from coal mines and unmined coalbeds, or over 8%—quite a success story for what was once a waste product and solely a "miner's curse". And this production is projected to increase as demand rises, as technology improves, and as mining companies cooperate with gas producers to utilize gas.

1.3 CMM: PROS AND CONS

CBM produced in conjunction with coal mining is called coal mine methane (CMM). Coal mines can simultaneously produce methane and consume it by generating electricity on site. This on-site capability is valuable because the mining operation needs electrical power to operate machinery and ventilation fans, coal cleaning plants, coal dryers to remove moisture, and other surface facilities. An underground mine's vent fans alone can consume 75% of the total electricity used at the site. Power generated from mine gas also can be fed to the grid that supplies electricity to the mine, selling the energy back to the power supplier. Such uses of mine gas can offset its cost of recovery in very gassy mines.

The US Department of Energy has sponsored several field tests since the 1970s to recover CMM and use it to generate power on site. Use of CBM to fuel combustion engines and gas turbines has been demonstrated. Also promising is the use of CBM in fuel cells.

One problem with CMM is that its quality varies, particularly if the gas has been mixed with ventilation air in an operating mine. Pipeline-grade natural gas must be at least 96% pure methane, so lower-quality mine gas must be upgraded for distribution by removing water and inert gases (CO_2, nitrogen, and oxygen).

The gassiest US coal mines are in the Appalachian coal basin, which stretches from Alabama to Pennsylvania. Some of these mines are recovering and using CBM. Annual methane recovery from US underground coal mines (not including methane tapped from unmined coal seams) has been about 25 US billion cubic feet.

Commercial success in this basin has included a large coal mine in Virginia and an Alabama mine. Both have sold over 40 million cubic feet of pipeline-quality CBM per day. Many smaller CBM projects in the country are in western Virginia, where pipeline-quality CBM is being sold.

Recovering methane from operating mines is not that easy. The coal industry has long seen methane as a problem, not particularly a resource, so a different viewpoint is required. Coalbeds are far more complex geologically than conventional natural gas reservoirs. Mine gas recovery adds cost in equipment, work force, services, and meeting additional regulatory requirements. Not all CBM is of pipeline quality. There can be questions of gas ownership and royalty rights. And there is the question of unproven economics of recovery. But improved technology for recovery, combined with potential utilization, and the need to meet future greenhouse gas regulations are making CBM recovery projects a commercial success.

1.4 PROFIT OPPORTUNITIES IN UNMINEABLE COAL

Today, coal is mined from coalbeds with high-volume mining machinery that feeds a virtual conveyor of road-rail-river transportation to our nation's power plants. This high-volume strategy holds down the cost of coal and electricity, but the reality of mineable coal is that about 60% of US coal is "unmineable", meaning that it is unprofitable to mine with present technology. It is unprofitable because of coal-seam thinness, poor, or inconsistent quality of the coal, or difficult mining conditions, such as depth.

But unmineable coal represents a vast, largely untapped methane resource. As demand for natural gas increases, CBM is growing more attractive as a fuel. More than 16,000 communities, many in the Appalachian area lie above coal seams that could produce methane. This fuel could be delivered locally, reducing the need for interstate gas transportation, hundreds of miles from distant gas fields. CBM has been gaining strong interest nationally, but the Southeast and the Appalachian areas are particularly promising markets, because of its large gas demand and the Appalachian basin's many gassy coal seams.

CBM is potentially capable of making seven eastern states at least partially "energy self-sufficient" in gas supply—Kentucky, Maryland, Ohio, Pennsylvania, Tennessee, Virginia, and West Virginia. In West Virginia's case, the state can be 100% self-sufficient, satisfying its entire natural gas demand from its own coalbeds. Furthermore, CBM can be locally competitive with conventionally produced pipeline gas from sandstone and shale reservoirs. In addition to recovering methane from unmineable coalbeds, the gas can be harvested from active underground mines as well.

1.5 CONVERGENCE OF CBM AND CO₂ SEQUESTRATION

International economic, environmental, and technological advances over the past decade have contributed toward the consideration of CBM recovery and CO_2 sequestration together. The idea is to geologically sequester CO_2, while at the same time recovering the methane already in them. The CO_2 would be injected via wells drilled into the coal, and the CO_2 would drive the methane out of the coal through the wells to the surface, where it would be collected. This "two-birds-with-one-stone" idea is feasible because bituminous coal stores twice the volume of CO_2 than it stores methane. The net result would be less CO_2 in the atmosphere, no significant new methane added to the atmosphere, and enhanced recovery of methane to help pay for the process.

What about the logistics and cost of this CBM/CO_2 strategy? Most US power plants are within 3–5 miles of a coalbed (not necessarily a suitable one). For a plant near a gassy coalbed (or multiple beds, for coal often occurs in multiple seams), pipeline length would be minimal to convey CO_2 from the plant into the coal, and to pipe recovered methane back to the power plant.

1.6 CBM'S FUTURE

Today, industry and academic research interest is running high because the methane recovery/CO_2 sequestration concept could be a "least-cost" option in the energy–economy–environment scenario. But much work still lies ahead. Candidate coalbeds must be targeted, and the potential methane resource must be determined for each bed. The feasibility of drilling CO_2 injection wells and methane recovery wells must be determined for each targeted bed. All environmental factors must be considered, including surface land use and water quality. Cost of production is a paramount consideration. Time frames must be established.

Better technology tomorrow could let us recover methane from coal deposits in regions that are not economic today. For example, the Appalachian coal basin currently accounts for about two-thirds of the CMM emissions in the United States—a large potential resource.

Properly developing our CBM resource can provide more clean energy, reduce our greenhouse gas contribution, and maintain a safe mining environment. Today, the "miner's curse" that haunted many coal mines has become an economic asset for our country.

The Origin of Coalbed Methane

Ling Gao[1], Maria Mastalerz[2], Arndt Schimmelmann[1]

[1]Department of Geological Sciences, Indiana University, Bloomington, IN, USA, [2]Indiana Geological Survey, Indiana University, Bloomington, IN, USA

2.1 INTRODUCTION

Coalbed methane (CBM) occurs as unconventional natural gas in coal seams. During the past 20 years, CBM has emerged as an important energy resource in the United States (Figure 2.1) and presently accounts for about 9% of total US natural gas production. Because its combustion releases no toxins, produces no ash, and emits less carbon dioxide per unit of energy than combustion of coal, oil, or even wood (U.S. Environmental Protection Agency), it is expected that CBM will remain an environmentally friendly component in our energy portfolio over the next decades.

Based on the large volume of coal in the US, conservative estimates suggest that there are more than 700×10^{12} cubic feet (i.e., 700 Tcf or 19.8×10^{12} m^3) of coalbed gas in US coal seams (Rice, 1997), of which 100 Tcf may be economically recoverable with today's technology—roughly equivalent to 5 year total natural gas use in the US at present consumption.

The recognition of coalbed gas as a largely untapped energy source in the 1980s (Rice and Claypool, 1981; Rice, 1983; Dugan and Williams, 1988) triggered many studies toward CBM exploration and production (Rice, 1993; Scott et al., 1994; Whiticar, 1994; Ayers, 2002; Faiz and Hendry, 2006; Strąpoć et al., 2007; Flores et al., 2008). Several coal basins in North America are already producing prolific CBM, for example, the Powder River Basin (Montgomery, 1999; Flores, 2004), San Juan Basin (Scott et al., 1994), and Black Warrior Basin (Pashin and McIntyre, 2003). Other basins such as the Illinois Basin (Tedesco, 2003; Faiz and Hendry, 2006), Forest City Basin (Tedesco, 1992; McIntosh et al., 2008), and Michigan Basin (Martini et al., 2003, 2008) are considered to be prospective sources of CBM in the future (Figure 2.2).

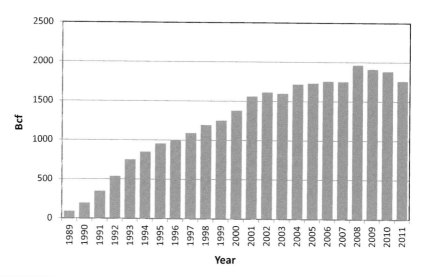

FIGURE 2.1   Annual coalbed methane production in the United States, rising from 91 Bcf in 1989 to 1966 Bcf in 2008. $1 \times 10^9 \, m^3 = 35.289 \times 10^9$ cubic feet $= 35.289$ Bcf. *Data from U.S. Department of Energy, Energy Information Administration (2007).*

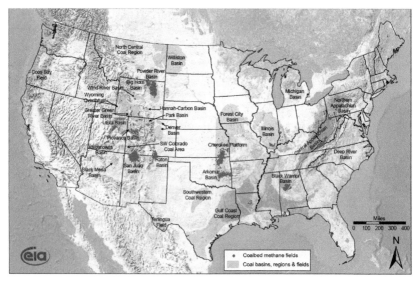

FIGURE 2.2   Basins producing coalbed methane in the lower 48 states in the US. *From U.S. Department of Energy, Energy Information Administration (2007).*

The objectives of this paper are to (1) characterize geochemical and microbial origins of coalbed gases, (2) assess geological and environmental factors controlling coalbed gas composition and generation, (3) evaluate petrological, chemical, and isotopic proxies that have been used to distinguish between types of coalbed gas origin, and (4) discuss implications of gas origin and generation for exploration and production.

2.2 ORIGINS OF CBM

CBM is technically defined as natural gas that can be recovered from coal seams. Most CBM has been produced in situ by microbial, thermal, or possibly catalytic degradation of organic material present in coal, although some allochthonous gas components may have migrated into coal seams from other strata. CBM is mainly composed of methane (CH_4) with variable additions of carbon dioxide (CO_2), elemental nitrogen (N_2), and heavier hydrocarbons, such as ethane (C_2H_6), and traces of propane (C_3H_8), and butanes (C_4H_{10}).

In general, coalbed gas is traditionally assumed to be either of biogenic, thermogenic, or mixed origin. Biogenic coalbed gas is generated by the breakdown of coal organic matter by methanogenic consortia of microorganisms at low temperature (usually less than 150 °F or 56 °C) (Scott et al., 1994). Many shallow and thermally immature coal seams contain biogenic CBM, such as the Powder River Basin. In contrast, thermogenic gas is produced from coal organic matter by chemical degradation and thermal cracking mainly above 100 °C, above the thermal threshold where microbial methanogenic activity becomes biochemically impossible (Hunt, 1979; Rice and Claypool, 1981). Thermogenic gas generation commences at the high-volatile bituminous coal rank with vitrinite reflectance values R_o between 0.6% and 0.8% (Scott et al., 1994). Thermogenic methane production from coal reaches a maximum at $R_o \sim 1.2\%$ and becomes negligible at $R_o \sim 3.0\%$ after the organic hydrogen pool in coal has been severely depleted and remaining organic carbon is largely present in condensed aromatic structures. Therefore, higher rank coal is expected to have generated more thermogenic CBM than relatively lower rank coal, which would translate into a higher CBM content if the gas has not been lost.

Knowledge of the origin of CBM is important to formulate an effective and successful CBM exploration strategy (Scott et al., 1994; Martini et al., 2008). A predominantly microbial gas play should target shallow coal seams at lower temperature close to basin margins, where the organic matter is likely less mature and larger fractures allow for faster gas extraction. In contrast, predominantly thermogenic coalbed gas accumulations likely occur in deeper, more thermally mature coal seams (Rice, 1993; Strąpoć et al., 2007; McIntosh et al., 2008) with a more restricted network of open fractures. Basins with coal seams hosting mixed (i.e., thermogenic and biogenic) CBM require complex exploration and production strategies depending on local geological and hydrological conditions. Knowledge of the origin of coalbed gas has become a prerequisite for successful CBM exploration (Rice, 1993; Scott et al., 1994; Whiticar, 1999; Faiz and Hendry, 2006; Flores et al., 2008; Dai et al., 2009).

2.2.1 Methods for Characterizing the Origin of CBM

Originally, coal rank was used to diagnose the origins of associated coalbed gases. Coals of low rank with vitrinite reflectance values, R_o <0.3% were considered to host biogenic gas whereas higher rank coals were assumed to contain thermogenic gases (Claypool and Kaplan, 1974). However, with the recognition of secondary biogenic gases (Rice, 1993; Scott et al., 1994), which overwhelmingly contribute to biogenic gas in coals with elevated vitrinite reflectance (R_o ~0.6–0.8%) (Russell, 1990; Scott et al., 1994), the chemical gas composition and the ratio of the abundance of CH_4 relative to higher hydrocarbons (i.e., "gas dryness") became indispensable proxies for evaluating the origin of coalbed gas (Stahl, 1973; Bernard, 1978; Schoell, 1980, 1983; Whiticar, 1996) (Table 2.1). However, these chemical compositional indices may vary significantly, depending on geological conditions, and thus sometimes result in ambiguous distinctions between biogenic and thermogenic coalbed gases.

Advances of stable isotope analytical techniques during the past 30 years have made it possible to exploit the carbon $^{13}C/^{12}C$ and

TABLE 2.1   Frequently Used Proxies for Distinction between Thermogenic and Biogenic Coalbed Gases

Parameters	Origin of Gas		Reference
	Thermogenic	Biogenic	
Vitrinite reflectance (R_o in %)	0.6–3.0%	0.3–0.8%	Rice (1993), Scott et al. (1994)
Hydrocarbon index [$CH_4/(C_2H_6+C_3H_8)$]	<20	>1000	Bernard (1978), Strąpoć et al. (2008a)
Gas wetness index[2] $C_{2+} = [(C_2H_6+C_3H_8+C_4H_{10}+C_5H_{12})/ (CH_4+C_2H_6+C_3H_8+C_4H_{10}+C_5H_{12})]$	>3%	<3%	Rice (1993)
CO_2 content	2–15 vol%	<5 vol%	Smith and Pallasser (1996)
$\delta^{13}C$ of methane (in ‰ vs Vienna Pee Dee Belemnite (VPDB)	>−50‰	<−55‰	Schoell (1980), Whiticar et al. (1986), Whiticar (1999)
δD of methane (in ‰ vs Vienna Standard Mean Ocean Water (VSMOW)	−275 to −100‰	−400 to −150‰	Whiticar (1999)
$\Delta^{13}C_{CO_2-CH_4}$[1]	<40‰	>60‰	Smith and Pallasser (1996)

[1] $\Delta^{13}C_{CO_2-CH_4} = \delta^{13}C_{CO_2} - \delta^{13}C_{CH_4}$.

[2] Gas wetness index is sometimes expressed as gas dryness ratio $C_1/(C_2+C_3)$, which is officially called hydrocarbon index. C_1, methane; C_2, ethane; C_3, propane.

hydrogen $^2H/^1H$ (or D/H) stable isotope ratios of individual gas components for classification of natural gases. Isotope ratios are commonly expressed in $\delta^{13}C$ and δD (or δ^2H) notations where more negative values indicate relative depletion in heavy isotopes ^{13}C and D (expressed in ‰ on isotopic scales anchored by VPDB and VSMOW) (Coplen, 1996). Based on a pioneering $\delta^{13}C$ data set on methanes, Schoell (Schoell, 1980) proposed that $\delta^{13}C_{CH_4}$ values below −55‰ indicated a biogenic source of natural gas whereas values greater than −55‰ implied a thermogenic origin (Table 2.1). The threshold value of −55‰ was later recognized to be an approximate guideline because biogenic methane can cover a wide range of $\delta^{13}C_{CH_4}$ values from −40 to −110‰ (Jenden and Kaplan, 1986; Whiticar et al., 1986) depending on the isotopic composition of the original organic substrate, the biochemical methanogenic pathways, and environmental factors (Rice, 1993; Whiticar, 1999; Valentine et al., 2004). Therefore, if $\delta^{13}C_{CH_4}$ values are in the range of −40 to −55‰, carbon isotopes alone cannot unequivocally characterize the proportion of biogenic versus thermogenic gas components. Fortunately, hydrogen isotope ratios expressed as δD values offer a second, entirely independent isotopic parameter for characterizing gas origins. In comparison with $\delta^{13}C_{CH_4}$ values, δD_{CH_4} values cover a much wider numerical range. Thermogenic gases express δD_{CH_4} values from approximately −275 to −100‰, and biogenic gases range from −400 to −150‰ (Whiticar, 1999). Owing to the considerable overlap in δD_{CH_4} values between some thermogenic and biogenic methanes (Figure 2.3(B)), δD_{CH_4} alone also offers limited analytical distinction between different origins of coalbed gases. However, combinations of both carbon and

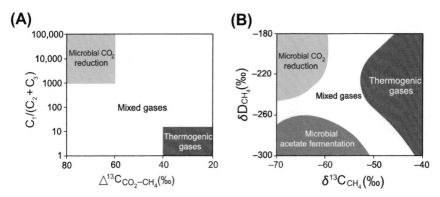

FIGURE 2.3 Origins of natural gases based on their chemical and compound-specific isotopic compositions. (A) Hydrocarbon index $C_1/(C_2+C_3) = [CH_4/(C_2H_6+C_3H_8)]$ and carbon isotopic difference between CO_2 and CH_4 $\Delta^{13}C_{CO_2-CH_4}$; (B) Carbon and hydrogen isotopic ratios of methane distinguish biogenic from thermogenic natural gases, as well as biogenic gases formed by the CO_2 reduction pathway from those formed by the acetate fermentation pathway. *Adapted from Whiticar (1999).*

hydrogen stable isotope ratios of gas components (e.g., methane, ethane, propane, CO_2) and chemical gas compositional data greatly enhance the diagnostic confidence (Whiticar, 1999; Strąpoć et al., 2007) (Figure 2.3).

2.2.2 Thermogenic Gases

2.2.2.1 Gas Generation and Gas Characteristics

Thermogenic coalbed gases are generated at elevated temperatures during coalification. The process of coalification encompasses physical and chemical changes that occur in coal shortly after deposition/burial and continue during thermal maturation (Figure 2.4). During coalification, the carbon in residual solid organic matter becomes progressively

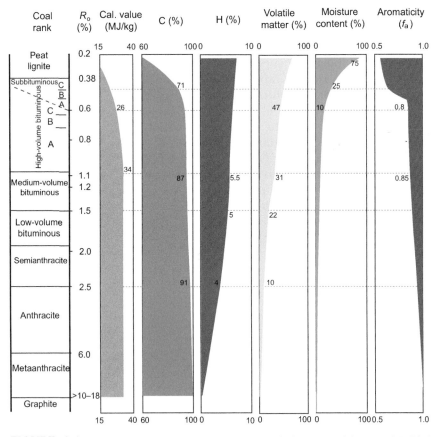

FIGURE 2.4 Physical and chemical changes in coal during coalification (Modified from Stach et al. (1982).); weight% of carbon, hydrogen, volatile matter, and moisture content are indicated on a dry, ash-free basis; cal. value = caloric value carbon; aromaticity $f_a = [(C/H) - H_{aliphatic}*/(H_{aliphatic}/C_{aliphatic})]/(C/H)$, where C/H is the atomic ratio of organic carbon/hydrogen, $H_{aliphatic}* = H_{aliphatic}/H_{total}$. Adapted from Brown and Ladner (1960).

more aromatic while hydrogen-rich aliphatic components with relatively low molecular weights (e.g., CH_4, C_2H_6) are liberated and expelled along with water and carbon dioxide during the generation of coalbed gases (Tissot and Welte, 1984).

In general, the formation of thermogenic gas can be divided into an early stage and a main stage. Early thermogenic gases are produced from hydrogen-rich coals of high-volatile bituminous C rank ($R_o \sim 0.6\%$) (Figure 2.5), and are commonly characterized by substantial amounts of ethane, propane, and other wet gas components (Scott et al., 1994). Important by-products CO_2 and H_2O are generated from breaking of heteroatomic bonds and the loss of organic functional groups via decarboxylation and dehydration. Carboxyl groups in coal are largely eliminated at a coal rank of ~0.7% vitrinite reflectance (Hunt, 1979; Cook and Struckmeyer, 1986) and the amount of CO_2 generated at even higher ranks is small (Faiz and Hendry, 2006).

With increasing temperatures and pressures, further coalification causes disproportionation of the overall pool of organic hydrogen as hydrogen-rich liquid and gaseous hydrocarbons are liberated at the main stage of gas generation ($R_o > 0.8\%$) leaving the residual solid organic matter hydrogen depleted and aromatized. The peak generation of liquid hydrocarbon and wet gas (i.e., ethane and higher hydrocarbon gases) occurs near $R_o \sim 1.1\%$ in proximity to the boundary between high-volatile bituminous B and medium-volatile bituminous ranks. At R_o values >1.2%,

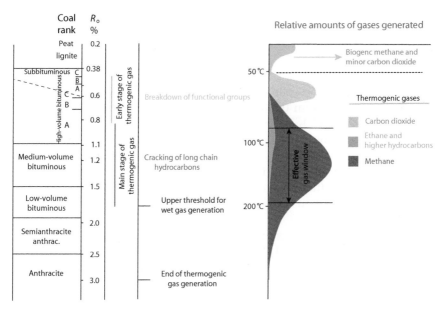

FIGURE 2.5 Generation of thermogenic gas from coal with increasing thermal maturity.

the generation of liquid hydrocarbons and wet gases rapidly diminishes. At the same time, the generation of CH_4 increases (Figure 2.5), because at higher temperatures and degrees of coalification, remaining aliphatic moieties in coal and previously formed hydrocarbons with more than two carbon atoms (C_{2+}) are thermally cracking to form additional CH_4. Up to a coal rank of R_o ~1.8%, methane production still occurs within the effective methane formation window, but methane generation decreases dramatically at higher vitrinite reflectance and effectively ceases at R_o ~3.0% (Figure 2.5).

Several studies empirically estimated the theoretical amounts of methane expected from coals during the entire coalification process (Jüntgen and Karweil, 1966; Meissner, 1984) and suggested that the methane yield is influenced by coal rank and composition. Based on the empirical relationship between vitrinite reflectance, R_o and the expected relative abundances of thermogenic gas components, Berner and Faber (Berner and Faber, 1988) subsequently established the following equations quantifying the generation of methane, ethane, and propane during coalification.

$$CH_4 = 9.1 \ln R_o + 93.1$$

$$C_2H_6 = -6.3 \ln R_o + 4.8$$

$$C_3H_8 = -2.9 \ln R_o + 1.9$$

Based on laboratory heating experiments, Tang et al. (Tang et al., 1991) suggested that large quantities of methane could be generated from coals with R_o values larger than 0.7% and concluded that the threshold of major thermogenic methane generation coincides with the rank of high-volatile A bituminous coal (R_o ~1.0%) (Figure 2.5).

In addition to methane, thermogenic gas also contains carbon dioxide, ethane, propane, butanes, nitrogen, and occasionally hydrogen sulfide. Early generated gas at R_o from 0.6% to 0.8% is relatively wet, usually with C_{2+} >3 vol% and a high CO_2 content of >10 vol%. The $CH_4/(C_2H_6+C_3H_8)$ volume ratio is typically <100 (Table 2.1). In contrast, thermogenic gas generated during the main stage at R_o >1.2% is dryer (i.e., containing less C_{2+}), since C_{2+} gases are susceptible to thermal cracking to form CH_4 (Figure 2.5). The fact that not all thermogenic gases are wet requires detailed geochemical analyses and cautious interpretations based on a variety of diagnostic indicators.

Thermogenic CH_4 is generally, but not universally, enriched in ^{13}C compared with biogenic CH_4 that covers a $\delta^{13}C$ range from −55 to −20‰. δD_{CH_4} values of thermogenic methanes can vary from approximately −275 to −100‰ and thus partially overlap with the hydrogen isotopic range of biogenic methane (Figure 2.3(B)). A clear distinction between biogenic and thermogenic gases was observed by comparing compound-specific δD

values of methane and ethane from comparable coal seams in southwestern Indiana and western Kentucky (Strąpoć et al., 2007). δD values of methane and ethane are similar in thermogenic gases from western Kentucky, but in predominantly biogenic gases from southwestern Indiana, δD_{CH_4} values are far lower than accompanying $\delta D_{C_2H_6}$ values (Figure 2.6). Small contributions of ethane and other heavier hydrocarbons in Indiana coalbed gases (Strąpoć et al., 2007) are attributed to the thermogenic pathway. At a vitrinite reflectance of ~0.60%, coal has already generated small amounts of early thermogenic gas. It was speculated that C_{2+} gases in dominantly biogenic gases from southwestern Indiana were partially biodegraded and became isotopically fractionated, especially propane. With increasing biodegradation of C_{2+} in thermogenic gas, the carbon isotopic signature of methane shifts toward values that are characteristic for biogenic gas (Larter and di Primio, 2005; Jones et al., 2008). Rather than using low gas wetness and high $\delta^{13}C_{CO_2}$ values of up to +20‰ (Larter and di Primio, 2005; Jones et al., 2008) as insufficient evidence leading to an erroneous biogenic diagnosis, it is imperative to carefully inspect dry gas for possible biodegradation of thermogenic gas.

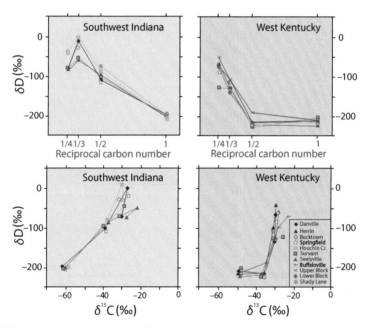

FIGURE 2.6 Compound-specific $\delta^{13}C$ and δD values of methane C_1 to straight-chain n-C_4 butane and combined butanes C_4 distinguish between biogenic coalbed gases from southwest Indiana and thermogenic coalbed gases from west Kentucky. *Adapted from Strąpoć et al. (2007).*

2.2.2.2 Controls on Thermogenic Gas Distribution and Producibility

Previous studies have shown that coalbed gas producibility is dependent on several critical factors, namely, depositional system and coal distribution, coal rank, tectonic and structural settings, gas content, permeability, hydrodynamics and hydrogeochemistry (Rice, 1993; Kaiser et al., 1994; Scott, 1999; Ayers, 2002). Understanding these geological and hydrological controls that affect the amount of gas adsorbed within coalbeds is critical to develop an effective and successful exploration strategy. The following

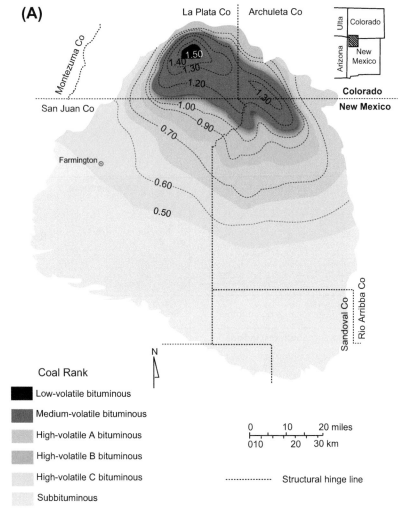

FIGURE 2.7   (A) Distribution of coal rank (vitrinite reflectance, R_o) in the Upper Cretaceous Fruitland Formation in San Juan Basin. (B) Geographic trends of coalbed gas in place $(1\,\mathrm{Bcf/miles^2} = 1.09 \times 10^7\,\mathrm{m^3/km^2})$. *Adapted from Scott et al. (1994).*

discussion of the controls on thermogenic gas distribution largely relies on the example of the San Juan Basin. This basin has been widely studied and is well known for its mature production of predominantly thermogenic coalbed gas (Rice, 1983; Scott et al., 1994; Whiticar, 1994; Formolo et al., 2008).

The San Juan Basin (Figure 2.7(A)) in northwestern New Mexico and southwestern Colorado is one of the most prolifically producing coalbed gas basins in the world and has been responsible for ~65% of the 2006 US coalbed gas production (EIA, 2007). Coalbed gas in the San Juan Basin has a predominantly thermogenic origin based on our understanding of the geological and hydrological characteristics and the large matrix of available analyses (e.g., gas dryness, isotopes, ranks of coals, biomarkers in source rocks) (Rice, 1983; Scott et al., 1994; Formolo et al., 2008).

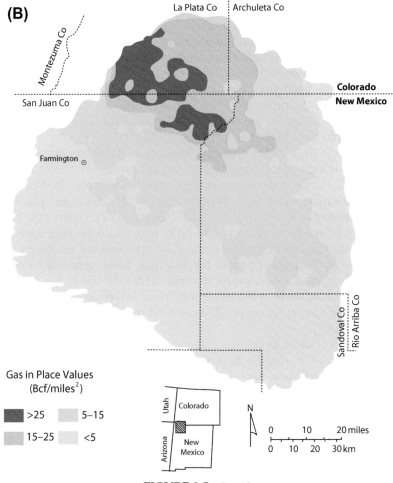

FIGURE 2.7 Cont'd

   Coalbeds are both sources and reservoirs for coalbed gas, and therefore the distribution of coal is important for gas producibility. Coal thickness and extent are closely related to tectonic, structural, and depositional settings, which strongly affect gas migration pathways and the distribution of gas (Rice, 1993; Scott, 2002). The regional heat flow influences coalification and the amount of thermogenic gas produced from coal. Thermogenic gas usually requires deep burial of coal in order to reach the required temperature for cracking reactions. For example, the Upper Cretaceous Fruitland coalbeds in the San Juan Basin are of moderate thickness, but their maximum depths of burial of ~4000 ft (~1219 m) make them the main CBM-producing interval (Table 2.2). Although most thermogenic CBM is produced during deep burial, some thermogenic gas can also be generated at relatively shallow depth if tectonic activities allow for abnormally high heat flow (Meissner, 1984).

   Coal rank is one of the most important factors controlling the generation of thermogenic gas. As discussed in Section 2.2, the onset for the generation of thermogenic gas in coal occurs at the high-volatile bituminous C rank (R_o ~0.6%), and coals must reach a certain threshold of thermal maturity (R_o between 0.8% and 1.0%) before large volumes of thermogenic gas can be generated. For example, thermal maturity in Fruitland coals expressed by vitrinite reflectance ranges from 0.4% to 1.5% (Table 2.2, Figure 2.7(A)), and the largest coalbed gas-in-place volumes coincide with the highest coal rank in the basin (Figure 2.7(B)).

   The producibility of CBM from coal depends on the ability of gas to flow along cleats and fractures in coal seams toward points of collection. Present-day in situ stress directions may significantly influence CBM producibility, for example, when S_{Hmax} is orthogonal to face cleats and fractures, gas permeability is lowered (Ayers, 2002; Scott, 2002). Uplift of a basin and erosion of overlying sedimentary rocks may cause the migration and possible loss of gas from shallow coalbeds. Late Cenozoic fracturing and uplift in the northernmost part of the San Juan Basin generated a structural hingeline. As a result, the highly fractured coalbeds in the northern and northwestern parts of the basin may have leaked gas, and therefore resulted in reduced amounts of CBM retained in these areas (Ayers, 2002; Zhou et al., 2005).

   Higher gas content in coal increases coalbed gas producibility. However, gas content is usually highly variable between and even within individual coalbeds, and depends on local hydrogeological and reservoir conditions (Scott and Kaiser, 1996). Gas contents in San Juan Basin coals vary from less than 150 scf/t (4.7 cm^3/g) in the southern and southwestern parts (blue, green regions in Figure 2.7(B)), increase to 200–400 scf/t (6.2–12.4 cm^3/g) in the central part (orange region in Figure 2.7(B)), and peak to >500 scf/t (>15.6 cm^3/g) in the north-central part of the basin (red region in Figure 2.7(B), (Ayers, 2002)).

TABLE 2.2 Comparison of Factors Controlling the Distribution and Producibility of Thermogenic versus Biogenic Gas in San Juan Basin and Powder River Basin, Respectively

Controls on Gas Distribution and Producibility	Thermogenic Gas in San Juan Basin	Biogenic Gas in Powder River Basin	Reference
Depositional systems and coal distribution	Deep (maximum 4000 ft; 1219 m); net coal thickness in belts 30–70 ft (9–21 m); individual coal thickness 6–30 ft (1.8–9 m)	Shallow (<2500 ft; 762 m); net coal thickness in belts 50 to >215 ft (15 to >65 m); individual coal thickness >60 ft (18 m)	Rice (1993), Ayers et al. (1994), Ayers (2002)
Coal rank	Vitrinite reflectance, R_o 0.4–1.5%	Vitrinite reflectance, R_o 0.3–0.4%	Rice (1993)
Tectonic and structural setting	Highly fractured in the northern and northwestern parts of the basin because of uplift; compact in the north-central part of the basin	Fairly well cleated Good aquifers	Tyler (1995) Ayers (2002)
Permeability (millidarcy, mD)	5–60 mD	10 to >1000 mD	Pratt et al. (1999), Ayers (2002)
Gas content	150 to >500 scf/t (4.7 to >15.6 cm³/g)	16 to 76 scf/t (0.5–2.4 cm³/g)	Ayers (2002)
Hydrodynamic	Artesian overpressure in the north-central part of the basin; underpressured in much of the rest of the basin	Normal to artesian pressure	Ayers (2002)
Hydrogeo-chemistry	Brackish formation water; bicarbonate in the north-central part of the basin; low chlorinity in the northwestern part of the basin; moderate to high TDS[1]	Fresh formation water (chlorinity <2 mol/L Cl⁻); very low SO_4^{2-} (<10 mmol/L); low TDS	Larson and Daddow (1984), Martini et al. (1998), Rice et al. (2008), Ayers (2002)

[1] *TDS (total dissolved solids): ions dissolved in water, such as Ca^{2+}, Na^+, Mg^{2+}, Fe^{2+}; HCO_3^{3-}, Cl^-, CO_3^{2-}, etc.*

Permeability and migration pathways of gas and water in coal are largely controlled by the abundance and geometry of open cleats and fractures in coalbeds. Elevated permeability facilitates the flow of desorbed gas and producibility of CBM, but very high permeability may result in enhanced water production. Produced formation water is often saline and may pose an environmental hazard if not reinjected underground. According to Lucia (Lucia, 1983) and Scott (Scott, 1995), the most productive wells in San Juan Basin have permeabilities from 0.5 to 100 mD. This permeability is within the range of optimal producibility (Table 2.2).

Hydrodynamic characteristics of a CBM field play a central role in reservoir performance and coalbed gas producibility. Gas formed in coal seams is originally trapped by the hydrostatic groundwater or formation water pressure. During CBM development, formation water is first pumped out of coal seams in order to (1) lower the hydrostatic pressure, (2) aid in the desorption of gas, (3) expand the free gas phase (i.e., bubbles), and (4) build a pressure gradient to drive the free gas to the extraction well. However, the long-term reduction of hydrostatic pressure may potentially result in a drawdown of aquifers and cause environmental issues. Basin hydrogeology therefore must be carefully monitored.

2.2.3 Biogenic Gases

2.2.3.1 Gas Generation and Characteristics

Biogenic coalbed gas is a microbial product formed at low temperatures during the fermentative decomposition of organic matter in peat and coal. Microorganisms generate primary biogenic gas (Rice, 1993) during the earliest burial of organic matter ranking from peat to subbituminous coal (R_o values usually <0.3%) (Claypool and Kaplan, 1974; Rice, 1993). The resulting methane is commonly produced as bubbles in the form of "swamp gas", or dissolved in water and expelled into surrounding sediments during compaction (Rice, 1993; Scott et al., 1994; Flores, 2004). Owing to shallow burial, primary biogenic gas cannot accumulate in economic deposits, except in cold regions where it may solidify as methane hydrates (i.e., clathrates) in permafrost (Bily and Dick, 1974; Collett, 2008).

Microbially generated methane is able to accumulate in sediments only after moderately deep burial of organic matter and placement of an overlying gastight seal. In many cases, coal had intermittently been buried to depths where the geothermal gradient caused heat sterilization of coal. Subsequent uplift caused cooling and may have allowed reinoculation of coal seams with methanogenic microbial consortia via introduction of meteoric waters into permeable coalbeds. The newly generated CBM is called secondary biogenic gas (Rice, 1993; Scott et al., 1994). Most preserved CBM is probably secondary gas, because late-stage biogenic gases are found in many coals with higher ranks (Rice, 1993; Scott et al., 1994; Whiticar, 1999).

Biogenic methane can be generated either by methyl-type fermentation or by carbon dioxide reduction. Both pathways differ biochemically and in terms of their stable isotope fractionations (Schoell, 1980; Whiticar et al., 1986; Rice, 1993). Figure 2.8 depicts the process of biogenic methanogenesis in four stages (Winfrey, 1984; Zinder, 1993; Faiz, 2004): (1) Fermentative

Biogenic gas generation

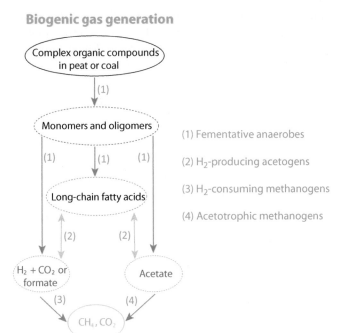

(1) Fementative anaerobes

(2) H_2-producing acetogens

(3) H_2-consuming methanogens

(4) Acetotrophic methanogens

FIGURE 2.8 Pathways of biogenic methanogenesis expressed in four distinct stages. Solid line ellipses identify original source material and end products; dashed line ellipses mark intermediate products. *(Modified from Zinder (1993).)*; double arrows in (2) stand for reversible reactions.

anaerobic microbes breakdown complex organic molecules in coal to simple organic acids and alcohols. (2) H_2-producing, acetogenic microbes convert fatty acids and alcohols into acetate, H_2, and CO_2 or formate under anoxic conditions. The third and fourth stages refer to alternative methanogenic pathways whereby either (3) H_2-consuming methanogens use available H_2 to convert formate or CO_2 to methane, or (4) acetotrophic methanogens utilize acetate to produce methane and CO_2. Based on free energy calculations by Strąpoć et al. (Strąpoć et al., 2008b), H_2-utilizing methanogenesis is favored during the terminal process of coal organic matter biodegradation over homoacetogenesis and acetoclastic methanogenesis. CO_2 reduction is generally the predominant pathway of methanogenesis after acetate fermentation ceases, although both pathways can occur simultaneously at certain conditions (Whiticar, 1999). Therefore, late-stage biogenic methane generation is mainly due to reduction of carbon dioxide.

Biogenically derived gas in coalbeds is predominantly composed of methane and contains only traces of C_{2+} hydrocarbons (i.e., dry gas). The CO_2 content is usually <5 vol%. This contrasts sharply with thermogenic gases that usually contain more C_{2+} hydrocarbons and CO_2. Biogenic CH_4 expresses wide ranges of δD_{CH_4} and $\delta^{13}C_{CH_4}$ values from $-400‰$ to $-150‰$ and from $-110‰$ to $-50‰$, respectively (Figure 2.3). During methyl-type fermentation, the hydrogen in methane is derived primarily from methyl groups of parental organic matter, whereas methanogenesis via CO_2 reduction relies entirely on hydrogen from formation water. Organic hydrogen tends to be strongly depleted in D relative to hydrogen in water. Therefore, methane from microbial CO_2 reduction tends to have higher δD_{CH_4} values ($-150‰$ to $-250‰$) in comparison to those relating to methyl-type fermentation ($-250‰$ to $-400‰$). In terms of carbon stable isotope ratios, $\delta^{13}C_{CH_4}$ values relating to CO_2 reduction tend to be more negative ($-55‰$ to $-110‰$) than those associated with methyl-type fermentation ($-40‰$ to $-70‰$) (Rice, 1993; Whiticar, 1999).

2.2.3.2 Controls on Distribution and Producibility of Biogenic Gas

Controlling factors for the generation, distribution, and producibility of biogenic gas in coalbeds are fundamentally similar to those affecting thermogenic gas, but their roles and contributions may differ (Table 2.2). The Powder River Basin of northeastern Wyoming and southwestern Montana is a good example of a microbial CBM system. The Tongue River Member of the Paleocene Fort Union Formation is currently the main CBM-producing formation. Geological and chemical characteristics of its coals suggest a biogenic pathway for CBM generation (Rice, 1993; Rice et al., 2008). Chemical and isotopic CBM analyses indicate a microbial, methyl-type fermentation pathway (Whiticar, 1999; Flores et al., 2008).

Coalbeds with active production of biogenic gas are typically located at shallower depths than coalbeds generating thermogenic gas because microbial methanogenesis is commonly limited to temperatures below 56 °C (Scott et al., 1994), with ~30 °C as the optimal temperature for microbial methanogenesis (Strąpoć et al., 2008b; Stams and Hansen, 1984; Green et al., 2008). For example, the predominant producer of biogenic gas in the Powder River Basin, the Paleocene Fort Union Formation, occurs at a depth of less than 2500 ft (762 m). Similarly in the Illinois Basin, an active microbial gas zone has been documented in Pennsylvanian coalbeds at a depth of less than 700 ft (213 m) (Strąpoć et al., 2008a,b).

The rank of coal limits the amount of biogenic gas that can be generated. A coal rank above R_o ~0.8% indicates that the original microbial communities had not been able to survive the high temperatures of coalification. Even if the coal had eventually cooled and become inoculated with a new methanogenic microbial consortium via infiltrating meteoric waters, the high-rank coal would be unable to produce significant quantities of biogenic gas because fermentative processes are less effective in degrading highly aromatic and cross-linked organic macromolecules (Strąpoć et al., 2008a,b). The rank of Paleocene Fort Union coalbeds is relatively low over the entire Powder River Basin, ranging from lignite to subbituminous

C (R_o values of 0.3–0.4%). These deposits represent the lowest rank coals from which commercial gas production has been developed in the US High-volatile bituminous C and B coals are the source of microbial gas in the Illinois Basin (Strąpoć et al., 2008a,b).

The location and geometry of folds and faults is dependent on tectonic and structural settings, which may strongly influence the recharge of meteoric water into coalbeds. The infiltration of formation waters not only lowers salinity and temperature of formation waters and can make the coalbeds habitable for microbial consortia, but also may reinoculate coal with methanogenic microbes (Ulrich and Bower, 2008). Therefore, tectonic and structural settings are critical for evaluating the possible generation of biogenic CBM. Although limited data are available to document cleat orientations in Fort Union coals along the eastern margin of Powder River Basin, cleats are generally trending east–northeastward, which is nearly perpendicular to the basin's axis (Glass, 1975; Tyler, 1995) and prevents compressive closure of cleats. These coals are fairly well fractured most likely because of their low ash and relatively high vitrinite contents, increasing the brittleness of the coals (Ayers, 2002). As a result, Fort Union coalbeds and associated interbedded sand units serve as good aquifers with access to freshwater recharge.

Gas content in coal is an important parameter controlling producibility of all coalbed gases, including biogenic gases. Gas content in biogenic CBM systems can vary significantly over short distances along coalbeds and even vertically across individual coal seams, as has been documented in the eastern part of the Illinois Basin, where gas content ranges from few to more than 150 scf/t (4.7 cm^3/g) (Drobniak et al., 2004; Strąpoć et al., 2008a). One underlying reason may be variance in maceral composition and available micro- and mesoporosity over short distances affecting the coal's ability to adsorb gas (Mastalerz et al., 2008). Gas content in the Powder River Basin is much lower than in the San Juan Basin (Table 2.2), but the large production of gas in the Powder River Basin suggests that low gas content is compensated by large coal volumes. This also suggests that the Powder River Basin may feature a significant contribution of free (i.e., not adsorbed) gas in its coal. Free gas seems to contribute significantly to total gas in low-rank coals (Bustin and Clarkson, 1999).

The very high permeability of Fort Union coals ranges between 10 mD and several darcys (Pratt et al., 1999, Table 2.2) and, combined with the coalbeds' thickness, provides for the quality of major Fort Union aquifers. The hydrochemistry of formation waters in coal seams is of critical importance for microbial methanogenesis, and consequently influences CBM distribution and producibility. Effective microbial CBM generation requires anoxic formation waters with relatively low chlorinity (<2 mol/L Cl$^-$) and low-sulfate concentrations (<10 mmol/L

SO_4^{2-}) (Rice and Claypool, 1981; Rice, 1993; Martini et al., 1998; Strąpoć et al., 2008a). Relatively fresh, low-sulfate formation waters serve as an important prerequisite for the economically important gas reservoirs in the Powder River Basin. Even if coal seams receive well-oxygenated meteoric water via recharge, the overwhelming abundance of strongly reducing organic matter and sulfide minerals in coal rapidly depletes free O_2 (Jin et al., 2010) and renders formation waters in coal seams anoxic.

2.2.4 The Hypothesis of Catalytically Generated Gas

Several large commercial gas reserves in low-rank coalbeds exhibit little or no evidence of microbially generated gas. Examples come from the Western Canadian Sedimentary Basin (e.g., Corbett area) (Mastalerz and Drobniak, 2007) and from India (Ramaswamy, 2003). These coalbed gas plays carry a chemical gas signature similar to that of thermogenic gases, but there is no evidence from thermal maturity measurements that temperatures have ever been high enough to generate thermogenic gases via cracking reactions. Such findings suggest an alternative pathway for nonbiogenic coalbed gas generation. Butala (Butala et al., 2000) proposed catalysis by transition metals (e.g., possibly nickel and iron) to explain low-temperature gas generation in bituminous coals, an idea consistent with experimental work by others in hydrocarbon gas formation (Mango, 1996). Mango's recent data strongly suggest that marine shales possess natural catalytic ability for converting parts of kerogen and bitumen to hydrocarbon gas at temperatures as low as 50°C (i.e., ~300°C below the required temperatures of pyrolytic cracking). Mango's proposed catalytic pathway with nonlinear kinetics differs fundamentally from high-temperature pyrolytic cracking reactions leading to thermogenic gases, but yields similar to wet gas generated in laboratory experiments at 50°C (Mango and Jarvie, 2009a,b). The catalytic hypothesis of gas generation can be logically extended from organic-rich marine shales to coals. Further research is needed before catalytic gas generation can become an accepted pathway for coalbed gas generation.

2.3 CONCLUSIONS

* The origins of coalbed gas can be biogenic, thermogenic, mixed, and possibly catalytic. Several compositional and isotopic gas parameters are applied to distinguish between the various pathways in order to help establishing a successful and effective CBM exploration strategy. This paper reviews coalbed gas generation pathways and summarizes compositional and isotopic characteristics of thermogenic and biogenic gases.

- Coalbed gas content and producibility are dependent on many factors that are often interrelated. Controlling factors for the generation, distribution, and producibility of biogenic gas in coalbeds are fundamentally similar to those affecting thermogenic gas and they include, among others, coal rank, tectonic and structural settings, coal permeability, hydrodynamics, and hydrogeochemistry of the formation waters. However, their role may drastically differ for different gas origins. For example, high rank favors thermogenic gas, whereas more biogenic gas is generally expected in low-rank coal. This paper compares these controlling factors using the San Juan Basin (dominantly thermogenic gas origin) and the Powder River Basin (biogenic origin) as examples.
- Exploration for a thermogenic gas play should focus on deeper coal zones with higher maturity. In contrast, exploration for a predominantly microbial gas play should focus on coal seams close to basin margins, where organic matter is less mature and larger fractures may facilitate increased gas producibility.

References

Ayers Jr, W.B., 2002. Coalbed gas systems, resources, and production and a review of contrasting cases from the San Juan and Powder River Basins. AAPG Bulletin 86, 1853–1890.

Ayers Jr, W.B., Ambrose, W.A., Yeh, J., 1994. Coalbed methane in the Fruitland Formation, San Juan Basin: depositional and structural controls on occurrence and resources. In: Ayers Jr, W.B., et al. (Ed.), Geological and Hydrologic Controls on the Occurrence and Producibility of Coalbed Methane, Fruitland Formation, San Juan Basin. Gas Research Institute Topical Report GRI-91/0072, Chicago, pp. 9–46.

Bernard, B.B., 1978. Light Hydrocarbons in Marine Sediments (Ph.D. thesis). Texas A & M University, 144 p.

Berner, U., Faber, E., 1988. Maturity related mixing model for methane, ethane and propane, based on carbon isotopes. Organic Geochemistry 13, 67–72.

Bily, C., Dick, J.W.L., 1974. Naturally occurring gas hydrates in the Mackenzie delta, N.W.T. Bulletin of Canadian Petroleum Geology 22, 340–352.

Brown, J.K., Ladner, W.R., 1960. A study of the hydrogen distribution in coal-like materials by high resolution N.M.R. spectroscopy, II. A comparison with infra-red measurement and the conversion to coal structure. Fuel 39, 87–96.

Bustin, R.M., Clarkson, C.R., 1999. Free gas storage in matrix porosity: a potentially significant coalbed resource in low rank coals. In: International Coalbed Methane Symposium, Tuscaloosa, Alabama, pp. 197–214.

Butala, S.J.M., Medina, J.C., Taylor, C.R.T.Q., Bowerbank, C.H., Lee, M.L., 2000. Mechanisms and kinetics of reactions leading to natural gas formation during coal maturation. Energy Fuel 14, 235–259.

Claypool, G.E., Kaplan, I.R., 1974. The origin and distribution of methane in marine sediments. In: Kaplan, I.R. (Ed.), Natural Gases in Marine Sediments. Plenum Press, New York, pp. 99–139.

Collett, T.S., 2008. Geologic and engineering controls on the production of permafrost-associated gas hydrates accumulations. In: Proceedings of the 6th International Conference on Gas Hydrates(Vancouver, Canada).

Cook, A.C., Struckmeyer, H., 1986. The role of coal as a source rock for oil. In: Glenie, R.C. (Ed.), Proceedings 2nd Southeastern Australia Oil Exploration Symposium, pp. 19–32.

Coplen, T.B., 1996. New guidelines for reporting stable hydrogen, carbon, and oxygen isotope-ratio data. Geochimica et Cosmochimica Acta 60, 3359–3360.

Dai, J., Ni, Y., Zou, C., Tao, S., Hu, G., Hu, A., Yang, C., Tao, X., 2009. Stable carbon isotope of alkane gases from the Xujiahe coal measures and implication for gas-source correlation in the Sichuan Basin, SW China. Organic Geochemistry 40, 638–646.

Drobniak, A., Mastalerz, M., Rupp, J., Eaton, N., 2004. Evaluation of coalbed gas potential of the Seelyville Coal Member, Indiana, USA. International Journal of Coal Geology 57, 265–282.

Dugan, T.A., Williams, B.L., 1988. History of gas produced from coal seams in the San Juan Basin. In: Fassett, J.E. (Ed.), Geology and Coalbed Methane Resources of the Northern San Juan Basin, Colorado and New Mexico. Rocky Mt. Assoc. Geologists, Denver, pp. 1–9.

Faiz, M.M., 2004. Microbial influences on coal seam gas reservoirs-a review. In: Proceedings of the Bac-Min Conference. The Australian Institute of Mining and Metallurgy Publication Series No. 6, pp. 133–142.

Faiz, M., Hendry, P., 2006. Significance of microbial activity in Australian coal bed methane reservoirs—a review. Bulletin of Canadian Petroleum Geology 54, 261–272.

Flores, R.M., 2004. Coalbed Methane in the Powder River Basin, Wyoming and Montana: An Assessment of the Tertiary-upper Cretaceous Coalbed Methane Total Petroleum System. U.S. Geological Survey Digital Data Series, DDS-69-C(Chapter 2), 56p.

Flores, R.M., Rice, C.A., Stricker, G.D., Warden, A., Ellis, M.S., 2008. Methanogenic pathways of coal-bed gas in the Powder River Basin, United States: the geologic factor. International Journal of Coal Geology 76, 52–75.

Formolo, M., Martini, A., Petsch, S., 2008. Biodegradation of sedimentary organic matter associated with coalbed methane in the Powder River and San Juan Basins, U.S.A. International Journal of Coal Geology 76, 86–97.

Glass, G.B., 1975. Analyses and measured sections of 53 Wyoming coal samples. Wyoming, Geological Survey, Report of Investigations 11, 219.

Green, M.S., Flanegan, K.C., Gilcrease, P.C., 2008. Characterization of a methanogenic consortium enriched from a coalbed methane well in the Powder River Basin, U.S.A. International Journal of Coal Geology 76, 34–45.

Hunt, J.M., 1979. Petroleum Geochemistry and Geology. W.H. Freeman and Co., San Francisco. 617 p.

Jenden, P.D., Kaplan, I.R., 1986. Comparison of microbial gases from the Scripps submarine canyon: implications for the origin of natural gas. Applied Geochemistry 1, 631–646.

Jin, H., Schimmelmann, A., Mastalerz, M., Pope, J., Moore, T.A., 2010. Coalbed gas desorption in canisters: consumption of trapped atmospheric oxygen and implications for measured gas quality. International Journal of Coal Geology 81, 64–72.

Jones, D.M., Head, I.M., Gray, N.D., et al., 2008. Crude-oil biodegradation via methanogenesis in subsurface petroleum reservoirs. Nature 451, 176–181.

Jüntgen, H., Karweil, J., 1966. Formation and storage of gas in bituminous coals, pt. 1 gas storage (English Trans.) Erdöl und Kohle-Erdgas-Petrochemie 19, 229–344.

Kaiser, W.R., Hamilton, D.S., Scott, A.R., Taylor, R., Finley, R.J., 1994. Geological and hydrological controls on the producibility of coalbed methane. Journal of the Geological Society of London 151, 417–420.

Larson, L.R., Daddow, R.L., 1984. Ground-Water-Quality Data from the Southern Powder River Basin, Northeastern Wyoming. U.S.G.S. Open-file Report 83-939. 56 p.

Larter, S., di Primio, R., 2005. Effects of biodegradation on oil and gas field PVT properties and the origin of oil rimmed gas accumulations. Organic Geochemistry 36, 299–310.

Lucia, J.L., 1983. Petrophysical parameters estimated from visual descriptions of carbonate rocks: a field classification of carbonate pore space. Journal of Petroleum Technology, 629–637.

Mango, F.D., 1996. Transition metal catalysis in the generation of natural gas. Organic Geochemistry 24, 977–984.

Mango, F.D., Jarvie, D.M., 2009a. Low-temperature gas from marine shales. Geochemical Transactions. 10. http://dx.doi.org/10.1186/1467-4866-10-3.

Mango, F.D., Jarvie, D.M., 2009b. Low-temperature gas from marine shales: wet gas to dry gas over experimental time. Geochemical Transactions. 10. http://dx.doi.org/10.1186/1467-4866-10-10.

Martini, A.M., Walter, L.M., Budai, J.M., Ku, T.C.M., Kaiser, C.J., Schoell, M., 1998. Genetic and temporal relations between formation waters and biogenic methane: Upper Devonian Antrim Shale, Michigan Basin, USA. Geochimica et Cosmochimica Acta 62, 1699–1720.

Martini, A.M., Walter, L.M., Ku, T.C.W., Budai, J.M., McIntosh, J.C., Schoell, M., 2003. Microbial production and modification of gases in sedimentary basins: geochemical case study from a Devonian Shale gas play, Michigan Basin. AAPG Bulletin 87, 1355–1375.

Martini, A.M., Walter, L.M., McIntosh, J.C., 2008. Identification of microbial and thermogenic gas components from Upper Devonian black shale cores, Illinois and Michigan Basins. AAPG Bulletin 92, 327–339.

Mastalerz, M., Drobniak, A. Depositional Setting and Coalbed Methane Potential in the Corbett Area, Alberta, Canada, Nexen, Calgary, unpublished report.

Mastalerz, M., Drobniak, A., Strąpoć, D., Solano Acosta, W., Rupp, J., 2008. Variations in pore characteristics in high volatile bituminous coals: implications for coalbed gas content. International Journal of Coal Geology 76, 205–216.

McIntosh, J.C., Martini, A.M., Petsch, S., Nüsslein, K., 2008. Biogeochemistry of the Forest City Basin coalbed methane play. International Journal of Coal Geology 76, 111–118.

Meissner, F.F., 1984. In: Woodward, J., Meissner, F.F., Clayton, J.L. (Eds.), Cretaceous and Lower Tertiary Coals as Sources for Gas Accumulations in the Rocky Mountain Area. Rocky Mt. Assoc. Geologists, Denver, Colorado, pp. 401–430.

Montgomery, S.L., 1999. Powder River Basin, Wyoming: an expanding coalbed methane (CBM) play. AAPG Bulletin 83, 1207–1222.

Pashin, J.C., McIntyre, M.R., 2003. Temperature-pressure conditions in coalbed methane reservoirs of the Black Warrior Basin: implications for carbon sequestration and enhanced coalbed methane recovery. International Journal of Coal Geology 54, 167–183.

Pratt, T.J., Mavor, M.J., De Bruin, R.P., 1999. Coal gas resource and production potential of subbituminous coal in the Powder River Basin. In: Proceedings of the 1999 Int. Coalbed Methane Symposium. University of Alabama College of Continuing Studies, pp. 23–34.

Ramaswamy, G.A., 2003. A field evidence for mineral-catalyzed formation of gas during coal maturation. Oil & Gas Journal 100, 32–36.

Rice, D.D., 1983. Relation of natural gas composition to thermal maturity and source rock type in San Juan Basin, northwestern New Mexico and southwestern Colorado. AAPG Bulletin 67, 1119–1281.

Rice, D.D., 1993. Composition and origins of coalbed gas. In: In: Law, B.E., Rice, D.D. (Eds.), Hydrocarbons from Coal, AAPG Studies in Geology, vol. 38. , pp. 159–185.

Rice, D.D., 1997. Coalbed Methane—An Untapped Energy Resource and an Environmental Concern: U.S. Geological Survey Fact Sheet FS–019–97. Available at: http://energy.usgs.gov/factsheets/Coalbed/coalmeth.html.

Rice, D.D., Claypool, G.E., 1981. Generation, accumulation, and resource potential of biogenic gas. AAPG Bulletin 65, 5–25.

Rice, C.A., Flores, R.M., Stricker, G.D., Ellis, M.S., 2008. Chemical and stable isotopic evidence for water/rock interaction and biogenic origin of coalbed methane, Fort Union Formation, Powder River Basin. International Journal of Coal Geology 76, 76–85.

Russell, N.J., 1990. The vitrinite reflectivity and thermal maturation of coal. In: Paterson, L. (Ed.), Methane Drainage from Coal. Commonwealth Scientific and Industrial Research Organization, Australia, pp. 19–26.

Schoell, M., 1980. The hydrogen and carbon isotopic composition of methane from natural gases of various origins. Geochimica et Cosmochimica Acta 44, 649–661.

Schoell, M., 1983. Genetic characterization of natural gases. AAPG Bulletin 67, 2225–2238.

Scott, A.R., 1999. Review of Key Hydrogeologic Factor Affecting Coalbed Methane Producibility and Resource Assessment, Okalahoma Geological Survey Open File Report of 6-99, 25 p.

Scott, A.R., 2002. Hydrogeologic factors affecting gas content distribution in coal beds. International Journal of Coal Geology 50, 363–387.

Scott, A.R., 1995. Factors affecting gas content distribution in coal beds. In: Kaiser, W.R., et al. (Ed.), Geology and Hydrology of Coalbed Methane Producibility in the United States: Analogs for the World, INTERGAS' 95 Short Course Notes. The University of Alabama, Tuscaloosa, Alabama, pp. 205–250.

Scott, A.R., Kaiser, W.R., 1996. Factors affecting gas-content distribution in coalbeds: a review (exp. Abs.). In: Expanded Abstracts Volume, Rocky Mountain Section Meeting: AAPG Geologists, pp. 101–106.

Scott, A.R., Kaiser, W.R., Ayers Jr, W.B., 1994. Thermogenic and secondary biogenic gases, San Juan Basin, Colorado and New Mexico—implications for coalbed gas producibility. AAPG Bulletin 78, 1186–1209.

Smith, J.W., Pallasser, R.J., 1996. Microbial origin of Australian coalbed methane. AAPG Bulletin 80, 891–897.

Stach, E., Mackowsky, M.T., Teichmüller, M., Taylor, G.H., Chandra, D., Teichmüller, R., 1982. Stach's Textbook of Coal Petrology, third ed. Gebrüder Borntraeger, Berlin, Stuttgart. Chapter 2, pp. 38–46.

Stahl, W., 1973. Carbon isotope ratios of German natural gases in comparison with isotope data of gaseous hydrocarbons from other parts of the world. In: Tissot, B., Bienner, F. (Eds.), Advances in Organic Geochemistry. Editions Technip, Paris, pp. 453–462.

Stams, A.J.M., Hansen, T.A., 1984. Fermentation of glutamate and other compounds by Acidaminobacter hydrogenoformans gen. nov. sp. nov., an obligate anaerobe isolated from black mud. Studies with pure cultures and mixed cultures with sulfate-reducing and methanogenic bacteria. Archives of Microbiology 137, 329–337.

Strąpoć, D., Mastalerz, M., Eble, C., Schimmelmann, A., 2007. Characterization of the origin of coalbed gases in southeastern Illinois Basin by compound-specific carbon and hydrogen stable isotope ratios. Organic Geochemistry 38, 267–287.

Strąpoć, D., Mastalerz, M., Schimmelmann, A., Drobniak, A., Hedges, S., 2008a. Variability of geochemical properties in a microbially-dominated coalbed gas system from the eastern margin of the Illinois Basin. International Journal of Coal Geology 76, 98–110.

Strąpoć, D., Picardal, F.W., Turich, C., Schaperdoth, I., Macalady, J.L., Lipp, J.S., Lin, Y., Ertefai, T.F., Schubotz, F., Hinrichs, K., Mastalerz, M., Schimmelmann, A., 2008b. Methane-producing microbial community in a coal bed of the Illinois Basin. Applied and Environmental Microbiology 74, 2424–2432.

Tang, Y., Jenden, P.D., Teerman, S.C., 1991. Thermogenic methane formation in low-rank coals-published models and results from laboratory pyrolysis of lignite. In: Manning, D.A.C. (Ed.), Organic Geochemistry—Advances and Applications in the Natural Environment. Manchester University Press, Manchester, England, pp. 329–331.

Tedesco, S.A., 1992. Coalbed methane potential assessed in Forest City Basin. Oil & Gas Journal 90, 68–72.

Tedesco, S.A., 2003. Positive factors dominate negatives for Illinois Basin coalbed methane. Oil & Gas Journal 101, 28–32.

Tissot, B.P., Welte, D.H., 1984. Petroleum Formation and Occurance. Springer-Verlag, New York. 699.

Tyler, R., 1995. Structural setting and coal fracture patterns of foreland basins: controls critical to coalbed methane producibility. In: Kaiser, W.R., Scott, A.R., Tyler, R. (Eds.), Geology and Hydrology of Coalbed Methane Producibility in the United States: Analogs for the World. University of Texas at Austin, Bureau of Economic Geology, Intergas' 95 Short Course Notes, pp. 1–50.

Ulrich, G., Bower, S., 2008. Active methanogenesis and acetate utilization in Powder River Basin coals, United States. International Journal of Coal Geology 76, 25–33.

U.S. Environmental Protection Agency. Available at: http://www.epa.gov/cleanenergy/energy-and-you/affect/natural-gas.html.

U.S Department of Energy, Energy Information Administration (EIA). http://www.eia.doe.gov/oil_gas/rpd/cbmusa2.pdf, 2007.

Valentine, D.J., Chidthaisong, A., Rice, A., Reeburgh, W.S., Tyler, S.C., 2004. Carbon and hydrogen isotope fractionation by moderately thermophilic methanogens. Geochimica et Cosmochimica Acta 68, 1571–1590.

Whiticar, M.J., 1994. Correlation of natural gases with their sources. In: Magoon, L., Dow, W. (Eds.), The Petroleum System—From Source to Trap. AAPG Memoir 60, Tulsa, pp. 261–284.

Whiticar, M.J., 1996. Stable isotope geochemistry of coals, humic kerogens and related natural gases. International Journal of Coal Geology 32, 191–215.

Whiticar, M.J., 1999. Carbon and hydrogen isotope systematic of bacterial formation and oxidation of methane. Chemical Geology 161, 291–314.

Whiticar, M.J., Faber, E., Schoell, M., 1986. Biogenic methane formation in marine and freshwater systems: CO_2 reduction vs. acetate fermentation-isotopic evidence. Geochimica et Cosmochimica Acta 50, 693–709.

Winfrey, M.R., 1984. Microbial production of methane. In: Altas, R.M. (Ed.), Petroleum Microbiology. Macmillan, New York, pp. 153–219.

Zhou, Z., Ballentine, C.J., Kipfer, R., Schoell, M., Thibodeaux, S., 2005. Noble gas tracing of groundwater/coalbed methane interaction in the San Juan Bain, USA. Geochimica et Cosmochimica Acta 69, 5413–5428.

Zinder, S.H., 1993. Physiological ecology of methanogens. In: Ferry, J.G. (Ed.), Methanogenesis. Chapman & Hall, New York, pp. 128–206.

Geology of North American Coalbed Methane Reservoirs

Jack C. Pashin

Devon Energy Chair of Basin Research, Boone Pickens School of Geology, Oklahoma State University, Stillwater, Oklahoma, USA

3.1 INTRODUCTION

Coalbed methane is an integral source of natural gas in North America, currently accounting for 9% of U.S. and 3% of Canadian dry natural gas production. A global expansion of the coalbed methane industry is underway, with major resources being developed in Australia, China, and India. The North American experience is helping guide this expansion and serves as a model for understanding the geologic controls on coalbed gas production. This chapter reviews our current understanding of the fundamental geologic controls on the distribution and producibility of coalbed methane. Numerous examples from the primary commercial basins in North America (Figure 3.1) highlight the variety of factors that are essential for understanding the geology of coalbed methane reservoirs.

Coal is classified as a continuous-type unconventional reservoir, and the basic mechanisms of gas storage and production are relatively simple. Coal is a microporous polymer, and the vast majority of the gas in coal is stored on pore surfaces in an adsorbed state (Figure 3.2); that is, the gas clings by Van der Waals forces to surfaces within the microporous structure. These forces, which solve charge imbalances on free surfaces, enable coal to store gases at much higher concentrations than would be predicted by ideal gas law under similar temperature–pressure conditions. Lowering reservoir pressure below the saturation point (i.e., the critical desorption pressure) causes gas to desorb from the microporous structure and flow toward the wellbore, where the gas can be produced, processed, and delivered to market. The concentration of gas and the ability to effectively depressurize the reservoir are controlled by a host of geologic factors,

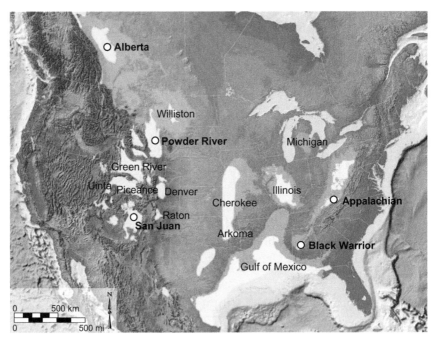

FIGURE 3.1 Generalized map of the coal basins of North America *(Source: National Energy Technology Laboratory)*. White circles denote major coalbed methane basins that are highlighted in this chapter.

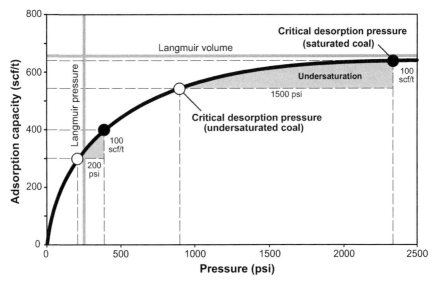

FIGURE 3.2 Idealized adsorption isotherm showing Langmuir parameters and the impact of isotherm geometry on the degree of depressurization required to reach critical desorption pressure and recover significant volumes of natural gas.

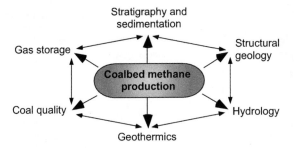

FIGURE 3.3   Basic geologic factors affecting the distribution and producibility of coalbed methane resources.

which operate from the molecular scale to the scale of a sedimentary basin. Indeed, a thorough understanding of the geology of a coalbed methane prospect is central to successfully identifying commercial potential and implementing the technologies that are suited to realizing that potential (e.g., Pashin et al., 1991; Ayers and Kaiser, 1994; Scott, 2002).

The geologic factors that are discussed in this chapter include stratigraphy, sedimentation, structural geology, hydrology, geothermics, coal quality, and gas storage (Figure 3.3). The following sections discuss the relevance of each of these factors to the characterization and development of coalbed methane resources. The examples cited within the text are from proven coalbed methane basins, including the Black Warrior, Appalachian, San Juan, Alberta, and Powder River basins. These examples demonstrate the impact different combinations of geologic variables have on the distribution of coalbed gas resources and the performance of coalbed methane reservoirs.

3.2 GEOLOGIC FACTORS

3.2.1 Stratigraphy and Sedimentation

Stratigraphy and sedimentation are key considerations when evaluating coalbed methane reservoirs because the thickness and continuity of coal seams, as well as the geometry and reservoir properties of the enveloping strata, are determined largely in the original sedimentary environment. Coal is fossilized peat, which is an organic-rich soil type that is characteristic of wetlands, such as swamps and marshes. The distribution of coalbed methane reservoirs through geologic time is therefore linked inextricably to the evolution of peat-forming wetland floras (Figure 3.4). The oldest known coaly material constitutes algal stringers resembling boghead coal in Precambrian stromatolitic rocks of the Canadian Shield, and the oldest known coal seams of significance formed in eastern Europe

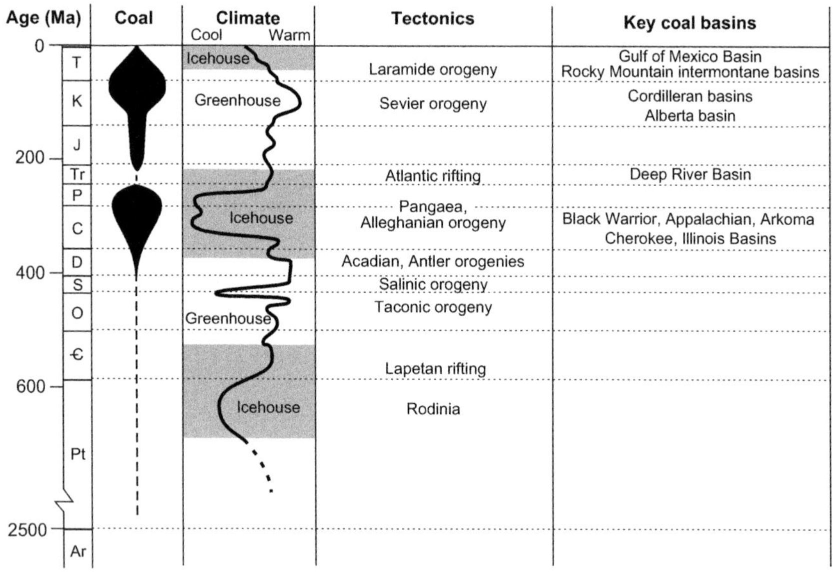

Age (Ma)	Coal	Climate		Tectonics	Key coal basins
		Cool	Warm		
0 — T		Icehouse		Laramide orogeny	Gulf of Mexico Basin Rocky Mountain intermontane basins
K		Greenhouse		Sevier orogeny	Cordilleran basins Alberta basin
J					
200 — Tr				Atlantic rifting	Deep River Basin
P				Pangaea, Alleghanian orogeny	
C		Icehouse			Black Warrior, Appalachian, Arkoma Cherokee, Illinois Basins
400 — D				Acadian, Antler orogenies	
S				Salinic orogeny	
O		Greenhouse		Taconic orogeny	
€				Lapetan rifting	
600 —		Icehouse		Rodinia	
Pt					
2500 — Ar					

FIGURE 3.4   Distribution of coal and coalbed methane through geologic time. *After Pashin (1998).*

during the Silurian, shortly after the first vascular plants (psilophytes) evolved (Diessel, 1992). Plants radiated through the Late Silurian and Devonian, and the gymnosperm floras (i.e., seed-bearing plants, including lycopods, calamites, and seed ferns) that dominated late Paleozoic wetland communities were established by Mississippian time.

Although Early Mississippian anthracite has been mined locally in the Appalachians of Virginia (Bartholomew and Brown, 1992), the oldest seams of major economic significance are in the Early Pennsylvanian Pocahontas Formation in the Appalachian foreland basin of southwestern Virginia. The Pocahontas Formation hosts a strategic reserve of metallurgical coal and is geologically the oldest formation in North America that has been developed for coalbed methane (Nolde and Spears, 1998; Milici et al., 2010). Pennsylvanian-age bituminous coal seams are productive in several eastern North American basins, including the Black Warrior and Arkoma foreland basins and the Cherokee and Illinois cratonic basins (Figure 3.1). Pennsylvanian coal-bearing strata were deposited near the paleoequator as the supercontinent Pangaea was assembled. Glaciation in the Southern Hemisphere drove high-frequency, high-magnitude changes of global sea level. This tectonic and climatic activity resulted in the deposition of thick, cyclic stratigraphic successions containing a great diversity of marine, coastal, and terrestrial deposits. A result was the distribution of numerous thin coal seams (thickness of ~1–12 ft) throughout the stratigraphic column (Figure 3.5). Permian through Jurassic strata contain little

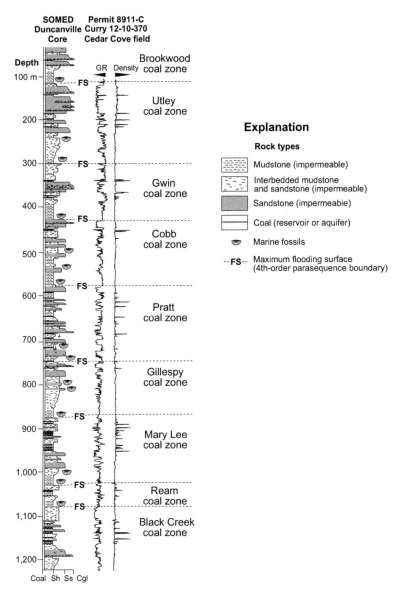

FIGURE 3.5 Stratigraphic column and geophysical well log of the upper Pottsville Formation in the Black Warrior Basin (*After Pashin and Hinkle (1997)*). Alternation of marine and nonmarine strata and distribution of numerous thin coal seams through a thick stratigraphic section is typical of Pennsylvanian-age coal-bearing successions in eastern North America. SOMED stands for the University of Alabama School of Mines and Energy Development.

economic coal in North America, save for a single-seam coalbed methane prospect in the Triassic-age Deep River Basin of North Carolina, which was deposited in a continental rift basin during the breakup of Pangaea (Robbins et al., 1988).

Cretaceous-age bituminous coal in the western interior of North America accounts for the majority of U.S. and Canadian coalbed methane production. Indeed, the prolific reservoirs of the San Juan Basin in Colorado and New Mexico account for more than 40% of U.S. coalbed gas production. Other basins with significant production from Cretaceous strata include the Raton, Uinta, and Alberta basins (Figure 3.1). The depositional setting of the Cretaceous coal-bearing strata in the western interior contrasts strongly with those deposited during the Pennsylvanian. By the Cretaceous, Pangaea had broken up, North America had drifted into the temperate zone, and wetlands were dominated by angiosperms (flowering plants). The Late Cretaceous marked maximum greenhouse conditions during which no ice sheets existed to drive high-frequency, high-magnitude sea-level change (Figure 3.4). Sedimentation took place along the western margin of a broad, subtropical continental seaway that was bounded on the west by the Sevier orogenic belt. The result was the deposition of thick coal seams (thickness of ~5–30 ft) behind wave-dominated shorelines (e.g., Fassett and Hinds, 1971; Ryer, 1981; Ambrose and Ayers, 2007) (Figure 3.6). Minor episodes of sea-level rise are thought to have caused vertical stacking of shoreline sandstone units, which was accompanied by the formation of extensive backshore wetland complexes. Whereas Pennsylvanian coalbed methane reservoirs are commonly distributed through thousands of feet of section, Cretaceous coalbed methane reservoirs tend to be concentrated in isolated coal zones less than 500 feet thick. Where multiple coal zones are present, only one zone tends to have the appropriate depth and reservoir quality to be commercialized in a given area.

Tertiary-age (Paleocene–Eocene) coal-bearing strata are widespread in the Powder River and Williston Basins of the western interior and in the Gulf of Mexico Basin. These basins contain giant resources of lignite and subbituminous coal. Coalbed methane development has been highly successful in the Powder River Basin of Wyoming, which currently accounts for about 25% of U.S. coalbed methane production. The Powder River is a large intermontaine basin that formed during the Laramide orogeny. The basin was initially occupied by a large, mud-floored lake, which is represented by the Lebo Shale Member of the Fort Union Formation (Figure 3.7). Infilling of the lake set the stage for accumulation of the exceptional coal seams of the Tongue River Member of the Fort Union Formation, which are locally thicker than 300 ft. One school of thought holds that the Tongue River coal seams accumulated in response to deltaic infilling of the basin (Ayers and Kaiser, 1984; Ayers, 2002), whereas another school favors the development of widespread peat swamps in the interfluves of a tributary system that drained the former lake bed (Flores, 1981, 1986; Flores and Bader, 1999). Regardless of origin, the heterogeneous assemblage of shale, sandstone, and coal bodies that constitute the Tongue River Member

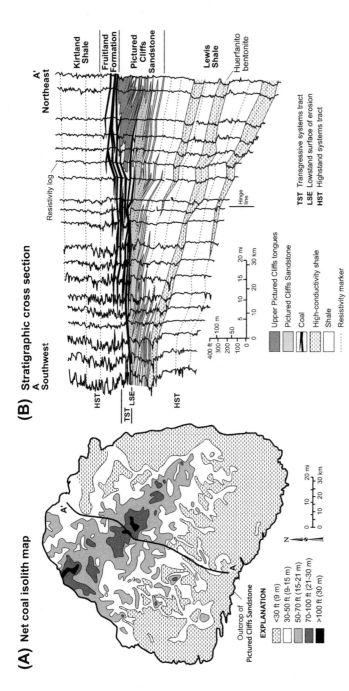

FIGURE 3.6 Depositional framework of the Fruitland Formation and associated strata in the San Juan Basin showing common characteristics of Cretaceous coal-bearing strata in the Cordilleran region (*After Ayers and Kaiser (1994)*). (A) Net coal isolith map showing development of thick coal seams in a northwest–southeast belt in the central part of the basin. (B) Stratigraphic cross section showing development of thick coal behind stacked shoreline deposits of the Pictured Cliffs Sandstone.

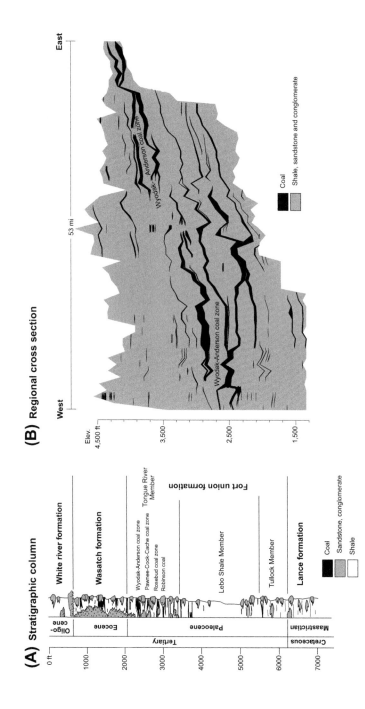

FIGURE 3.7 Depositional framework of the Fort Union Formation and associated strata in the Powder River Basin in Wyoming (*After Flores* (*2004*)). (A) Stratigraphic section of Upper Cretaceous and Tertiary strata. (B) Stratigraphic cross section through the central part of the basin showing complex geometry of exceptionally thick coal seams in the Tongue River Member of the Fort Union Formation.

underscores the variable thickness and continuity of reservoir facies must be considered when assessing and developing coalbed methane resources.

3.2.2 Structural Geology

The structural framework of a sedimentary basin begins forming at the time of sedimentation and continues evolving throughout the regional burial and unroofing history. Folding and faulting modify the original depositional framework and are fundamental controls on the shape and continuity of coalbed methane reservoirs. Natural fracture networks, moreover, are a major part of the natural plumbing that facilitates commercial flow rates in coal and determines the degree to which coal seams and the intervening strata are hydraulically confined (e.g., Laubach et al., 1998; Pashin et al., 2004).

North American sedimentary basins have a diverse tectonic history, and the coal-bearing strata contained therein are sensitive recorders of the dynamic compressional and extensional stresses associated with basin formation. Compressional structures in coalbed methane reservoirs include a variety of folds and thrust faults that formed during the development of the Appalachian–Ouachita orogenic belt in eastern North America and the Cordilleran orogenic belt in the west (e.g., Pashin and Groshong, 1998; Riese et al., 2005). Large-scale folds bring reservoir coal seams to the surface at basin margins and thus have a major impact on basin hydrodynamics, which will be discussed later in this chapter. Basin interiors vary considerably in structural complexity. Some basins, like most of the Alberta Basin, are structurally simple, consisting of gently dipping strata with little internal deformation. In basins containing small-scale folds, productivity sweet spots are in places associated with deformation in structural hinges (Pashin and Groshong, 1998; Pashin, 2009) (Figure 3.8). Extensional faults in the interiors of sedimentary basins, by contrast, form horst and graben and half-graben systems that partition coalbed methane reservoirs into structural blocks with different production characteristics.

Fracture networks are a central source of heterogeneity in coal-bearing strata and include cleat systems in coal, joint systems in the enveloping shale, sandstone, and carbonate beds, and shear fractures associated with fault zones (Figure 3.9). Cleat and joint systems are orthogonal fracture systems composed of systematic fractures and cross fractures. These types of fractures exhibit no displacement parallel to the fracture plane; they form under normal stress and are commonly referred to as opening-mode fractures. Systematic fractures propagate parallel to the maximum horizontal stress; they tend to be planar and have great lateral continuity. Cross fractures, by contrast, are connective structures that commonly curve, have irregular surfaces, and terminate at systematic fractures. In coal, the systematic fractures are called face cleats, whereas the cross fractures are

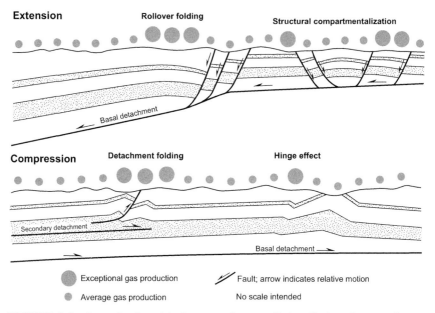

FIGURE 3.8   Generalized model of structurally controlled coalbed methane production in the Black Warrior Basin. *After Pashin et al. (1995).*

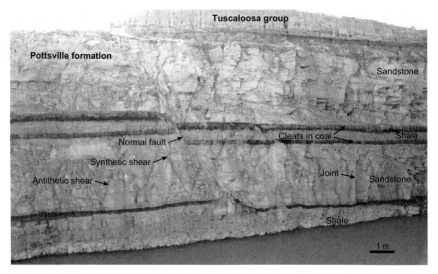

FIGURE 3.9   Mine highwalls in the Brookwood coal zone of the Black Warrior Basin showing the complexity of fracture architecture in coal-bearing strata (*After Pashin et al. (2004)*). Coal seams contain cleat systems, shale and sandstone contain vertical joints with different spacing in each bed, and crossing shear fractures are developed in sandstone adjacent to a listric normal fault that flattens in the shale near the base of the highwall.

called butt cleats (Figure 3.10). Systematic fractures commonly maintain consistent orientation across large parts of sedimentary basins and even continents (Engelder and Whitaker, 2006). Localized orthogonal fracture networks associated with folds and faults are commonly superimposed on the regional networks.

Cleat systems are strata-bound fractures; that is, they are restricted to the host bed. They commonly exhibit a geometric hierarchy, with primary or master cleats cutting complete beds or benches of coal, secondary cleats cutting part of a bed or bench, and tertiary cleats being isolated in vitrain bands (i.e., the bright, glossy bands in coal) (Laubach et al., 1998) (Figure 3.10). Cleat spacing typically ranges from a decimeter to less than a millimeter and generally decreases as the thermal maturity of the coal increases (Ting, 1977). This dependence of cleat spacing on thermal maturity suggests that internal stresses associated with the devolatization and shrinkage of organic matter and changes of pore pressure during burial, heating, hydrocarbon generation, and explusion is an important mechanism of cleat formation.

Joints in shale, sandstone, and carbonate resemble cleats geometrically, but spacing tends to be greater by least an order of magnitude (1–10m). Whereas devolatization and matrix shrinkage are important cleating mechanisms, tectonic stress and pore pressure are important jointing mechanisms in siliciclastic and carbonate rocks (Pollard and Aydin, 1988). Different rock types have different mechanical properties, and so

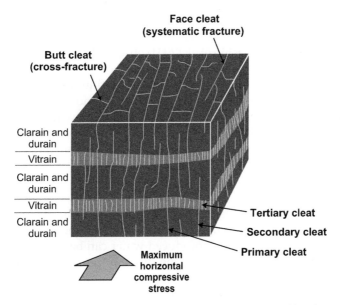

FIGURE 3.10 Block diagram showing relationship of cleating to coal banding and maximum horizontal compressive stress at time of fracturing.

interlayering of those rock types is a source of heterogeneity that facilitates the development of strata-bound joint networks, as well as transstratal structural elements. Accordingly, joints in brittle sandstone and carbonate beds tend to be spaced more closely than those in more ductile shale beds. In addition, joint spacing in a given rock type tends to increase with bed thickness. The vertical stacking of strata-bound and transstratal fracture networks with variable spacing and orientation is an important source of reservoir heterogeneity in coal-bearing strata that is difficult to detect in the subsurface but can have a strong effect on the performance of coalbed methane reservoirs (Pashin et al., 2004).

Faults constitute fundamental reservoir discontinuities, especially where they cut across bedding (Figures 3.8 and 3.10). Whereas joints and cleats are opening-mode fractures, dislocation of strata along fault panes and many of the associated fractures indicates formation under shear stress. Faults are classified on the basis of geometry and relative displacement. Common types of fault systems in coal-bearing strata include normal faults, high-angle reverse faults, low-angle thrust faults, oblique-slip faults, and strike-slip faults. Fault zones can be complex zones of deformation, with intensely deformed gouge distributed along the fault plane, and swarms of dipping shear fractures extending from the gouge zone into the formation (Pashin, 1998). Shear fractures dipping with the fault are called synthetic shears, whereas those dipping opposite the fault are called antithetic shears.

Faults can act as reservoir seals or as conduits for cross-formational flow depending on how sealing and transmissive rock types are juxtaposed by the fault, as well as the internal architecture and degree of mineral cementation in the fault zone. Pashin et al. (2004) observed that wells intersecting faults in the Black Warrior Basin tend to produce little water or gas. This apparently owes to several factors. First, discontinuities may be present that limit the drainage area of nearby wells. Second, hydraulic fracturing may be ineffective along transmissive fault zones, which can intercept completion fluid. Finally, fault zones can be cemented with calcite and other minerals, which occlude fracture porosity in coal and adjacent strata.

3.2.3 Hydrology

Water production is the principal mechanism by which coal is depressurized, and so understanding the hydrology of coalbed methane reservoirs and the ways in which coproduced water can be managed is every bit as important as understanding the distribution and producibility of gas in coal. Coalbed methane, moreover, is commonly referred to as a hydrodynamic gas play because basin hydrology is a major determinant of reservoir pressure, water chemistry, and gas chemistry (e.g., Pashin et al., 1991; Ayers and Kaiser, 1994; Scott, 2002; Pashin, 2007). This section reviews permeability, water chemistry, and reservoir pressure and presents basic hydrodynamic models of North American coalbed methane reservoirs.

3.2.3.1 Porosity and Permeability

Coal is a stress-sensitive rock type in comparison to siliciclastic and carbonate rocks. The permeability of coal decreases exponentially with depth as overburden stress compresses the coal and closes the cleat system (McKee et al., 1988). Near the surface, coal can have Darcy-class permeability and total porosity exceeding 5% (Gan et al., 1975). In the Powder River Basin, for example, the permeability of many shallow coalbed methane reservoirs is greater than 3 darcys. At depths of 2000 feet or greater, by contrast, many coal seams have permeability on the order of 1 mD and porosity lower than 0.1%. The Black Warrior Basin is an excellent example where a strong permeability-depth gradient influences reservoir performance (McKee et al., 1988; Pashin, 2010). It should be noted, however, that local variation of permeability is a major source of reservoir heterogeneity in coal. Accordingly, permeability within a given coal seam commonly varies by more than an order of magnitude at a given depth. And given favorable tectonic stress conditions, as in the San Juan Basin, some coal seams support hundreds of mD of permeability at depths beyond 3000 ft (Ayers and Kaiser, 1994; Riese et al., 2005). Geologic structure can have a strong effect on permeability. In the Frying Pan Anticline of the Appalachian Basin in southwestern Virginia, the highest ultimate recovery values for vertical wells are in shallow, permeable coal at the crest of the structure (Figure 3.11). Here, positive structural curvature helps maintain open cleats that are filled with free gas. Low recovery values along the flanks reflect low permeability that appears to be related to increased overburden stress, weak structural curvature, and water-saturated cleat systems.

Permeability in coalbed methane reservoirs is a transient reservoir property. As gas is produced, coal matrix shrinks, thereby widening cleat apertures and improving permeability (Harpalani and Chen, 1995). In some cases, matrix shrinkage may result in order-of-magnitude increases of permeability over the life of a well. However, geomechanical models indicate that the evolution of permeability through time can be highly variable, reflecting a delicate interplay among matrix shrinkage, overburden stress, tectonic stress, Young's modulus of elasticity, and Poisson's ratio (Palmer and Mansoori, 1998).

3.2.3.2 Water Chemistry

Water that is produced from coalbed methane reservoirs ranges from nearly potable freshwater to hypersaline brine (Van Voast, 2003). Disposing of that water in an environmentally responsible manner is a central concern in coalbed methane development. This is especially true in

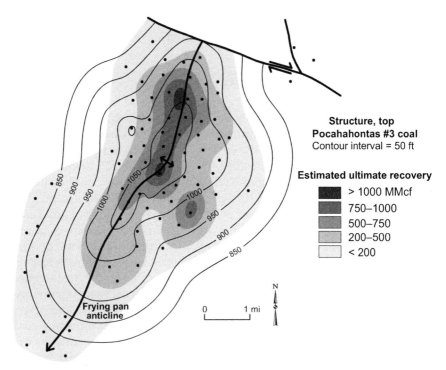

FIGURE 3.11 Structural contour map of the Frying Pan anticline in the Appalachian Basin of southwestern Virginia showing relationship of estimated ultimate recovery of vertical coalbed methane wells to geologic structure *(Source: Marshall Miller and Associates).* Low stress in shallow reservoirs is interpreted to favor gas recovery in the crestal region of the anticline.

the Powder River and Black Warrior basins, where subsurface disposal options are limited, and so the coalbed methane industry is dependent on stream discharge. Produced waters from coalbed methane reservoirs are dominated by sodium bicarbonate and sodium chloride water types. Bicarbonate waters in commercial coalbed methane reservoirs typically have total dissolved solids (TDS) content between 300 and 3000 mg/L, whereas chloride waters commonly have TDS content between 3000 and 60,000 mg/L.

Low TDS bicarbonate water forms freshwater plumes that are fed by meteoric recharge along basin margins (Figure 3.12). The best known plumes of this type are fed along the structurally upturned margins of the Black Warrior and San Juan basins, where reservoir coal seams come to the surface (Ayers and Kaiser, 1994; Pashin, 2007). An exceptional recharge system exists in the Uinta Basin, where freshwater percolates downward along fault zones into coal seams (Stark and Cook, 2009). In the shallow subsurface, calcium magnesium sulfate and bicarbonate water passes into

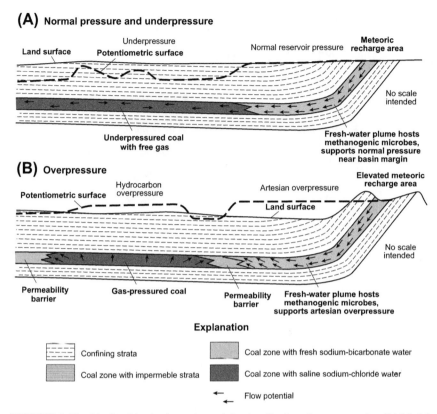

FIGURE 3.12 Idealized hydrodynamic models of coalbed methane reservoirs. (A) Model for normally pressured and underpressured reservoirs. (B) Model for overpressured reservoirs. *After Pashin (2008).*

sodium bicarbonate water, and the interface between the sulfate and bicarbonate waters is an important zone of microbial methanogenesis. Cation exchange with clay minerals and carbonate precipitation cause formation water to be depleted in calcium and magnesium and enriched in sodium, and microbial processes favor reduction of sulfate into sulfide and enrichment of formation water with bicarbonate (Lee, 1981; Van Voast, 2003).

Microbial methanogenesis also is thought to occur in the interiors of some basins, where organic compounds in coal and formation water appear to support microbial communities, even where formation water is saline. In the Powder River Basin, for example, coal is of insufficient thermal maturity to have generated thermogenic hydrocarbons, and geochemical data indicate that all produced gas is effectively of biogenic origin (Flores, 2008). Methanogenesis occurs along two major metabolic pathways: methyl-type fermentation and CO_2 reduction (Figure 3.13). These pathways can be distinguished by analyzing carbon and deuterium

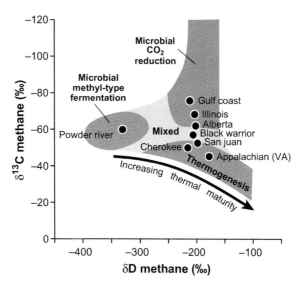

FIGURE 3.13 Cross plot showing generalized stable isotopic composition of coalbed methane from selected North American basins. *Diagram after Whiticar (1994); analyses compiled from numerous sources.*

isotopes in methane (Whiticar, 1994, 1996). Most coalbed methane reservoirs plot along a corridor between thermogenic gas generation with microbial CO_2 reduction. A prominent exception may be in the shallow reaches of the Powder River Basin, where low-rank coal may support methyl-type fermentation (Flores, 2008). However, questions about isotopic fractionation processes in coal and formation water has made this interpretation uncertain (Vinson et al., 2012). The metabolic pathway can have an influence on impurities associated with coalbed gas. For example, methyl-type fermentation can generate significant quantities of CO_2 along with methane, whereas CO_2 is consumed as methane is generated along the CO_2 reduction pathway.

3.2.3.3 *Reservoir Pressure*

The pressure regime in coalbed methane reservoirs is yet another source of heterogeneity that must be accounted for during exploration and development (Figure 3.14). Pressure in sedimentary basins has two major components: lithostatic pressure, which is the pressure caused by the weight of the overburden (gradient ~1.0 psi/ft) and hydrostatic pressure, which is an opposing pressure caused by reservoir fluid (i.e., pore pressure). Many reservoirs have normal hydrostatic pressure gradients of about 0.43 psi/ft, under which the water level in the coalbed methane reservoir will rise to the land surface. In the Black Warrior Basin, for example, normal reservoir pressure is associated with freshwater plumes, which are supported by a

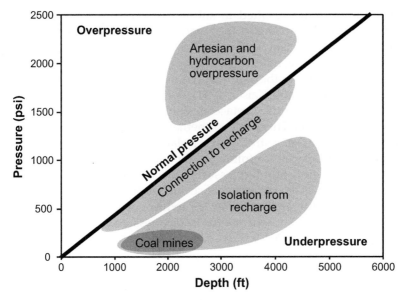

FIGURE 3.14   Pressure–depth diagram showing common pressure regimes in North American coalbed methane reservoirs.

structurally upturned recharge area with a similar elevation to the interior of the basin (Pashin and McIntyre, 2003) (Figure 3.12).

Abnormal reservoir pressure includes overpressure, in which water will rise above the land surface, and underpressure, in which reservoir pressure is insufficient for water to rise to the surface (Figures 3.12 and 3.14). Overpressure takes two basic forms: artesian overpressure and hydrocarbon overpressure. Artesian overpressure is driven by a recharge area that is elevated above the basin interior. A famous example of artesian overpressure is in the San Juan Basin, where an elevated recharge area in the Hogback monocline contributes to overpressure in an elongate fairway, which is the most productive coalbed methane trend in the world (Kaiser et al., 1994). Hydrocarbon overpressure is typically developed in geologically young strata that are enveloped by a low-permeability barrier that prevents hydrocarbons from migrating. An example of hydrocarbon overpressure is in the Cameo coal zone of the Piceance Basin, where large volumes of thermogenic hydrocarbons have been generated in a low-permeability setting near the center of the basin (Scott, 2002).

Underpressure results from numerous processes ranging from isolation from recharge, thermal contraction and fracturing of rocks in geologically old sedimentary basins, as well as anthropogenic factors, such as the fluid withdrawals associated with wells and coal mines (Kaiser, 1993). Zones of mine-related pressure depletion have been documented in the Appalachian region. In the Black Warrior Basin, for example, deep longwall mining has resulted in substantial pressure sinks within the otherwise normally pressured freshwater plumes (Pashin and McIntyre, 2003; Pashin, 2007). Basinward of the freshwater plumes is a large area of natural underpressure, reflecting isolation from recharge. A similar area of underpressure exists in the San Juan Basin south of the region of artesian overpressure (Ayers and Kaiser, 1994). A remarkable example of under-pressuring is found in the Cretaceous-age Horseshoe Canyon coal zone of the Alberta Basin. This coalbed methane play is developed within a vast expanse of flat-lying strata that is isolated from recharge, and melting of about 3 km of glacial ice following the Pleistocene is thought to have contributed to regional underpressuring (Bachu and Michael, 2003). Low water production is characteristic of severely underpressured coal seams (Pashin, 2007, 2010), and no water is produced from a large part of the Horseshoe Canyon play, where hydrostatic pressure gradients are as low as 0.04 psi/ft (Gentzis, 2010).

3.2.4 Geothermics

Burial and unroofing during the development of a sedimentary basin results in significant changes of reservoir temperature that affect the generation, expulsion, and retention of hydrocarbons (Waples, 1980). These changes have a strong impact on the ability of coal to generate and retain commercial quantities of gas. Generation of large volumes of thermogenic gas is thought to commence as organic matter reaches a temperature of about 100 °C. In coal, the lower boundary of the thermogenic gas window corresponds approximately with a vitrinite reflectance (R_o) of 0.8% (Jüntgen and Klein, 1975). Closed system pyrolysis experiments suggest that coal can generate 4 to 10 times more methane than can be retained in coal matrix (Zhang et al., 2008). The excess gas will be expelled through adjacent strata if coal is poorly confined, will migrate along bedding if coal is well confined, and will contribute to overpressuring and cleating if the coal is confined and has low permeability.

The adsorption capacity of coal is dependent not only on pressure, but on temperature as well. Indeed, if other factors are held constant, the adsorption capacity of coal increases substantially as reservoir temperature decreases (Yang and Saunders, 1985; Zhou et al., 2000). Coal is a relatively weak sorbent at the temperatures associated with thermogenic hydrocarbon generation. Hence, basins that have been cooled substantially during episodes of regional uplift and erosion can be expected to contain coal that is undersaturated with gas unless gas migrates in from another source, is resorbed from fractures, or is formed by late-stage microbial methanogenesis (Scott et al., 1994; Scott, 2002).

A recent modeling study reveals the complexity of the relationship among reservoir temperature, gas saturation, and regional unroofing

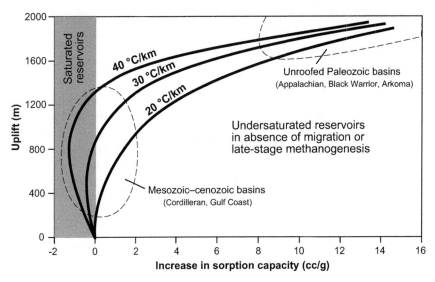

FIGURE 3.15 Relationship among uplift, geothermal gradient, adsorption capacity, and gas saturation in North American coalbed methane reservoirs. *After Bustin and Bustin (2008).*

(Bustin and Bustin, 2008). Results indicate that, in basins where geothermal gradient is lower than 20 °C/km, regional uplift results in significant undersaturation early in the unroofing history (Fig. 3.15). In basins with geothermal gradient higher than 30 °C/km, early uplift can actually decrease adsorption capacity, which favors the maintenance of gas-saturated reservoir conditions. Regardless of geothermal gradient, basins from which more than 1.5 km of overburden has been eroded are predicted to be undersaturated with thermogenic gas in the absence of migration from other areas. Many Paleozoic coal basins, such as the Black Warrior, Arkoma, and Appalachian basins, have had more than 2 km of overburden removed and therefore are expected to be significantly undersaturated. By contrast, geologically young coal basins, such as the Mesozoic basins of the Cordilleran region, have not been unroofed to the extent of the Paleozoic basins. Hence, the parts of these basins that have entered the thermogenic gas window are predicted to have gas-saturated or nearly saturated reservoir conditions.

3.2.5 Coal Quality

Coal quality encompasses a range of type, grade, and rank parameters. These parameters describe composition, purity, and thermal maturity, respectively, and govern the composition and quantity of the gases that may be generated, retained, and transmitted by coal. Coal can be defined simply as any rock containing more than 50% organic matter by weight.

Other organic-rich rock types, such as black shale, can contribute to coalbed gas production, but the vast majority of the gas, whether of thermogenic or biogenic origin, comes from true coal.

3.2.5.1 Type

In terms of type, North American coal seams are predominantly bright banded and are composed mainly of vitrain and clarain (bright bands) with lesser amounts of durain and fusain (dull bands). Macerals are the basic organic constituents of coal and are divided into three major groups: vitrinite, inertinite, and liptinite. Vitrinite macerals, which are the vitrified remains of woody material, typically constitute 60–90% by weight of the coal seams in North America that are productive of natural gas. Vitrinite macerals appear gray in reflected light. The remaining fraction of the coal is dominated by intertinite macerals, which are the oxidized plant and fungal remains that form by processes ranging from atmospheric exposure to catastrophic swamp fires. Inertinite macerals appear white in reflected light. Liptinite macerals, which are derived from the waxy organic matter derived from leaves, cuticles, palynomorphs, and algae, typically form less than 5% of North American coal seams. These macerals are distinctly fluorescent and appear dark gray in reflected light.

Maceral types differ in the capacity to generate, store, and transmit hydrocarbons. The major maceral groups are analogous to kerogen, which is the dispersed organic matter in sedimentary rocks that is insoluble in aqueous alkaline solvents and common organic solvents. Kerogen and coal macerals are classified as types I through IV organic matter based on chemical composition, specifically the proportions of carbon, hydrogen, and oxygen (van Krevelen, 1961; Cornelius, 1978) (Figure 3.16). Type I organic matter is rich in hydrogen and is highly oil prone. Alginite, which is a liptinite maceral derived from algae, is the signature Type I maceral. Alginite is not a major constituent of North American coal seams but is a very important constituent of many Australian seams, which are proving to be important coalbed methane reservoirs. Type II organic matter has a lower hydrogen–carbon ratio than Type I and includes most of the liptinite macerals, such as sporinite and cutinite, which are derived from waxy plant parts, such as palynomorphs, leaves, and cuticles. Type II organic matter is oil prone but yields less oil than Type I organic matter. Type III organic matter is derived from humic material, including wood, and is represented by the macerals of the vitrinite group. Type III organic matter has a hydrogen–carbon ratio lower than 1.00 and a broad range of oxygen–carbon ratios that decrease as coal progresses from lignite through anthracite rank. Vitrinite group macerals are typically gas prone and also store large volumes of natural gas (Mastalerz et al., 2004). Perhydrous vitrinite, which contains more than 5.5% hydrogen by weight and is common in Mesozoic–Cenozoic coal may be oil prone (Bertrand, 1984; Smith and

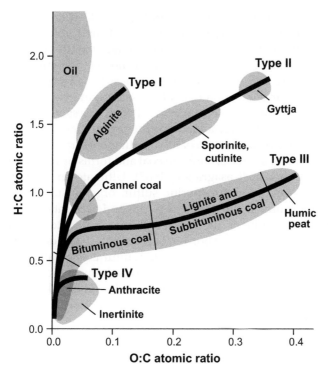

FIGURE 3.16 Van Krevelen diagram showing hydrogen–carbon and oxygen–carbon ratios of organic matter in coal. *Modified from Cornelius (1978).*

Cook, 1984), although the oil-generative potential of coal is controversial (Wilkins and George, 2002). Type IV organic matter, which includes the inertinite macerals, has low hydrogen–carbon and oxygen–carbon ratios compared to the other types of organic matter. Type IV organic matter has no potential to generate hydrocarbons but does have sufficient porosity and internal surface area to store significant volumes of gas (Crosdale et al., 1998).

3.2.5.2 Grade

Whereas type refers broadly to coal composition, grade refers to purity. The grade parameters that are most relevant to coalbed methane reservoirs are ash content, sulfur content, and mineral matter content (Spears and Caswell, 1986; Spears, 1987; Faraj et al., 1996). Ash is the residue that remains after coal is combusted, and the weight% is determined to provide an approximation of the inorganic content of the coal. Mineral matter content is also expressed in terms of weight% and can be estimated from ash content using the Parr formula, which accounts for the moisture lost from minerals, such as clay, during combustion. A diverse suite of minerals are common in coal and typically include quartz, feldspar, various clay minerals, carbonate minerals, and pyrite. Detrital mineral matter deposited with peat during events like floods and volcanic ash falls is thus part of the coal matrix. Cleat-filling mineral matter is also common and includes an array of carbonate, silicate, and sulfide minerals. Mineral matter dilutes organic matter and thus reduces the adsorption capacity of coal. For reservoir evaluation of coals, assessing the volumetric percentage of mineral matter is more important than assessing weight%. Scott et al. (1995) proposed basic methodology for evaluating the impact of mineral matter and ash content on coalbed methane reserves.

Pyrite is a major source of sulfur in coal and forms principally by microbial sulfate reduction (Casagrande, 1987). Ultimate analyses commonly report three forms of sulfur: pyritic sulfur, organic sulfur, and sulfatic sulfur. Pyritic sulfur is the sulfur in pyrite and accounts for most sulfur in coal. Organic sulfur is the fraction that is bound to organic matter, whereas sulfatic sulfur is the portion of the sulfur that is derived by oxidation and weathering. Accordingly, pyritic sulfur is of greatest relevance to the evaluation of coalbed methane reservoirs. Indeed, the high density of pyrite relative to the other minerals in coal should be taken into account when estimating mineral matter volume, particularly in coal with high sulfur content.

3.2.5.3 Rank

Thermal maturity affects numerous fundamental coal properties, including gas capacity and geomechanical properties, thus coal rank is an extremely important coal quality parameter for the evaluation of coalbed methane reservoirs. Vitrinite reflectance, volatile matter (dry, ash free), and moisture (ash free) are the most commonly used measures of coal rank in the coalbed methane industry. Rank parameters track the coalification process from lignite through metaanthracite. Moreover, the type and quantity of hydrocarbons generated and expelled from coal are closely related to thermal maturity and can be gauged by determining coal rank.

The transformation of peat to lignite is complete when moisture falls below 75%, carbon content (dry, ash free) is greater than ~60%, and free cellulose is absent. Lignin and the remaining cellulosic substances are transformed to humic compounds as coal progresses through lignite and subbituminous rank. Below bituminous rank, the principal volatile compounds generated are water and carbon dioxide, methane generation is dominantly biogenic, and the vast majority of the gas is expelled by compaction (Levine, 1993). Oil, furthermore, can begin to be generated from liptinite at lignite rank ($R_o \sim 0.35$) (Paterson et al., 1997).

Devolatization emerges as the principal factor driving volumetric changes as coal approaches bituminous rank. Moisture decreases from 15 to 1% as coal progresses from high volatile C bituminous rank to medium volatile bituminous rank. The lower boundary of the thermogenic oil window corresponds with the subbituminous–bituminous transition ($R_o \sim 0.50$), and peak oil generation corresponds with high volatile B and A bituminous rank ($R_o \sim 0.70$–1.10) (Hunt, 1979). Most oil generation occurs at temperatures between 50 and 150 °C, and large volumes of carbon dioxide also can be generated. Major thermogenic generation of gaseous hydrocarbons apparently begins at the boundary between high volatile B and A bituminous rank ($R_o \sim 0.80$) (Jüntgen and Klein, 1975). Most methane is generated between temperatures of 100 and 225 °C, and significant volumes of nitrogen also can be generated. Thermal cracking of hydrocarbons is an important process in the thermogenic gas window, and hydrocarbon generation is thought to be effectively complete at anthracite rank ($R_o > 3.00$).

As mentioned previously, coal has the capacity to generate multiple times more methane than can be held in matrix and is thus an important source rock for natural gas. Although coal contains significant quantities of oil-prone organic matter, the capacity of bright-banded coal to expel oil is uncertain (Wilkins and George, 2002). Indeed, the aromatic network of coal has capacity to store significant amounts of oil, and it is possible that much of this oil is cracked to gas as it is expelled. Indeed, vitrinite commonly fluoresces and has suppressed reflectance within the oil generation window, indicating retention of oil in coal matrix. This oil apparently occludes gas storage, thereby adversely affecting the ability of coal to store and transmit natural gas. This phenomenon is particularly common in Cretaceous and Tertiary coal seams, and in the southern part of the San Juan Basin, coal seams within the oil window typically exhibit poor reservoir performance (Clayton et al., 1991; Meek and Levine, 2006). By contrast, production of gas from coal in the oil window has proven successful in the underpressured Horseshoe Canyon play of western Canada (Gentzis, 2010).

Cleating of coal is commonly rank dependent, reflecting progressive devolatization as well as changing mechanical properties during thermal maturation (Ting, 1977; Law, 1993). Lignite and subbituminous coal are commonly weakly cleated, whereas cleat spacing in medium volatile bituminous and low volatile bituminous is typically on the order of centimeters or millimeters. Pashin et al. (1999), for example, observed that reservoir quality cleat systems in the Black Warrior Basin are restricted primarily to coal within the thermogenic gas window, suggesting that gas generation was an important cleating mechanism. In North American basins, anthracite generally lacks cleats, has conchoidal fracture, and appears glassy, suggesting that cleats have annealed. By contrast, some Chinese anthracite is finely cleated (Su et al., 2005), establishing that cleat systems can be preserved at elevated rank.

Coal rank also has a strong influence on the mechanical properties of coal. The strength and brittleness of coal is dependent on rank and coal

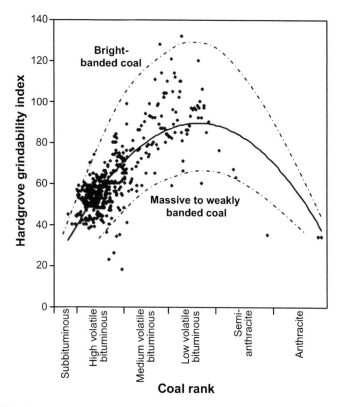

FIGURE 3.17   Relationship of Hardgrove grindability index to coal rank and coal type. *After Esterle (2008).*

composition, and the Hardgrove grindability index a commonly employed proxy for these properties (Hower and Wild, 1988; Chelgani et al., 2008). In general, the grindability index of coal is greatest in medium and low volatile bituminous rank (Esterle, 2008) (Figure 3.17). Accordingly, hydrofracturing appears to be most effective in coal seams with high grindability index. In the San Juan Basin, moreover, cavity completions have proven most effective in medium volatile bituminous to low volatile bituminous coal (Young et al., 1994).

3.2.6 Gas Storage

Gas may be stored in coal in adsorbed and free states, depending on pressure–temperature conditions and gas saturation. Considering the low porosity of coal under typical reservoir conditions, the vast majority of the gas is stored in an adsorbed state. Adsorption capacity is characterized by the Langmuir isotherm, which depicts how gas capacity changes

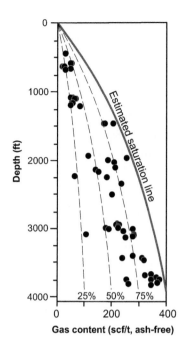

FIGURE 3.18 Plot of gas content vs depth showing major variability of gas content and estimated gas saturation in a single core from the Black Warrior Basin. *After Levine and Telle (1989).*

with pressure at constant temperature (Figure 3.2). Langmuir parameters, specifically Langmuir volume and Langmuir pressure, are important for characterizing total gas capacity and the shape of adsorption isotherms. Langmuir volume is the total adsorption capacity of a substance at infinite pressure, whereas Langmuir pressure is the pressure at which storage capacity equals one half of Langmuir volume. Langmuir volume describes the height of the isotherm, whereas Langmuir pressure provides important information on shape. All adsorption isotherms have steep slope near the origin that flattens as pressure increases. Isotherms with low values of Langmuir pressure tend to be flat at elevated pressure, whereas those with high values tend to maintain slope at elevated pressure.

The adsorption capacity of coal is influenced by a number of factors, including maceral composition, ash content, and rank. Gas capacity in bright-banded coal seams is influenced mainly by the vitrinite and intertinite group macerals (Crosdale et al., 1998; Mastalerz et al., 2004). Minerals dilute the organic matter in coal, thus gas capacity correlates negatively with mineral matter and ash content. Rank parameters, including moisture, volatile matter, and vitrinite reflectance correlate strongly with adsorption capacity (Joubert et al., 1973, 1974; Kim, 1977). As a rule, adsorption capacity increases with coal rank.

Gas content and saturation can vary considerably among coal seams, and most seams are undersaturated (Figure 3.18). The critical desorption pressure is the maximum pressure at which coal matrix can be saturated with gas (Figure 3.2). Above this pressure, coal is undersaturated and stored almost exclusively in an adsorbed state. Below this pressure, coal becomes supersaturated with gas, which will therefore desorb from matrix and flow into open pores and cleats. Free storage in saturated coal has been identified in extremely underpressured reservoirs in the Alberta and Black Warrior basins (Gentzis, 2010; Pashin and McIntyre, 2003). In the Powder River Basin, significant accumulations of free gas have been identified in the crestal regions of anticlines, indicating a conventional component to hydrocarbon trapping (Ayers, 2002), and similar relationships have been observed in anticlinal fields in the Appalachian Basin (e.g., Figure 3.11).

Knowing the shape of an isotherm is important for analyzing the producibility of coalbed gas (Figure 3.2). At elevated pressure, where the slope of the isotherm is low, a modest undersaturation of only 100 scf/t may require a major lowering of reservoir pressure to reach the critical desorption pressure. At low pressure, where the slope of the isotherm is steep, by comparison, a similar undersaturation may require only a modest lowering of reservoir pressure for coal to reach the saturation point. Accordingly, successful recovery of a large coalbed gas resource requires a lowering of reservoir pressure into the steep part of the adsorption isotherm. Indeed, successful coalbed methane production can be favored in parts of basins where native reservoir conditions are in the steep region of the isotherm, such as in the northeastern part of the Black Warrior Basin (Pashin, 2010). Highly pressured reservoirs can necessitate selective completion of coal seams with high gas saturation, and a large quantity of water may need to be produced to lower a significant reservoir volume below the critical desorption pressure. In this situation, deep coal seams with low permeability are particularly problematic. Alternatively, where permeability is high and hydrodynamic conditions are favorable, like in the exceptionally productive fairway of the San Juan Basin, even highly overpressured reservoirs can be depressurized rapidly and brought into the steep region of the isotherm (Meek and Levine, 2006).

3.3 SUMMARY AND CONCLUSIONS

North American coalbed methane reservoirs are geologically diverse, with each productive basin exhibiting a distinctive mix of depositional, tectonic, hydrodynamic, petrologic, and petrophysical characteristics. Coalbed methane resources are concentrated in Pennsylvanian, Cretaceous, and Tertiary strata in which reservoir architecture reflects a nexus of the sedimentologic, tectonic, climatic, and biological factors that prevailed during each time period. Folds, faults, and fractures record the tectonic processes that occurred during burial and subsequent exhumation of the host basins and are fundamental controls on the geometry, continuity, and transmissivity of coal.

Coal seams can be considered as hydrodynamic reservoirs that respond to myriad geologic processes. Porosity and permeability in coal are highly stress-sensitive, and so reservoir properties vary considerably among basins depending on overburden stress, tectonic stress, and hydraulic stress. Recharge at basin margins has a strong effect on water chemistry and reservoir pressure. Water chemistry ranges from fresh to hypersaline, which can necessitate varied water management strategies within productive basins. Reservoirs are underpressured, normally pressured, or overpressured depending on basin hydrodynamics and geothermics. The regional hydrodynamic framework and geothermal history, moreover, determine the distribution and intensity of biogenic and thermogenic gas charge in coal.

Most reservoirs are bright-banded coal that is dominated by vitrinite, or gas-prone type III organic matter. Gas capacity is inversely correlated to the volume of mineral matter in coal, which includes clay, quartz, pyrite, and carbonate. Rank of the major coalbed methane reservoirs ranges from subbituminous to low volatile bituminous. Subbituminous to high volatile B bituminous coal is submature with respect to major thermogenic gas generation and thus contains mainly microbial gas, whereas higher rank coal contains thermogenic gas that is in many basins supplemented by late-stage biogenic gas. Importantly, adsorption capacity tends to increase substantially with coal rank. The mechanical properties of coal, which may determine the effectiveness of hydraulic fracture treatments, also are influenced by rank.

References

Ambrose, W.A., Ayers Jr, W.B., 2007. Geologic controls on transgressive-regressive cycles in the upper Pictured Cliffs Sandstone and coal geometry in the lower Fruitland Formation, northern San Juan Basin, New Mexico and Colorado. American Association of Petroleum Geologists Bulletin 91, 1099–1122.

Ayers Jr, W.B., 2002. Coalbed gas systems, resources, and production and a review of contrasting cases from the San Juan and Powder River Basins. American Association of Petroleum Geologists Bulletin 86, 1853–1890.

Ayers Jr, W.B., Kaiser, W.R., 1984. Lacustrine interdeltaic coal in the Fort Union Formation (Paleocene), Powder River Basin, Wyoming and Montana, U.S.A. In: Armani, R.A., Flores, R.M. (Eds.), Sedimentology of Coal and Coal-bearing Sequences: International Association of Sedimentologists Special Publication, vol. 7. pp. 61–84.

Ayers Jr, W.B., Kaiser, W.A. (Eds.), 1994. Coalbed Methane in the Upper Cretaceous Fruitland Formation, San Juan Basin, Colorado and New Mexico, vol. 218. Texas Bureau of Economic Geology Report of Investigations. 216 p.

Bachu, S., Michael, K., 2003. Possible controls of hydrogeological and stress regimes on the producibility of coalbed methane in Upper Cretaceous-Tertiary strata of the Alberta Basin, Canada. American Association of Petroleum Geologists Bulletin 87, 1729–1754.

Bartholomew, M.J., Brown, K.E., 1992. The Valley Coalfield (Mississippian age) in Montgomery and Pulaski Counties, Virginia, vol. 124, Virginia Division of Mineral Resources Publication. 33 p.

Bertrand, G., 1984. Geochemical and petrographic characterization of humic coals considered as possible petroleum source rocks. Organic Geochemistry 6, 481–488.

Bustin, A.M.M., Bustin, R.M., 2008. Coal reservoir saturation: impact of temperature and pressure. American Association of Petroleum Geologists Bulletin 92, 77–86.

Casagrande, D.J., 1987. Sulphur in peat and coal. In: Scott, A.C. (Ed.), Coal and coal-bearing Strata: Recent Advances, vol. 32. Geological Society Special Publication, London, pp. 87–105.

Chelgani, S.C., Hower, J.C., Jorjani, E., Mesroghli, Sh, Bagherieh, A.H., 2008. Prediction of coal grindability based on petrography, proximate and ultimate analysis using multiple regression and artificial neural networks. Fuel Processing Technology 89, 13–20.

Clayton, J.L., Rice, D.D., Michael, G.E., 1991. Oil-generating coals of the San Juan Basin, New Mexico and Colorado, U.S.A. Organic Geochemistry 17, 735–742.

Cornelius, C.D., 1978. Muttergesteinfazies als Parameter der Erdölbildung. Erdördgas Zeitschrift 3, 90–94.

Crosdale, P.J., Beamish, B.B., Valix, M., 1998. Coalbed methane sorption related to coal composition. International Journal of Coal Geology 35, 147–158.

Diessel, C.F.K., 1992. Coal-bearing Depositional Systems. Springer-Verlag, Berlin. 721 p.

Engelder, T., Whitaker, A., 2006. Early jointing in coal and black shale: evidence for an Appalachian-wide stress field as a prelude to the Alleghanian orogen. Geology 34, 581–584.

Esterle, J.S., 2008. Mining and beneficiation. In: Ruiz, I.S., Crelling, J.C. (Eds.), Applied Coal Petrology—The Role of Petrology in Coal Utilization. Elsevier, Amsterdam, pp. 61–83.

Faraj, S.M., Fielding, C.R., MacKinnon, I.D.R., 1996. Cleat mineralization of Upper Permian Baralaba/Rangal coal measures, Bowen Basin, Australia. Geological Society of London Special Publication 109, 151–164.

Fassett, J.E., Hinds, J.S., 1971. Geology and fuel resources of the Fruitland formation and Kirtland Shale of the San Juan Basin, New Mexico and Colorado. U.S. Geological Survey Professional Paper 676, 76.

Flores, R.M., 1981. Coal deposition in fluvial paleoenvironments of the Paleocene Tongue River Member of the Fort Union Formation, Powder River area, Powder River Basin, Wyoming and Montana. Society of Economic Paleontologists and Mineralogists Special Publication 31, 169–190.

Flores, R.M., 1986. Styles of coal deposition in Tertiary alluvial deposits, Powder River Basin, Montana and Wyoming. Geological Society of America Special Paper 210, 79–104.

Flores, R.M., 2004. Coalbed Methane in the Powder River Basin, Wyoming and Montana: An Assessment of the Tertiary-upper Cretaceous Coalbed Methane Total Petroleum System. U.S. Geological Survey Digital Data Series DDS. 69-C, 56 p.

Flores, R.M. (Ed.), 2008. Microbes, Methanogenesis, and Microbial Gas in Coal. International Journal of Coal Geology, vol. 76. 185 p.

Flores, R.M., Bader, L.R., 1999. Fort Union coal in the Powder River Basin, Wyoming and Montana—A synthesis. U.S. Geological Survey Professional Paper 1625-A, PS1–PS71.

Gan, H., Nandi, S.P., Walker Jr., P.L., 1975. Nature of porosity in American coals. Fuel 51, 272–277.

Gentzis, T., 2010. Coalbed natural gas activity in western Canada: the emergence of major unconventional gas industry in an established conventional province. In: Reddy, K.J. (Ed.), Coalbed Natural Gas, Energy and Environment. Nova Science Publishers, New York, pp. 377–399.

Harpalani, S., Chen, G., 1995. Influence of gas production induced volumetric strain on permeability of coal. Geotechnical and Geological Engineering 15, 303–325.

Hower, J.C., Wild, G.D., 1988. Relationship between Hardgrove grindability index and petrographic composition for high-volatile bituminous coals from Kentucky. Journal of Coal Quality 7, 122–126.

Hunt, J.M., 1979. Petroleum Geochemistry and Geology. Freeman, San Francisco. 617 p.

Joubert, J.L., Grein, C.T., Bienstock, D., 1973. Sorption of methane in moist coals. Fuel 52, 181–185.

Joubert, J.L., Grein, C.T., Bienstock, D., 1974. Effects of moisture on the methane capacity of American coals. Fuel 53, 186–191.

Jüntgen, H., Klein, J., 1975. Entstehung von Erdgas aus kohligen Sedimenten. Erdöl und Kohle, Erdgas Petrochem. Ergängsband 1, 52–69.

Kaiser, W.R., 1993. Abnormal pressure in coal basins of the western United States. In: 1993 International Coalbed Methane Symposium Proceedings, vol. 1, University of Alabama, Tuscaloosa, Alabama, pp. 173–186.

Kaiser, W.R., Hamilton, D.S., Scott, A.R., Tyler, R., 1994. Geological and hydrological controls on the producibility of coalbed methane. Journal of the Geological Society of London 151, 417–420.

Kim, A.G., 1977. Estimating methane content of bituminous coals from adsorption data. U.S. Bureau of Mines Report of Investigations 2845, 22 p.

Laubach, S.E., Marrett, R., Olson, J.E., Scott, A.R., 1998. Characteristics and origins of coal cleat: a review. International Journal of Coal Geology 35, 175–207.

Law, B.E., 1993. The relation between coal rank and cleat spacing: implications for the prediction of permeability in coal. In: International Coalbed Methane Symposium Proceedings. University of Alabama College of Continuing Studies, Tuscaloosa, Alabama, pp. 435–442.

Lee, R.W., 1981. Geochemistry of water in the Fort Union formation of the Powder River Basin, southeastern Montana. U.S. Geological Survey Water-Supply Paper 2076, 17.

Levine, J.R., 1993. Coalification: the evolution of coal as a source rock and reservoir rock for oil and gas. American Association of Petroleum Geologists Studies in Geology 38, 39–77.

Levine, J.R., Telle, W.R., 1989. A Coalbed Methane Resource Evaluation in Southern Tuscaloosa County, Alabama. University of Alabama, Tuscaloosa, Alabama. School of Mines and Energy Development Research Report 89-1, 90 p.

Mastalerz, M., Gluskoter, H., Rupp, J., 2004. Carbon dioxide and methane sorption in high volatile bituminous coals from Indiana, USA. International Journal of Coal Geology 60, 43–55.

McKee, C.R., Bumb, A.C., Koenig, R.A., March 1988. Stress-dependent permeability and porosity of coal and other geologic formations. Society of Petroleum Engineers Formation Evaluation, 81–91.

Meek, R.H., Levine, J.R., 2006. Delineation of four "type producing areas" (TPA's) in the Fruitland coal bed gas field, New Mexico and Colorado, using production history data. American Association of Petroleum Geologists Search and Discovery. Article 20034, (unpaginated).

Milici, R.C., Hatch, J.R., Pawlewicz, M.J., 2010. Coalbed methane resources of the Appalachian Basin, eastern USA. International Journal of Coal Geology 82, 160–174.

Nolde, J.E., Spears, D., 1998. A preliminary assessment of in place coal bed methane resources in the Virginia portion of the central Appalachian Basin. International Journal of Coal Geology 38, 115–136.

Palmer, I., Mansoori, J., 1998. How permeability depends on stress and pore pressure in coalbeds: a new model. Society of Petroleum Engineers Reservoir Evaluation and Engineering, Paper SPE 52607, 539–544.

Pashin, J.C., 1998. Stratigraphy and structure of coalbed methane reservoirs in the United States: an overview. International Journal of Coal Geology 35, 207–238.

Pashin, J.C., 2007. Hydrodynamics of coalbed methane reservoirs in the Black Warrior Basin: key to understanding reservoir performance and environmental issues. Applied Geochemistry 22, 2257–2272.

Pashin, J.C., 2008. Coal as a petroleum source rock and reservoir rock. In: Ruiz, I.S., Crelling, J.C. (Eds.), Applied Coal Petrology—The Role of Petrology in Coal Utilization. Elsevier, Amsterdam, pp. 227–262.

Pashin, J.C., 2009. Shale gas plays of the southern Appalachian thrust belt. In: College of Continuing Studies, International Coalbed & Shale Gas Symposium Proceedings, Paper, vol. 0907, University of Alabama, Tuscaloosa, Alabama. 14 p.

Pashin, J.C., 2010. Variable gas saturation in coalbed methane reservoirs of the Black Warrior Basin: implications for exploration and production. International Journal of Coal Geology 82, 135–146.

Pashin, J.C., Carroll, R.E., Hatch, J.R., Goldhaber, M.B., 1999. Mechanical and thermal control of cleating and shearing in coal: examples from the Alabama coalbed methane fields, USA. In: Mastalerz, M., Glikson, M., Golding, S. (Eds.), Coalbed Methane: Scientific, Environmental and Economic Evaluation. Kluwer Academic Publishers, Dordrecht, Netherlands, pp. 305–327.

Pashin, J.C., Groshong Jr., R.H., 1998. Structural control of coalbed methane production in Alabama. International Journal of Coal Geology 38, 89–113.

Pashin, J.C., Groshong Jr, R.H., Wang, Saiwei, 1995. Thin-skinned structures influence gas production in Alabama coalbed methane fields. In: InterGas '95 Proceedings. University of Alabama, Tuscaloosa, Alabama, pp. 39–52.

Pashin, J.C., Jin, G., Payton, J.W., 2004. Three-dimensional computer models of natural and induced fractures in coalbed methane reservoirs of the Black Warrior Basin. Alabama Geological Survey Bulletin 174, 62.

Pashin, J.C., Hinkle, Frank, 1997. Coalbed methane in Alabama. Alabama Geological Survey Circular 192, 71.

Pashin, J.C., McIntyre, M.R., 2003. Temperature-pressure conditions in coalbed methane reservoirs of the Black Warrior Basin, Alabama, U.S.A: implications for carbon sequestration and enhanced coalbed methane recovery. International Journal of Coal Geology 54, 167–183.

Pashin, J.C., Ward II, W.E., Winston, R.B., Chandler, R.V., Bolin, D.E., Richter, K.E., Osborne, W.E., Sarnecki, J.C., 1991. Regional analysis of the Black Creek-Cobb coalbed-methane target interval, Black Warrior Basin, Alabama. Alabama Geological Survey Bulletin 145, 127.

Paterson, D.W., Bachtiar, A., Bates, J.A., Moon, J.A., Surdam, R.C., 1997. Petroleum system of the Kutei Basin, Kalimantan, Indonesia. In: Howes, J.V., Noble, R.A. (Eds.), Petroleum Systems of SE Asia and Australasia. Indonesian Petroleum Association, pp. 709–726. Report 97-OR-35.

Pollard, D.D., Aydin, A., 1988. Progress in understanding jointing over the last century. Geological Society of America Bulletin 100, 1181–1204.

Riese, W.C., Perlman, W.L., Snyder, G.T., 2005. New insights on the hydrocarbon system of the Fruitland Formation coal beds, northern San Juan Basin, Colorado and New Mexico, USA. Geological Society of America Special Paper 387, 73–111.

Robbins, E.I., Wilkes, G.P., Textoris, D.A., 1988. Coals of the Newark rift system. In: Manspeizer, W. (Ed.), Triassic-Jurassic Rifting: Continental Breakup and the Origin of the Atlantic Ocean and Passive Margins, Part B. Elsevier, Amsterdam, pp. 649–678.

Ryer, T.A., 1981. Deltaic coals of Ferron Sandstone Member of Mancos Shale: predictive model for Cretaceous coal-bearing strata of Western Interior. American Association of Petroleum Geologists Bulletin 65, 2323–2340.

Scott, A.R., 2002. Hydrogeologic factors affecting gas content distribution in coal beds. International Journal of Coal Geology 50, 363–387.

Scott, A.R., Kaiser, W.R., Ayers Jr., W.B., 1994. Thermogenic and secondary biogenic gases, San Juan Basin, Colorado and New Mexico—Implications for coalbed gas producibility. American Association of Petroleum Geologists Bulletin 78, 1186–1209.

Scott, A.R., Zhou, N., Levine, J.R., 1995. A modified approach to estimating coal and gas resources: example from the Sand Wash Basin, Colorado. American Association of Petroleum Geologists Bulletin 79, 1320–1336.

Smith, G.C., Cook, A.C., 1984. Petroleum occurrence in the Gippsland Basin and its relationship to rank and organic matter type. Australian Petroleum Exploration Association Journal 24, 196–216.

Spears, D.A., 1987. Mineral matter in coals, with special reference to the Pennine Coalfields. Geological Society (London) Special Publication 32, 171–185.

Spears, D.A., Caswell, S.A., 1986. Mineral matter in coals: cleat minerals and their origin in some coals from the English Midlands. International Journal of Coal Geology 6, 107–125.

Stark, T.J., Cook, C.W., 2009. Factors controlling coalbed methane production from helper, drunkards wash and buzzard bench fields, carbon and Emery Counties, Utah. American Association of Petroleum Geologists Search and Discovery. Article 90090, (unpaginated).

Su, X., Lin, X., Zhao, M., Song, Y., Liu, S., 2005. The upper Paleozoic coalbed methane system in the Qinshui Basin, China. American Association of Petroleum Geologists Bulletin 89, 81–100.

Ting, F.T.C., November 1977. Origin and spacing of cleats in coal beds. Journal of Pressure Vessel Technology, 624–626.

van Krevelen, D.W., 1961. Coal. Elsevier, Amsterdam. 362 p.

Van Voast, W.A., 2003. Geochemical signature of formation waters associated with coalbed methane. American Association of Petroleum Geologists Bulletin 87, 667–676.

Vinson, D.S., McIntosh, J.C., Ritter, D.J., Blair, N.E., Martini, A.M., 2012. Carbon isotope modeling of methanic coal biodegradation: metabolic pathways, mass balance, and the role of sulfate reduction, Powder River Basin, USA. Geological Society of America Special Programs 44 (no. 7), 466.

Waples, D.W., 1980. Time and temperature in petroleum formation. American Association of Petroleum Geologists Bulletin 64, 916–926.

Whiticar, M.J., 1994. Correlation of natural gases with their sources. American Association of Petroleum Geologists Memoir 60, 261–283.

Whiticar, M.J., 1996. Stable isotope geochemistry of coals, humic kerogens and related natural gases. International Journal of Coal Geology 32, 191–215.

Wilkins, R.W.T., George, S.C., 2002. Coal as a source rock for oil: a review. International Journal of Coal Geology 50, 317–361.

Yang, R.T., Saunders, J.T., 1985. Adsorption of gases on coals and heat-treated coals at elevated temperature and pressure. Fuel 64, 616–620.

Young, G.B.C., Kelso, B.S., Paul, G.W., 1994. Understanding cavity well performance. Society of Petroleum Engineers Paper 28579, 16.

Zhang, E., Hill, R.J., Katz, B.J., Tang, Y., 2008. Modeling of gas generation from the Cameo coal zone in the Piceance Basin, Colorado. American Association of Petroleum Geologists Bulletin 92, 1077–1106.

Zhou, L., Zhou, Y., Li, M., Chen, P., Wang, Y., 2000. Experimental modeling study of the adsorption of supercritical methane on a high surface activated carbon. Langmuir 16, 5955–5959.

Evaluation of Coalbed Methane Reservoirs

Kashy Aminian[1], Gary Rodvelt[2]

[1]West Virginia University, Morgantown, West Virginia, USA,
[2]Global Technical Services, Halliburton Energy Services, Canonsburg,
Pennsylvania, USA

4.1 INTRODUCTION

Coal, unlike conventional gas reservoirs, is both the reservoir rock and the source rock for methane. While much of the gas generated during coalification migrated out of the coal seam, significant quantities remain in coal that can be produced from a seam by understanding and evaluating the unique properties of coal. Coal is a heterogeneous and anisotropic porous media which is characterized by two distinct porosity systems (dual porosity): macropores and micropores. The macropores, also known as cleats, constitute the natural fractures common to all coal seams. Micropores, or the matrix, contain the vast majority of the gas. This unique coal characteristic has resulted in classification of coalbed methane (CBM) as an "unconventional" gas resource. The characteristics of CBM reservoirs differ from conventional gas reservoir in several areas. Table 4.1 summarizes the major difference between CBM and conventional gas reservoirs. The attributes that a reservoir engineer must know to forecast production from a coal include gas content, adsorption isotherm, sorption time, permeability and permeability anisotropy, relative permeability relationships, and the reservoir pressure. To maximize gas recovery, the reservoir engineer must also determine the optimum drilling pattern, hydraulic-fracture design, and operating scheme. Testing and simulation modeling can provide preliminary predictions that can be verified with production history matching once adequate production time has passed. The CBM characteristics and testing

TABLE 4.1   Comparison of Coalbed Methane (CBM) and Conventional Gas
Reservoir Characteristics

Characteristic	Conventional Reservoirs	CBM Reservoirs
Gas generation	Gas is generated in the source rock and then migrates into the reservoir	Gas is generated and trapped within the coal
Structure	Randomly-spaced fractures	Uniformly-spaced cleats
Gas storage mechanism	Compression	Adsorption
Transport mechanism	Pressure gradient	Concentration gradient and pressure gradient
Production behavior	Gas rate starts high and declines Little or no water production initially Gas Water Ratio (GWR) decrease with time	Gas rate increases and then declines The production is initially water GWR increases with time
Mechanical properties	Young modules ~10^6 Pore compressibility ~10^{-6}	Young modules ~10^5 Pore compressibility ~10^{-4}

techniques to determine key reservoir parameters needed for developing a CBM prospect will be discussed in the following sections.

4.1.1 Coal Structure and Gas Storage

Coal is an organic-rich rock with a micropore structure (with pores 5–10 Å in size). This micropore structure is often modeled as cubes, much like a set of building blocks stacked on each other, with natural fractures or cleats that run orthogonal to each other as illustrated in Figure 4.1. The cleats are an artifact of the coalification process and any tectonic stresses that have been applied. Cleats can be spaced as close as 0.1–1 inch apart. Regional joint sets can also occur during maturation process. Typical coal porosity ranges from 0.5% to 5%, with 1–2% being the most common for current coals with commercial production. This porosity is the "open space" within the cleats themselves. Any water and "free" gas will be stored within the cleat porosity at reservoir pressure in equilibrium with the micropore concentration. The cleats and the joints make up the in situ permeability of the system. Face cleats are the dominant cleats generally aligned in the direction of the maximum principal stress. Face-cleat permeability is almost always higher than the butt-cleat permeability.

Gas in the coal can be present as "free" gas within the macropores (cleats) or as an adsorbed layer on the internal surfaces of the coal micropore. The micropore of coal has immense capacity for methane storage. At pressures below 1000 psia, coal can store far more gas in adsorbed state than conventional reservoirs can by compression. The porosity of the cleat

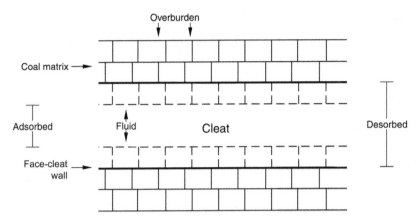

FIGURE 4.1 Artist's representation of coal-seam structure.

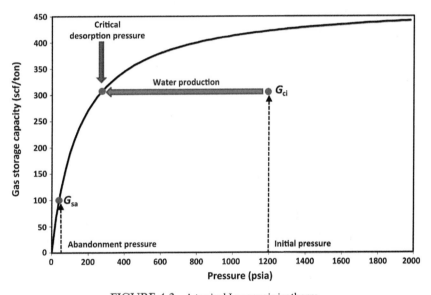

FIGURE 4.2 A typical Langmuir isotherm.

system is small and if any "free" gas is present, it would account for an insignificant portion of the gas stored in the coal. The vast majority of the gas in coals is stored by adsorption in the coal matrix. As a result, pressure–volume relationship is defined by the sorption isotherm. A sorption isotherm relates the gas storage capacity of a coal to pressure and depends on the rank, temperature, and the moisture content of the coal. The sorption isotherm can be used to predict the volume of gas that will be released from the coal as the reservoir pressure is lowered. A typical sorption isotherm illustrated in Figure 4.2. A common assumption is that the relationship between gas storage capacity and pressure can be described by an equation originally presented by Langmuir (Langmuir, 1916) and when it is modified to account for ash and moisture contents of the coal can be presented as follows:

$$G_S = \left(1 - f_a - f_m\right) \frac{V_L P}{P_L + P} \tag{4.1}$$

As can be seen from Figure 4.2, the sorbed gas eventually reaches a maximum value, which is represented by Langmuir volume constant (V_L). Langmuir pressure constant (P_L) represents the pressure at which gas storage capacity equals one half of the maximum storage capacity. The values of P_L and V_L for a particular coal are determined by laboratory isotherm testing.

The isotherm defines the maximum gas adsorptive capacity as a function of pressure. The gas content of the coal could, however, be less than this maximum value at the initial reservoir conditions. The gas content of the coal can be evaluated by desorption testing. When the initial content is below the equilibrium with the isotherm, as illustrated in Figure 4.2, no free gas will be present and the cleats will be filled with water. Gas desorption initiates once the pressure in the cleat system is lowered to "critical desorption pressure" normally by water production. Below critical desorption pressure, the gas content will be the maximum value (as determined by the isotherm) down to the abandonment pressure.

4.1.2 Gas Transport and Production

Most CBM reservoirs initially produce only water because the cleats are filled with water. Typically, water must be produced continuously from the coal seams to reduce the reservoir pressure and release the gas. The cost to treat and dispose the produced water can be a critical factor in the economics of a CBM project. Once the pressure in the cleat system is lowered to the "critical desorption pressure", gas will desorb from the matrix. Critical desorption pressure as illustrated on Figure 4.2, is the pressure on the sorption isotherm that corresponds to the initial gas content. As the desorption process continues, a free methane gas saturation builds up within the cleat system. Once the gas saturation exceeds the critical gas saturation, the desorbed gas will flow along with water through cleat system to the production well.

Gas desorption from the matrix surface in turn causes molecular diffusion to occur within the coal matrix. Diffusion through the coal can happen by bulk diffusion in large diameter pores, by Knudsen-type diffusion in small capillaries, or by surface diffusion along the surface of the micropore like a liquid. Figure 4.1 shows micropores, cleats, and the coal matrix that gas molecules must pass through en route to the hydraulic fracture or wellbore. The diffusion through the coal matrix is controlled by concentration gradient and can be described by Fick's Law (Rogers, 1994):

$$q_{gm} = 2.697\sigma D\rho_c V_c \left(\overline{G_c} - G_s\right) \tag{4.2}$$

Diffusivity and shape factor are usually combined into one parameter referred to as sorption time as follows:

$$\tau = \frac{1}{\sigma D} \tag{4.3}$$

Sorption time, τ, is the time required to desorb 63.2% of the initial gas volume. The sorption time characterizes the diffusion effects and generally is determined from desorption test results. Different coals will have different sorption times, which can be a major factor in forecasting the production rate. Table 4.2 depicts some representative sorption times for various coals in the United States.

The two-phase flow in the cleat system can be adequately represented by Darcy's law. Cleat system porosity, permeability, and relative permeability control the fluid flow within the cleat system. As desorption process continues, gas saturation within the cleat system increases and flow of methane becomes increasingly more dominant. Thus, the water production decline rapidly until the gas rate reaches the peak value and water saturation approaches the irreducible water saturation. This is the point for highest gas permeability—usually 6–18 months into production and maximum gas desorption (depending on spacing). Once this condition is reached, readmitting fluid into the cleats is not advisable because the energy to unload it has been lost. The implications are that a re-frac of an old well should be done with high-quality nitrogen foam to limit fluid loading and maintain maximum relative gas permeability. The typical production behavior of a CBM reservoir is illustrated in Figure 4.3. After the peak gas rate production is reached, the CBM reservoir behavior becomes similar to a conventional gas reservoir.

TABLE 4.2 Sorption Times for Various Coals in the US

Coal	Sorption Time
Fort Union, sub	<1 day (Hughes and Logan, 1990)
Fruitland, mvb	<1 day (Hughes and Logan, 1990)
Pennsylvanian age	>80 days (Hughes and Logan, 1990)
Fruitland (NW San Juan Basin)	4.1 h (Mavor and McBane, 1991)
Northern Appalachian	100–900 days (Mavor and McBane, 1989)
Central Appalachian	1–3 days (Mavor and McBane, 1989)
Warrior	3–5 days (Mavor and McBane, 1989)

After Rogers (1994).

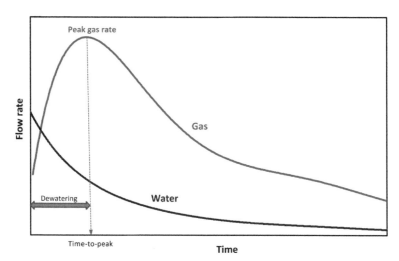

FIGURE 4.3   Typical production history of a coalbed methane reservoir.

4.1.3 Mechanical Properties

Several mechanical properties of coal are significantly different from most reservoir rock as summarized in Table 4.1. Coal is relatively compressible compared to the rock in many conventional reservoirs. Thus, the permeability of coal is more stress dependent than most reservoir rocks. The orientation and magnitude of stress can strongly influence CBM recovery. Permeability and porosity are functions of the net stress in the system. Permeability in the coal reservoirs can change by three mechanisms: Klinkenberg effect, matrix shrinkage, and effective stress changes. The Klinkenberg effect and matrix shrinkage increase permeability while effective stress reduces permeability.

Flow of gas through the cleats is modeled using Darcy's law, which assumes that the gas layer closest to the cleat walls does not move. As pressure declines, slippage does occur, as described by the Klinkenberg effect, giving a higher flow rate than calculated by Darcy's law (Patching, 1965). Because most coals are lower in pressure than conventional reservoirs, this effect is more important. An increase in permeability also occurs when the matrix shrinks as a result of gas desorption (Harpalani and Schraufnagel, 1990). The cleat apertures increase as gas volume is withdrawn, improving the porosity and permeability of the cleat system (Seidle and Huitt, 1995). The coal matrix shrinkage effects generally become significant at low pressures (Palmer and Mansoori, 1996). This could lead to better deliverability during late stages of production.

Finally, as the pressure is drawn down to promote desorption, the net vertical stress increases because the stress caused by overburden pressure

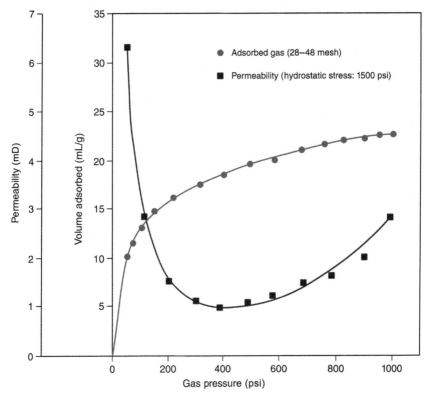

FIGURE 4.4 Permeability changes with production.

remains unchanged by production. In the absence of other factors, porosity and permeability will decrease as pore pressure drops (Collins, 1989). Figure 4.4 shows representative results on effective permeability from the combination of matrix shrinkage, net stress, and the Klinkenberg effect on in situ permeability. The effect of the overburden pressure on permeability is very pronounced—as much as a 10-fold reduction for every thousand feet of burial. This has been documented from data collected in the San Juan, Black Warrior, and Piceance basins (McKee et al., 1984) and is illustrated in Figure 4.5. Alternative methods, besides conventional application of hydraulic fracturing, will be needed to produce from deeper coal seams.

The friable, cleated nature of coal affects the success of hydraulic fracturing treatments, and in certain cases allows for cavitation techniques to dramatically increase production. Strength of the coal reaches a minimum where cleating is most developed. As a result, obtaining competent core samples from coals with well-developed cleat system is not possible. Therefore, porosity and permeability, and relative permeability of

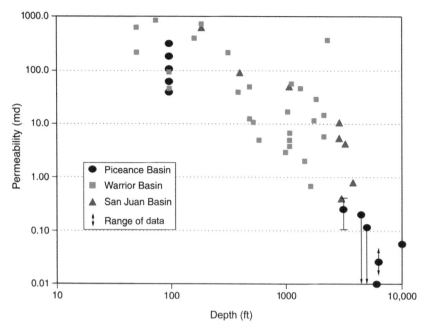

FIGURE 4.5   Effect of overburden pressure on coal permeability.

the fracture system cannot be accurately determined from core analysis. Properties of the fracture system are usually determined from well testing and/or history matching with a reservoir simulator.

4.2 PROSPECT EVALUATION

The key parameters for evaluation of CBM prospects are the gas resources, reserves, and deliverability. Table 4.3 lists the key required data for analysis of CBM reservoirs and their primary sources. The vast majority of the gas in CBM is stored in the low-permeability coal matrix by sorption. Therefore, properties of the coal matrix have the greatest effect on the estimates of the volume of gas-in-place and gas recovery. Gas content and storage capacity are the key parameters for determination of the gas resources and reserves both of which must be measured directly from core samples. Core data acquisition and analysis is an indispensable step in evaluating any CBM project. The flow of fluids (gas and water) to the wellbore in CBM reservoirs takes place through the natural fracture system (cleats) because coal matrix practically has no permeability. Therefore, properties of the natural fracture system have the greatest effect on gas and water production rates. Accurate estimate of deliverability requires accurate evaluation of the coal natural fracture system flow properties.

TABLE 4.3   Required Data and Their Sources for Analysis of Coalbed Methane Reservoirs

Property	Source
Storage capacity	Core measurements
Gas content	Core measurements
Diffusivity	Core measurements
Pore volume compressibility	Core measurements
Gross thickness	Well logs
Effective thickness	Well logs
Pressure	Well tests
Absolute permeability	Well tests
Relative permeability	Simulation
Porosity	Simulation
Fluid properties	Composition and correlations
Gas composition	Produced and desorbed gas
Drainage volume	Geologic studies

The natural fracture system permeability and relative permeability are the key parameters, which most influence the deliverability. Natural fracture system permeability can be accurately estimated by well testing.

4.2.1 Gas-in-Place

The following equation is generally used to estimate the total initial adsorbed gas in a CBM reservoir:

$$G = 1359.7Ah\overline{\rho_c}\overline{G_c} \tag{4.4}$$

Coal matrix properties can be reliably obtained from the interpretation and integration of the core and log data. However, numerous challenges are involved in the acquisition and analysis of data to determine the various parameters in Eqn (4.4) that can complicate accurate determination of the initial gas-in-place (Nelson, 1999). Therefore to minimize the errors, the established protocols (Mavor et al., 1996) must be followed when assessing the gas-in-place in coalbed reservoirs. The parameters in Eqn (4.4) will be discussed in the following sections.

4.2.1.1 Gas Content

The gas content is the standard volume of gas per unit weight of coal or rock and usually is reported in units of standard cubic feet per ton (scf/ton). This section provides a brief overview of gas content determination. Detailed information on the theory and procedures for determining the gas content of coal samples can be found elsewhere (McLennan et al., 1995). The desorbed gas is estimated by "Direct Method" or canister desorption test (Diamond and Levine, 1981). The canister desorption test is conducted by placing a freshly cut conventional core sample in a sealed container (canister) and measuring the amount of gas released as a function of time. This test, if conducted properly, usually provides reliable estimate of the gas content (Mavor et al., 1994). Since the temperature has a significant impact on diffusion rates, the test must be conducted at the reservoir temperature. In addition to the desorbed gas volume measured during the canister desorption test, "lost gas" and "residual gas" volumes must be evaluated to determine the total desorbed gas volume. "Lost gas" is the volume of the gas that desorbs from the sample during the recovery process before the core sample can be sealed in a desorption canister. "Residual gas" is the gas that remains sorbed on the sample at the conclusion of the canister desorption test.

Lost gas volume is estimated by analyzing the data obtained during the canister desorption tests. This analysis method is based on the solution, which suggests that the cumulative desorbed gas is proportional to the square root of the elapsed time since the inception of gas desorption. The inception of gas desorption is referred to as "time zero", which usually occurs during the core recovery process. Proper estimation of the "time zero" has a significant impact on the accuracy of the estimated lost gas volume (McLennan et al., 1995). In addition to the lost gas volume, the sorption time can be obtained from the Direct Method of analysis. Lost gas volume usually is the greatest source of error in the total gas content estimate (Nelson, 1999). Pressure coring can eliminate the lost gas but because of difficulties involved in operating the specialized equipment and the added expenses, its use has been limited to research studies. The residual gas volume is estimated by crushing the sample to smaller than a 60-mesh grain size and measuring the gas volume released at the reservoir temperature.

The total gas volume is the summation of the lost gas, desorbed gas, and residual gas volumes. The total gas volume is reported at standard temperature and pressure, and it reflects the total estimated gas volume in the coal at initial reservoir pressure and temperature. The measured gas volume is divided by the mass of the sample, and then converting to units of scf/ton. The basis for reporting gas content depends on the manner in which the sample mass is determined. Commonly used bases for reporting coal gas content are the air-dry basis; the dry, ash-free basis; the dry, mineral-matter-free basis; the "pure coal" basis; and the in situ basis. "Air-dry" implies that all extraneous material is removed from the sample and that the entire sample has been allowed to dry to a constant equilibrium weight in a laboratory climate. Because the gas is stored by sorption only in the organic matrix of the coal, the air-dry weight must be corrected for "noncoal" components such as residual moisture and ash. If the correction to gas content is based on individual sample composition, the gas content is referred to as the dry, ash-free basis or the dry, mineral-matter-free basis.

Coal composition is not uniform throughout the reservoir; therefore, gas content is not constant. To obtain statistically reliable estimate of in situ gas content, the gas content from multiple samples having a broad range of compositions must be measured. An inverse linear relationship between the gas content and the "noncoal" fraction has been shown to exist (McLennan et al., 1995). Consequently, a linear regression analysis is performed to establish the relationship between total gas content and the "noncoal" fraction. The "pure coal" gas content is estimated by extrapolating the regression line to pure coal (zero ash and moisture content). "Pure coal" gas content estimates can be used to compare gas contents from different geographic or geologic location. The linear relationship between the total gas content and the "noncoal" fraction can then be used to determine in situ gas content for any amount of ash present in the formation (Mavor and Nelson, 1997). Since the bulk density can be correlated with ash content, the average reservoir density in conjunction with equilibrium moisture contents can be used to determine average in situ gas content. The average in situ gas content is the value that must be used for gas-in-place calculations.

The term "coalbed methane" is commonly used even though often the gas produced from coal contains significant amounts of carbon dioxide, nitrogen, water, and heavier molecular weight hydrocarbons. Therefore, the composition of desorbed gas should be determined during desorption test to measure the total methane content which would be different from the total gas content (Saulsberry et al., 1996).

4.2.1.2 Storage Capacity

To reliably estimate gas reserves for a CBM reservoir, gas desorption behavior in addition to gas content must be determined. The most common model in use for coal is the Langmuir isotherm, which was presented in Eqn (4.1). If the initial reservoir conditions (pressure and gas content), desorption isotherm, and the abandonment pressure are available, the recovery factor at the economic limit can be estimated using the following equation:

$$R_f = \frac{\overline{G_{ci}} - G_{sa}}{\overline{G_{ci}}} \tag{4.5}$$

Gas reserves can be calculated by multiplying gas-in-place by the estimated recovery factor at the economic limit.

Determining accurate sorption isotherms requires a properly prepared sample and correct laboratory procedures. This section provides a brief overview of sorption isotherms determination. Detailed information on the theory and procedures for determining of the sorption isotherms can be found elsewhere (McLennan et al., 1995). The isotherm is determined by grinding the coal (core sample) to a fine mesh and systematically measuring the amount of gas that the coal can store at various pressures. The test results are used to determine Langmuir parameters P_L and V_L. It is critical to use the correct moisture content and perform the isotherm tests at reservoir temperature. Accepted practice is to perform sorption isotherm measurements at the equilibrium moisture content. The equilibrium moisture content, which is defined by American Society for Testing and Materials standards, is assumed to be the same as the in situ moisture content. The measured isotherm data must be normalized to a dry, ash-free basis because it is important to have the sorption isotherm and gas content values on the same basis. If significant quantities of other gases besides methane are present in the coal, isotherms for the other gases need to be determined in addition to a pure methane isotherm.

4.2.1.3 In situ Density

The correct in situ density should be estimated from open-hole density log data. A common practice of using a value of 1.32–1.36 g/cm³ for the average in situ density can lead to erroneous results (Nelson, 1999). In the absence of well log data, in situ density can be estimated based on the density of the ash, moisture, and organic (pure coal) fractions by the following Equation:

$$\frac{1}{\rho_c} = \frac{f_a}{\rho_a} + \frac{1 - (f_a + f_w)}{\rho_o} + \frac{f_w}{\rho_w} \tag{4.6}$$

4.2.1.4 Coal Thickness and Areal Extent

For the reservoir engineer to calculate gas-in-place, the reservoir must be defined. The reservoir can be a single seam or multiple seams that have a measured thickness and can be mapped over some distance to comprise an acreage position. Lease holdings are measured in acres of surface area that can be extrapolated across the coal-seam surface to give areal extent. Thicknesses are generally assumed constant until control wells are drilled to accurately measure the thickness.

Coal thickness is obtained by drilling a well and either measuring the thickness of coal core or evaluating the hole with a wireline log. Gross coal thickness usually can be determined accurately with wireline logs. Open-hole density logs generally provide the most reliable and cost-effective estimates of gross thickness in newly drilled wells. The gross reservoir thickness is commonly computed by summing the thicknesses of the intervals having densities less than a cut-off value generally equal to the coal ash density. However, using a too low value for density cut-off can lead to erroneous estimates. Determining net thickness is more complicated because it requires evaluating how much of the gross coal thickness actually contributes to production. Resistivity logs, well tests, production logs, or zonal isolation tests can be used to estimate the net thickness.

The reservoir area is usually estimated based on the well spacing if the coal seam can be assumed to be laterally continuous. Structural and stratigraphic variations throughout the reservoir determine the three dimensional distribution of the coal. Therefore, geological evaluations can provide clues about coal seam continuity and its other pertinent characteristics. However, it may be difficult to identify localized stratigraphic variations. Three-dimensional seismic data can be also used to determine CBM reservoir geometry.

Pattern development can take advantage of the greater permeability in the face-cleat direction by spacing wells further apart along the face-cleat direction. By the same reasoning, the lower-permeability butt-cleat system requires spacing to be closer along the butt-cleat direction. It is advantageous to have wells spaced so that they "interfere" with the offset wells. Drawing pressure down from two directions, lowers the reservoir pressure within the drainage pattern and will expedite coal dewatering which leads to quicker gas desorption. Permeability, well-stimulation effects, and well spacing all affect interference. Higher permeability means wells can "see" each other sooner and will promote desorption. Wells closer together accomplish the same thing. Longer fracture lengths allow wells to "see" each other and promote greater interference. Figures 4.6–4.8 show the effects of spacing, permeability, and hydraulic-fracture length, respectively. Anything that the operator can do to promote greater interference with offset wells will promote greater desorption and lead to quicker and

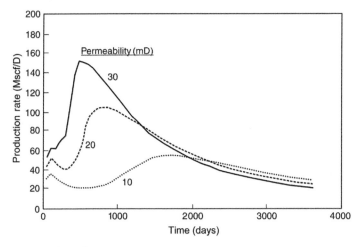

FIGURE 4.6 Production rates for various permeabilities.

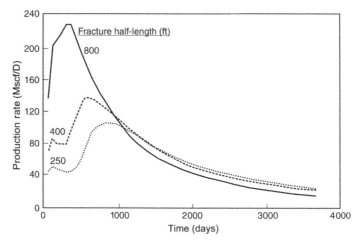

FIGURE 4.7   Production rates for various fracture half-lengths.

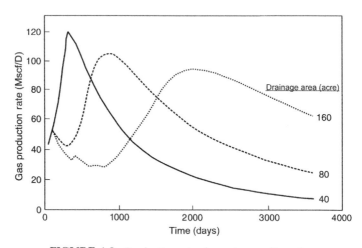

FIGURE 4.8   Production rates for various well spacing.

higher gas production. The economics will dictate what the optimum spacing and pattern should be, given the different scenarios.

4.2.2 Deliverability

Absolute permeability and relative permeability are two of the most important natural fracture system flow properties that affect gas and water production rates. Pressure/rate transient testing is the only reliable way to accurately estimate in situ natural fracture system permeability.

The results of well test interpretations need to be combined with geological information, wireline log data, core analysis results, and fluid properties to predict future production rates under a variety of operating conditions. The technology for designing, conducting, and interpreting oil and natural gas well tests can be applied to CBM reservoirs. However, characteristics of CBM reservoirs must be considered in the interpretation of the well test results. These characteristics include dual porosity system, two-phase flow, stress dependent properties of coal, and multiple seams. Even though it is possible to obtain estimates of cleat system permeability from test data measured during two-phase flow, the results would be highly dependent on the relative permeability of the relationship assumed. Therefore, it is advantageous to conduct tests under single-phase flow conditions. In an undersaturated CBM reservoir, such as the one illustrated in Figure 4.1, the initial reservoir pressure is above critical desorption pressure and the natural fracture system (cleats) is saturated with water. If a test is conducted by injecting water, the pressure in natural fracture system remains above than critical desorption pressure and single-phase flow conditions will prevail during the test. Further, the reservoir behaves as a single porosity system since the coal matrix is not affected during the test. This allows the use of single porosity models to interpret the test results.

In situ permeability can be determined from a flow test through injection of fluid in slug, tank, or forced-injection test followed by capture of the pressure-falloff data that can be analyzed for permeability and reservoir pressure. In most cases, reservoir pressure can be captured with bottomhole gauges before the injection of fluids if allowed to set for 24 h in a static condition. Slug testing is a simple test of instantaneously introducing a given volume of fluid to the reservoir and then measuring the pressure response as the fluid level reaches equilibrium. Tools might be required to isolate an interval. A bottomhole-pressure gauge, as illustrated in Figure 4.9, is used to capture the pressure response. Because the volume is fixed, no special metering equipment is required for flow rate measurement. Drawbacks to this test include long capture times for low-permeability seams to reach equilibrium, short radius of investigation, and constant monitoring over the long periods to effectively capture data needed for analysis. The test should be done early in the life of the well while water saturation is near 100% to simplify analysis.

An inexpensive procedure for conducting injection/falloff test is "tank test" that reduces the injection pressure. The procedure is based on using gravity drainage of water from a tank rather than using pumping equipment to inject water as illustrated in Figure 4.10. The tank test can be performed in hydrostatically low-pressured reservoirs where the pressure gradient is less than 0.4 psi/ft. Tank measurements must be taken over time to build a rate profile for test analysis. It might be necessary to breakdown perforations

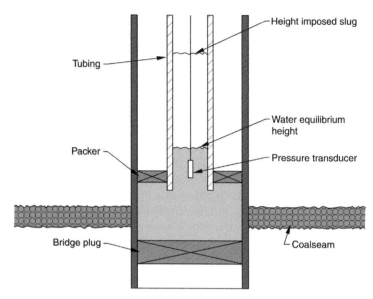

Tubing

Height imposed slug

Water equilibrium height

Packer

Pressure transducer

Bridge plug

Coalseam

FIGURE 4.9   Slug test showing pressure of transducer on the bottom.

so that fluid entry can take place. The volume might be limited to prevent extensive fracturing of the coal. Once again, unless a long test is performed, the radius of investigation is limited. The shut-in must be long enough to reach radial/pseudoradial flow to calculate pore pressure and permeability.

A matrix injection/falloff test (MIFOT) is performed at a rate well below the fracturing rate, typically about 0.1 gal/min. It is important to inject at low rates to avoid fracturing the coal and to minimize permeability changes due to stress effects. Typical injection time is 8 h followed by at least 48-h shut-in period. Longer injections and shut-in times might be required for lower-permeability coals if a greater radius of investigation is required. The objective of the injection period is to raise the injection pressure 15–25% above initial injection pressure (typically 100–200 psi) and then observe the pressure wave dissipate. This test will typically see 200–400 ft out into the coal, which will normally cover the extent of coal to be treated with a hydraulic fracture. A downhole shut-in valve, as illustrated in Figure 4.11, is recommended for data-quality purposes. The best practices for capturing quality data with this test for CBM applications has been documented elsewhere (Rodvelt and Oestreich, 2008).

In reservoirs that primarily produce gas, it may be more desirable to inject gas for injection/falloff test or conduct drawdown/buildup tests. Accurate estimates of the natural fracture system effective permeability can be obtained from multiphase flow interpretation methods if saturation changes are small. This would occur if a reservoir is producing at slowly changing gas–water ratios. When fluid saturations change rapidly,

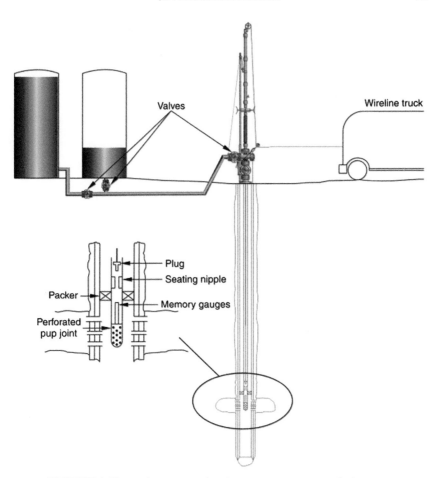

FIGURE 4.10 Tank test setup showing memory gauges on the bottom.

such as when the reservoir pressure is near the critical desorption pressure, accurate test analysis will be difficult.

The primary objective of well testing in CBM reservoir is to obtain estimates of the natural fracture system pressure, permeability, and skin factor. However, well tests can also be conducted to obtain estimates of the induced-fracture properties. Diagnostic fracture injection/falloff testing is performed at a rate well above fracturing rate; typically about 2–3 bbl/min. Typical injection time is 10 min followed by at least a 24-h shut-in. This test also benefits from a downhole shut-in valve, especially where fluid levels fall below ground level. Although the radius of investigation is less than the MIFOT (i.e., 100–200 ft), it does allow calculation of the minimum and maximum horizontal stresses as well as fracture leak-off characteristics (Ramurthy et al., 2002). This is invaluable early in a project to aid in effective fracture design for successful stimulation treatment.

Pressure transient packer gauges test assembly
with downhole shut-in
Slimhole configuration

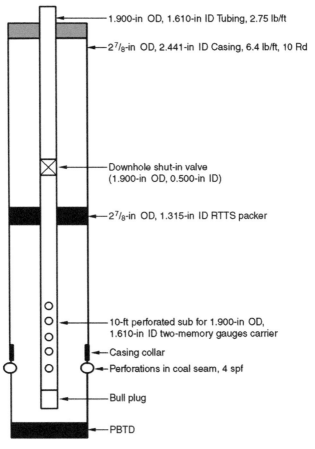

FIGURE 4.11   MIFOT bottomhole assembly showing downhole shut-in valve.

Relative permeability is one of the key parameters in determining the deliverability of CBM reservoirs. Recorded measurements of relative permeability in coal (Puri et al., 1991) are sparse despite the importance of relative permeability. It is generally accepted that laboratory measurements on core samples do not provide reliable relative permeability relationships for CBM reservoir because it is difficult to obtain competent core samples from highly fractured section of the reservoir. History matching is the only practical method to obtain realistic relative permeability values. History matching involves using a reservoir simulator to reproduce gas and water production data. Simulators have been developed specifically to account for the complex characteristics of coal. When production has not been initiated or when production history is limited, relative permeability has to be assumed.

4.3 PRODUCTION FORECASTING

CBM production behavior is complex and difficult to predict or analyze especially at the early stages of the recovery. This is because gas production from CBM reservoirs is governed by complex interaction of single-phase gas diffusion through micropore (matrix) system and two-phase gas and water flow through macropore (cleat) system that are coupled through desorption process. Therefore, the conventional reservoir engineering techniques cannot be used to predict CBM production behavior. The best tool to predict performance CBM reservoirs is a numerical reservoir simulator that incorporates the unique flow and storage characteristics of CBM reservoirs and accounts for various mechanisms that control CBM production. In addition, history matching with simulator is one of the key tools for determining reservoir parameters that are often difficult to obtain by other techniques.

Once the reservoir engineer has acquired the gas content, sorption isotherm, sorption time, thickness, areal extent, reservoir pressure, permeability, and a set of relative permeability curves, simulation can be performed to forecast gas and water production based on production limits and a given backpressure profile. Attention to acquiring quality data early will enable the operator to make accurate decisions using the reservoir simulator results. After a well has been completed, the operator should immediately begin gathering flow rate, pressure, and volume data to history match with the simulator. This permits early validation of the reservoir model or the need to make changes to it before a large investment has been made. Completion effectiveness, well spacing, and pattern development can all be varied within the simulator program. The end result is an optimum development strategy that could lead to economic recovery of the CBM.

Production decline analysis can be used once a CBM well is through the dewatering phase and has begun a normal decline similar to a conventional well. The time it takes to reach a normal decline can vary depending on the interference setup with the offset wells. Typically, one would like to see production of six months or more on a normal decline curve. Figure 4.12 depicts a Deerlick Creek well with exponential decline. Depending on the daily operating costs, an abandonment rate can be established for the field to then allow extrapolation of the decline curve to that time point and rate. The cumulative gas can be calculated and an Estimated Ultimate Recovery (EUR) established for the wells. The reservoir engineer can also

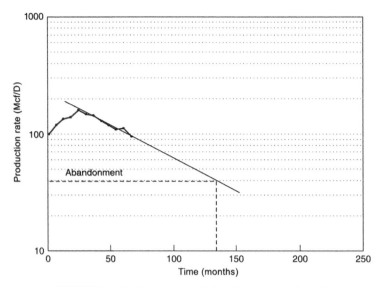

FIGURE 4.12   Exponential decline, Deerlick Creek well.

check decline rates to see if abnormally high declines are present, indicating well damage and/or a possible workover candidate. Decline rates greater than 25% were found to be candidates for a workover in a Virginia CBM field (Rodvelt and Moyers, 2009).

4.4 ENHANCED CBM RECOVERY

Enhanced recovery of CBM has been done with at least two methods: nitrogen injection and CO_2 flooding. Both processes use the method of partial pressure reduction to liberate methane. This means that nitrogen or CO_2 can be injected at reservoir pressure and liberate methane without a reduction in reservoir pressure. The coal has a greater affinity for storage of CO_2 than methane. Nitrogen will liberate some methane as it passes through the coal, but is not greatly adsorbed by the coal like CO_2. Nearly twice as much CO_2 can be stored as methane in most coals. This is the basis for storing CO_2 in coal as a way of sequestering it for removal from the atmosphere and preventing global warming. Much work is being done to develop storage reservoirs in coals, in deep saline, and old oil fields for this purpose. Introduction of CO_2 will cause permeability to decrease as the affinity for it is high causing swelling of the matrix. This swelling of the matrix and subsequent reduction in permeability must be accounted for when considering enhanced CBM production from CO_2 flooding or setting up a sequestration project.

4.5 APPENDIX: SOME CONSIDERATIONS FOR A DATA AND INFORMATION GATHERING PROGRAM TO EVALUATE THE COALBED METHANE POTENTIAL OF A COAL RESERVE

David C. Uhrin, Coalbed Methane Consulting

Assessment of the coalbed methane (CBM) content of a coal reserve arose from the need of the mining industry to design better ventilation systems. More recent developments have been driven by the gas industry's commercial production of CBM. The observations and findings on the nature of CBM resources in the Northern and Central Appalachian coal basins can pertain to coal basins elsewhere.

Continuous coring: Coal strata are best evaluated using narrow diameter, continuous coring procedures, and constitute a basic exploration tool of the coal industry. It is based on technology proven over decades of use and it is reliable and cost-effective.

Continuous coring provides intact, physical evidence that confirms the presence, thickness, and nature of the coal seam. Not only does it provide desorption samples, but it provides samples for chemical and physical properties tests, including petrographic analyses. It also provides information on the nature of adjacent enclosing strata that can be critical to well and stimulation design or mine planning. It is especially important that continuous coring provides for full recovery of coal samples in frontier areas, where the depths and thickness of the beds may be poorly known or highly variable, making conventional gas well rig spot-coring difficult and less successful.

Early on, some gas companies operating in the Northern and Central Appalachian coal basins realized the value of using narrow diameter, dedicated coring rigs to obtain coal samples. These core holes are often referred to as "slim holes" in the gas industry. Maximum coal depths in the Northern and Central Appalachian coal basins typically are 1500–2500 ft or less, although some of the rigs are easily capable of drilling to 3000 ft or more. A drilling company that routinely cores coal strata at the expected project depths and has demonstrated experience in the planned coring area should conduct the coring operation.

Beyond being used to establish the presence of coal and to provide samples for desorption testing and routine coal analyses, core holes can be used to conduct a variety of tests. Geophysical logs and various other tests such as permeability tests, which can be performed in the larger diameter holes produced by conventional gas well rigs, can also be conducted in narrow diameter core holes with the proper size tools.

Since the continuous coring rig is dedicated solely to coring, the operations are basic and straightforward. The cores are cut using water; in some areas, "mud" may be needed to control caving/squeezing strata. Coring rigs are readily available in the Appalachians and routinely achieve 100% core recovery. In order to provide some perspective on the application of continuous coring in the coal industry, consider that routine coal industry practice for proper characterization of a coal reserve is the use of core data at a minimum of 6–25 core holes per square mile (1 core per 25–100 acres).

Regulatory requirements for drilling core holes vary significantly from state to state, so it is important to contact the appropriate regulatory agency in the state of interest before proceeding with any exploration plans. For example, in Pennsylvania, a Notice of Intent to Explore must be filed. Once the notice is filed, if no problems or objections are noted, drilling can proceed; no permit is required. In West Virginia, a permit application must be submitted and a permit issued before drilling can proceed; a 4–8 week review period is common before a permit is issued.

In the Northern and Central Appalachian Coal basins, Nominal Cross-section (NX) size core holes are routinely drilled. NX cores are approximately two inches in diameter and produce an approximate three-inch borehole. Larger diameter cores can be drilled, but cost rises significantly as the diameter increases. A typical rig is truck mounted (approximately 13 tons) and the drilling pad requires relatively little site disturbance (as small as 40 ft × 100 ft). A crew of two, the driller and helper, conduct all operations including conduct of a mud program, if needed. A third person is needed during mobilization, for certain testing procedures or when hauling water is required. Once a larger diameter surface casing is set (approximately 20 ft depth), coring is conducted using 10-ft rods, usually in 15–20 ft runs. In recent years, even smaller, track mounted rigs are used in coring operations. Cores are drilled using 15 or 20 ft coring barrels and the core is retrieved with a wireline. Short runs are made when coal thickness and/or stratigraphic position is not well established, so as to insure the entire target coalbed is included in the core barrel. Typical time to reach the surface once the barrel has started up the hole would be about 7 min from a depth of 1200 ft.

When the coring barrel arrives at the surface, the core should be pushed out into a core tray, with no need for a core barrel liner. The core is quickly examined, described, and samples of the complete seam are placed into the desorption canisters. Samples should be canistered as quickly as possible and are routinely canistered within 5–10 min after reaching the surface, depending on such factors as need for detailed core description, condition of core, or weather conditions. A typical time to reach the surface once the core barrel has started up the hole would be about 7 min from a depth of 1200 ft. Split core barrels, that lay open the entire core at once for examination, can be used. The drilling rates are much slower with a split barrel. Experienced rig operators can accurately determine when a coal is first encountered, the coal thickness, and the thickness of coal partings, during the coring operation. This alerts personnel performing desorption tests well in advance of the arrival of the core at the surface, allowing for positioning of the appropriate number of canisters and other supplies. Rigs are typically operated dawn to dusk (10–14 h). Experience over the years has shown that efforts to reduce the number of coring days per hole by operating around the clock are neither usually cost-effective nor time saving. The gas industry had often shied from use of dedicated coring rigs as too costly and time consuming. Yet the gas industry has learned that it solves many of the problems inherent in coring from a conventional drilling rig. For example, it addresses much of the problem of estimating the lost gas component by significantly reducing the time for recovery of samples. Drillers experienced with the rock types in the basin being explored are much more efficient and usually produce better core recovery. Actual drilling costs will vary with the specific sites chosen for coring locations and will depend on numerous factors such as site access, preparation, and reclamation; nearness of available water sources; drilling rates and total depth of hole; hole stability problems; and need for mud programs. In the Northern and Central Appalachian coal basins, it may be necessary to drill through mined strata and/or broken strata; a larger diameter hole can be reamed through the mined or broken strata and cased temporarily with larger diameter rods during the coring operation. Drilling experience limited to foundation coring or "hard rock" coring is not acceptable for this task.

In the Northern and Central Appalachian coal basins, when CBM production is being considered for pipeline injection, the use of narrow diameter continuous cores from dedicated rigs, and a thoroughly designed coring/testing program, precludes the need to rush to judgment when the well drilling stage is reached. The variety of information, such as desorption and permeability data, coal characteristics and geologic data that can be gathered from a well-planned, on-going coring program, can be correlated with production data to direct successful assessment and development of a CBM reserve.

Geophysical logging: Related to the coring operation is the need for accurate and reliable geophysical logs. It is especially important to run a "coal suite of logs" at the proper resolution. Drilling conditions may dictate a need to run the logs through the drill steel to assure the safe retrieval of the geophysical sonde. A detailed discussion of geophysical logging is provided in this publication.

Geologist log: In addition to the driller's log and geophysical logs, a detailed geologic log should be kept. Reliable correlations of the deeper coals in the Appalachian and Illinois coal basins have been hampered by the lack of deep core information. State geological surveys are quite valuable for conducting core logging and assisting with correlations of the coals. Geologic surveys can also provide reliable determination of when coring has gone below the lower-most coal bearing horizons. They are quite useful in assisting on deep holes at "frontier depths" and will keep project data confidential through signed agreements, when legally possible. Core logs provide more reliable information about bed thickness, depth, and detail than do oil or gas well driller's logs. E-logs from oil or gas wells do provide accurate bed depths and total thicknesses, and therefore are more reliable than gas well driller's logs, although internal noncoal partings are not discernible as in a core. Often not all of the coals noted in the core logs are noted in the gas well driller's logs so that no conclusion can be drawn from the nonreporting of a coal in a gas well driller's log.

Desorption testing: When properly and reliably conducted, direct method desorption tests are an effective method for initial indication of whether gas is present in sufficient quantities for production by a gas company and/or provides data for mine ventilation designs. From a mine-planning stand-point, it is important to note that coal samples are generally "safe" (from deterioration) for coal quality analyses, following completion of desorption testing in the closed environment of desorption canisters. Desorption data are also vital for safety design of cleaning and storage facilities, as well as for safe transportation of the coal. The methane contents of coal seams can vary greatly throughout a given coal reserve and it is critical to the proper characterization of the reserve to use correct testing procedures and properly designed and constructed equipment. A review of the US Bureau of Mines (USBM) desorption database includes at least 1200 separate files (Diamond et al., 1986). These data are publicly available.

The principle of desorption testing is relatively simple, but it must be done in a manner that is easily repeated and provides reproducible and defensible results. When variable methane contents show up, in order to use the data with confidence for correlations and planning, there must be confidence that the variations are due to natural factors such as coal makeup, structure, and permeability or enclosing strata and not due to poor procedures and/or poor equipment design or equipment failure. Canister design and construction must be such that there is minimal chance of failures. A 4-mil polyethylene liner is a practical option used to isolate the samples from the canisters and for ease of handling samples when testing is completed. Four mil polyethylene sleeves are slipped over the core samples before being placed in the desorption canisters, which helps to maintain the stratigraphic integrity of the samples. Canisters should be routinely checked for wear and should be duration-pressure tested prior to use.

Multiple options for performing gas content testing are available to the user (American Society for Testing Materials, 2010). The manomeric, direct method test has a proven track record in the Northern and Central Appalachian Basin. Careful measurement of temperature and atmospheric pressure are made at initial sealing and each time a volume of desorbed gas is measured from each canister. Desorbed volumes measured are corrected to express the volumes at standard temperature and pressure (STP) conditions, using basic gas law principles. Using this method, pressure–volume–temperature relationships must be applied not only to measured desorbed gas volumes, but to canister headspace as well. Provision must be made for handling and review of the enormous number of calculations involved to properly conduct desorption testing for each coal sample canistered, including such aspects as calculation for headspace or canister water vapor pressure at testing temperature (coal samples are generally wet and gases are measured/collected over water).

Desorbed gas volumes are generally measured in the field for approximately 2 h immediately following canistering of samples. It is extremely important to maintain a constant temperature during this phase of testing and laboratory temperature should be maintained at a constant temperature throughout the testing procedure. Estimates of gas lost (gas released from the time the coal is first cored until it is canistered) prior to sealing the cores in the canisters are based on the readings taken during the first 2 h and are calculated by regression analysis using the "lost gas" method developed by the USBM. A Gas Research Institute report discusses three methods for calculating lost gas. While no method is completely satisfactory, the USBM method for the Northern and Central Appalachian basin coals has provided consistent results.

Owing to the great variability in coal characteristics that may occur for a given seam, the entire coal core should be tested. Quick sealing, cam locking desorption canisters provided significant advantages in the field. It is recommended that all obvious coal material (megascopic) be placed into a canister. Any distinct intervals of easily separated, noncoal material showing can be placed in a separate canister(s) to monitor gas liberation. Strata adjacent to the coals encountered that show evidence of gas, especially zones of coal, interlaminated with bone, shale, or other impure lithologies, should also be sealed in canisters. If no significant volumes of desorbed gas are measured for the noncoal intervals, these tests can be terminated. Owing to stratigraphic correlation considerations, it is generally best to identify individual seams by their depth and not by specific seam names.

The thicknesses and stratigraphic positions of coals that are encountered at greater depths often can be quite variable and difficult to predict, particularly in frontier areas. Owing to the uncertainty of the number of coals, coal thicknesses and intervals between the coals, a two-person team is generally used for fieldwork. This ensures tests are conducted without compromising the quality of the results. When desorption data are gathered for mine planning, the author modifies the typical procedure of taking readings every 15 min for the first 2 h (for lost gas estimation) and takes readings at shorter intervals. This provides better characterization of the potential for variability in methane liberation during the mining cycle, from when the coal is cut at the face and continuing through transport out of the mine.

Testing should continue until consistently low volumes are measured over at least a 30-day period. A minimum of approximately 120 days is allotted for desorption testing. Preliminary desorption results are periodically provided prior to final determination of the methane content estimates. Core samples are removed from the canisters at completion of desorption testing and core samples are air dried at room temperature before final weighing of each sample. It is typical to note a 5% or more weight loss owing to loss of water weight.

Residual gas testing: After desorption testing is completed, residual gas tests are sometimes conducted. The core sample is crushed to a fine powder in a sealed canister and the volume of gas "desorbed" is measured. Upon completion and opening of the canister, any obvious oversize particles (i.e., greater than approximately one quarter inch) are removed and not included in the estimate of residual gas per unit weight. Care must be taken not to prematurely terminate desorption testing, especially of coals for which very little other desorption data are available for a particular coal reserve or for frontier depth coals. Many Northern and some Central Appalachian coal basin coals desorb very slowly, reaching a "steady" curve during testing that may last for years. It is important to recognize this character, especially when testing reserves for which little or no reliable desorption data are available.

Gas composition analyses: Gas from each canister can be sampled during desorption testing for chemical analyses, especially if recovery of the gas is planned for utilization as natural gas. Gas samples are analyzed by gas chromatography. Most gas desorbing from virgin coals is primarily methane, generally with minor amounts of other higher hydrocarbons. Careful monitoring of the quality of the gas desorbed from coals is important. If potentially deleterious gases to energy utilization such as CO_2 are detected from any particular coal sample, additional gas samples should be analyzed from the canisters of such samples during the course of desorption testing. CO_2 has a very different water solubility than methane.

Coal characterization tests: Standard coal characterization tests, routinely used by the coal industry, including petrographic analyses, should be considered, beginning with the early stages of a project. To aid in comparison of gas content estimates, coal quality analyses can be used to normalize the estimates by removing the moisture and ash contents and reported on a moisture-ash-free basis. This is especially important when using methane contents for estimating the in-place gas resource for a given coal resource.

Slug testing: In situ permeability estimation based on the slug test has been used extensively in CBM work in Alabama and Australia, and is considered a valuable tool in project assessments and gas reservoir simulations. Slug tests offer an opportunity to correlate permeability data with geologic features, including linears, sandstone channels, and clay veins, and with coal characteristics. Depending on the permeability of the coal being tested, the slug test may last from 8h to as long as several days or more. A programmed data recorder is used to manage data collection for computer analysis and reduces the need for personnel to be continually on site during data collection. Testing requires installation of nitrogen inflated straddle packers on a string of 20 ft collared joints. Collared joints are most readily available, but flush joint casing is recommended. Packers/string installation is handled by the coring or service company. Packers are installed using the coring rig initially, and are usually moved up/down the hole to additional selected zones with a water well rig or other small rig.

Presenting the results: Desorption results are presented in a summary table reporting the estimated desorbed gas content, lost gas content, residual gas content, and total gas content of each sample in cubic feet per ton ("raw coal" basis recommended) at STP of 60 °F and 1 atm (dry gas basis). In addition, tables reporting the volumes of gas actually measured throughout the desorption of each sample, including the date, time, elapsed time, temperature, and barometric pressure at the time of each measurement, the corrected volumes, and the cumulative totals after each reading can be included in a final report to the client. Basic desorption plots of the cumulative gas contents (not including lost gas) for each sample corrected to STP, vs cumulative time illustrate how the gas desorbed from the samples, an important addition to a report. Composite plots, comparing samples from the same seam, different seams, and/or individual core holes are also informative, and should all be plotted at the same scale for proper comparability.

Lastly, separate tables reporting the results of chemical composition analyses and Btu testing for desorbed gas, slug tests, or any other such tests or analyses conducted, can be included in the final report.

CONCLUSION

Many of the elements discussed in this section are often ignored in CBM recovery efforts. Interpreting the results of desorption tests and slug tests is complicated by various factors. A given coal reserve has its own peculiar character related to structure (e.g., folds; linears), coal characteristics (e.g., chemical makeup; macerals) and other geologic features (e.g., discontinuities; clay veins) that control concentration and release of methane. A coal reserve's CBM character is the same whether producing methane by ventilating it for mining or by removing it for gas production. The biggest constraint on interpreting and using desorption information is most often the limited CBM database available for a given coal reserve. Due to the variable nature of most coal reserves, a large number of core holes are usually drilled for characterization for mine design and coal quality delineation. Proper characterization of a CBM reserve likewise needs to be based on a significant number of core holes, desorption tests, and other tests such as permeability analysis. Combining CBM testing with a traditional core drilling and analysis program will build a CBM database suitable to characterize the gas production potential of a coal reserve. This approach can result in successful and profitable production of CBM in the Northern and Central Appalachian coal basins, as well as in other coal basins. Such a comprehensive database will provide for correlations between desorption and in situ permeability information, water data, coal characteristics, geologic factors, and actual production/ventilation data. A plan should be developed that leads to understanding coal as a distinct natural gas reservoir, not merely just another conventional reservoir.

NOMENCLATURE

A Reservoir area, acres
D Matrix diffusivity constant, (per second)
f_a Ash content, fraction
f_m Moisture content, fraction
$\underline{G}$ Gas-in-place, scf
$\overline{G}_C$ Average gas content, scf/ton
G_{ci} Initial gas content, scf/ton (dry ash-free)
G_s Gas storage capacity, scf/ton
G_{sa} Gas storage capacity at abandonment pressure, scf/ton (dry ash-free)
h Thickness, ft
P Pressure, psia
P_L Langmuir pressure constant, psia
q_{gm} Gas diffusion rate, 1000 cubic feet (MCF)/day
V_L Langmuir volume constant, scf/ton
σ Matrix shape factor, dimensionless
ρ_a Ash density, g/cc
$\underline{\rho_c}$ Matrix density, g/cc
$\overline{\rho}_c$ Average matrix density, g/cc
ρ_o Pure cool density, g/cc
ρ_w Moisture density, g/cc

References

American Society for Testing Materials, 2010. D7569–10: Standard Practice for Determination of Gas Content of Coal-Direct Desorption Method. vol. 5.06.
Collins, R.E., 1989. New Theory for Gas Adsorption and Transport in Coal. Paper presented at the International CBM Symposium, Tuscaloosa, Alabama.
Diamond, W.P., Levine, J.R., 1981. Direct Method Determination of the Gas Content of Coal: Procedures and Results, Report of Investigations 8515. United States Department of the Interior, Bureau of Mines, Washington, D.C.
Diamond, W.P., LaScola, J.C., Hyman, D.M., 1986. Results of Direct-Method Determination of the Gas Content of US Coalbeds. USBM IC9067.
Harpalani, S., Schraufnagel, A.R., 1990. Influence of matrix shrinkage and compressibility on gas production from coalbed methane reservoirs. Paper SPE 20729. In: Proceedings of SPE Annual Technical Conference and Exhibition.

Hughes, B.D., Logan, T.L., (May 1990). How to Design a CBM Well, Petroleum Engineering, International, 16.

Langmuir, I., 1916. The constitution and fundamental properties of solids and liquids. Journal of American Chemical Society, 38(11), pp 2221–2295.

Mavor, M.J., McBane, R.A., (June 1989). Western Cretaceous Coalseam Project, Quarterly Review of Methane from Coalseams Technology, 6(3,4,24).

Mavor, M.J., McBane, R.A., (November 1991). Quarterly Review of Methane from Coalseams Technology, 9(1), 19–23.

Mavor, M.J., Nelson, C.R., 1997. Coalbed Reservoir Gas-in-Place Analysis. Gas Research Institute Report No. GRI-97/0263, Chicago, Illinois.

Mavor, M.J., Pratt, T.J., Britton, R.N., 1994. Improved Methodology for Determining Total Gas Content, vol. I, Canister Gas Desorption Data Summary, Gas Research Institute Report No. GRI-93/0410, Chicago, Illinois.

Mavor, M.J., Pratt, T.J., Nelson, C.R., Casey, T.A., 1996. "Improved gas-in-place determination for coal gas reservoirs. Paper SPE 35623. In: Proceedings of SPE Gas Technology Symposium.

McKee, C.R., Bumb, A.C., Bell, G.J., 1984. Effects of Stress-Dependent Permeability on Methane Production from Deep Coalseams. Paper SPE 12858 presented at the Unconventional Gas Recovery Symposium, Pittsburgh, Pennsylvania.

McLennan, J.D., Schafer, P.S., Pratt, T.J., 1995. A Guide to Determining Coalbed Gas Content. Gas Research Institute Report No. GRI-94/0396, Chicago, Illinois.

Nelson, C.R., 1999. Effects of coalbed reservoir property analysis methods on gas-in-place estimates. Paper SPE 57443. In: Proceedings of SPE Eastern Regional Conference.

Palmer, I., Mansoori, J., 1996. How permeability depends on stress and pore pressure in coalbeds: a new model. Paper SPE 36737. In: Proceedings of the SPE Annual Technical Conference and Exhibition.

Patching, T.H., 1965. Variations in Permeability of Coal. Paper presented at the Rock Mechanics Symposium. University of Toronto.

Puri, R., Evanoff, J.C., Brugler, M.L., 1991. Measurement of Coal Cleat Porosity and Relative Permeability Characteristics. Paper SPE 21491 Presented at the SPE Gas Technology Symposium, Houston.

Ramurthy, M., Marjerisson, D.M., Daves, S.B., 2002. Diagnostic Fracture Injection Test in Coals to Determine Pore Pressure and Permeability. Paper SPE 75701 presented at the Gas Technology Symposium, Calgary, Alberta, Canada.

Rodvelt, G., Moyers, W.L., 2009. Case History: Recompletions in a Virginia Coalbed-Methane Field Yield Additional Gas. Paper SPE 125458 presented at the Eastern Regional Meeting, Charleston, West Virginia, USA.

Rodvelt, G., Oestreich, R., 2008. Best Practices for Obtaining Quality Permeability Data with CBM Matrix Injection-Falloff Testing. Paper 0827 presented at the International Coalbed and Shale Gas Symposium, Tuscaloosa, Alabama, USA.

Rogers, R.E., 1994. Coalbed Methane: Principles and Practice. Prentice-Hall, Inc.

Saulsberry, J.L., Schafer, P.S., Schraufnagel, R.A. (Eds.), 1996. A Guide to Coalbed Methane Reservoir Engineering. Gas Research Institute Report GRI-94/0397, Chicago, Illinois.

Seidle, P.J., Huitt, G.L., 1995. Experimental measurement of coal matrix shrinkage due to gas desorption and implications for cleat permeability increases. Paper SPE 30010. In: Proceedings of the International Meeting on Petroleum Engineering.

Wireline Logs for Coalbed Evaluation

Todd Sutton

Schlumberger, Pittsburgh, PA, USA

5.1 BASIC COALBED LOG EVALUATION

Most basic log delineation of coalbeds involves running the caliper—density-gamma ray measurement. Some operators prefer to supplement that with one or more of either the compensated neutron, photoelectric factor (PEF), or resistivity logs. More precise vertical coalbed determination comes with acquiring high-resolution data. In "dry" coalbed regions drilled on air, audio, and temperature log measurement are used to identify gas entry, and thereby qualitative permeability. The 3⅜″ Platform Express and new 2¼″ Multi Express (MEX) have provided all of these essential coalbed measurements. See Figure 5.1 for a typical coalbed methane (CBM) log presentation.

The standard Borehole Compensated Sonic tools have been used to identify coals and tie-in to surface seismic, also. Resistivity measurements offer a look at the presence of an invasion profile (fluid-filled boreholes), thereby indicating qualitative permeability. The Array Induction Tool with its five depths of investigation as well as micrologs have successfully identified invasion in coalbeds. This drilling fluid invasion suggests cleating.

In some cased holes, coalbed identification is acquired with pulsed neutron tools, which provide a gamma ray, neutron porosity, sigma measurement, plus near/far count rate qualitative gas identification. The Reservoir Saturation Tool (RST) has answered the need for cased hole evaluation. See Figure 5.2 for the basic log values for coal identification.

In addition to providing improved slim and 3″ core hole access, the new MEX is notable for it's efficiency in obtaining high-quality, comprehensive data in all formations. The lightweight, small-diameter, compact logging string can log at speeds as high as 4500 ft/h [1372 m/h] recording

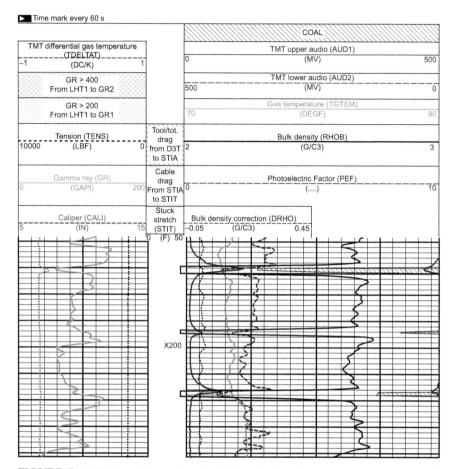

FIGURE 5.1   Platform Express in coal with low density plus gas entry temperature and audio response in air-filled borehole.

Coal type	Density (gm/cc)	CNL porosity	Sonic (usec/ft)	PEF(barn/cm²)	Sigma (capture unit)
Anthracite	1.47	37	105	.16	8.7
Bituminous	1.24	60+	120	.17	14
Lignite	1.19	52	160	.20	13

FIGURE 5.2   Though the formation presence of water and varying salinity influences log response, pure log responses in various types of coal are as follows. PEF (photoelectric factor), CNL, Compensated Neutron Log.

all sensors simultaneously. This newest tool can be run in real-time or memory mode, and offers through drill pipe access for vertical and horizontal wellbores.

The MEX platform is uniquely and fully characterized for CBM wells to account for the characteristically low density and low PEF of CBM formations. Standard logging tools are not characterized for the low density of coal and PEF's lower than 0.9. In addition, the unique articulated configuration of the Litho-Density* pad improves pad application to reduce the effects of hole rugosity, which is oftentimes associated with coalbeds (Herron and Herron, 1996).

5.2 ADVANCED COAL ANALYSIS

As stated above, the simplest log method of qualitative analysis is to use the bulk density to approximate ash content, which correlates to coal rank. The addition of the compensated neutron, PEF, and gamma ray complements local correlations. This technique can be complicated by the tendency of coalbeds to washout and thus creating uncertainty on the absolute measurement. In addition, the coal components, particularly ash can vary, which adds to parameter selection uncertainty.

An established alternative coal evaluation method is based on elemental analysis from neutron-induced gamma ray spectroscopy. The advantage is that the preponderance of formation signal arises from the formation elements, not the borehole in the case of washout; plus gamma ray spectroscopy yields discrete elemental signatures that more precisely define mineralogy such as ash content. Both the Elemental Capture Sonde (ECS) and RST have been used successfully for this quantitative coal analysis technique (Ahmed et al., 1991).

Neutron-induced gamma ray spectroscopy tools emit high-energy neutrons that are principally slowed down and then captured by the formation elements. Upon capture, the formation element emits a gamma ray at a unique energy level that is characteristic of that element. A gamma ray detector detects the energy level and also the count rate, called the energy spectrum, which reflects the type and quantity of element.

Figure 5.3 represents the Coal Advisor process, which begins by calculating the relative gamma ray yield by comparing the measured spectrum with the theoretical spectrum of each element. A mathematical inversion provides a percentage of the principal elements such as silicon, calcium, iron, sulfur, and hydrogen. The yields obtained at well site are only relative measures, because the varying borehole environment affects the total signal. Additional information is needed to gain the absolute elemental concentrations.

This is accomplished by using the oxide closure model, which states that dry rock consists of a set of oxides, the sum of whose concentrations must equal unity. The relative oxide yield allows total yield calculation, and hence the factor needed to convert to unity. This normalization factor then converts each relative yield to a dry weight elemental concentration.

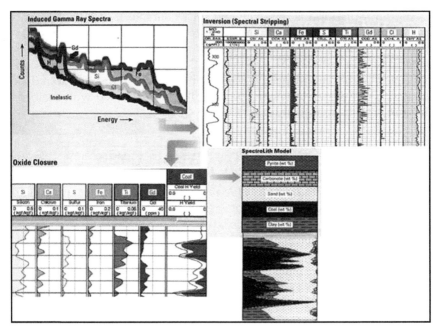

FIGURE 5.3   Mineralogy interpretation steps—from gamma ray spectra to SpectroLith model.

The final step in coal analysis is to transform elemental concentration to mineral concentrations via the SpectroLith model by using empirical correlations based on more than 400 core samples from varied clastic environments. The major advantage of this method is that it is virtually automatic and does not depend heavily on user intervention.

Results are presented as dry weight percentages of clay, coal, the aggregate of quartz—feldspar—mica, and other minerals such as pyrite and siderite. Coals are easily identified by their high hydrogen concentration. The challenge to quantify amounts of fixed carbon, volatile material, and moisture is accomplished with two assumptions. First, other sources of hydrogen in coal such as water in cleats, formation moisture, clay bound water, and borehole water (unless drilled on air), tend to have a consistent background and thus are subtracted to yield the actual hydrogen concentration in coal. Second, for a given area or formation, hydrogen coal concentration can be sufficiently consistent to allow for a conversion to coal percentage.

Total ash content is obtained from its components: quartz, clay, carbonates, and pyrite. Fixed carbon and volatile material can be estimated from ash content correlations. As mentioned, many correlations have been established for specific areas or formations. ECS Coal mineralogy is further enhanced with other log information with the ELAN computation. Detailed ash description from the ECS, combined with the density, PEF, and neutron data, which

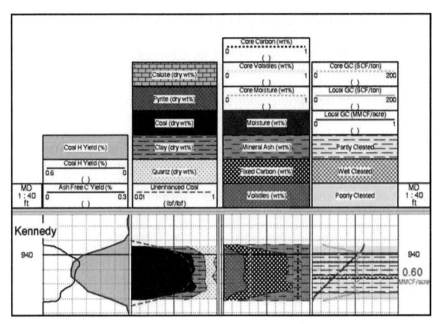

FIGURE 5.4 Coal Advisor example using openhole logs and Elemental Capture Spectroscopy Tool.

identifies fixed carbon from volatile matter; all contributes to a more precise coal analysis. See a finalized Coal Advisor in Figure 5.4.

Cleating can be identified with the presence of calcite and pyrite, which suggests formation water flow causing secondary mineralization. High quantities of calcite and pyrite may indicate filled cleats or low-grade coal. Quartz and clay have been observed in cleats, also. Large quantities of such minerals plus a large total ash content indicate a lower ranked coal. This coal type will have fewer cleats since less water and volatile matter was lost during coalification. By observation, well-cleated coal has between 2 and 7% calcite and 0.5–5% pyrite. Poorly cleated coals have total ash above 45%, clay above 25%, and quartz above 10%. Partly cleated coals will have percentages that run in between, but cutoffs should be tempered by local production experience.

In summary, neutron-induced gamma ray spectroscopy as with the ECS, in combination with other logs provides a continuous record of the major variables needed for coal seam evaluation and surrounding formations. Cleat grading indicates permeability and ash content lends to coal grading and gas content. Similar elemental analysis can be done in cased holes not only with the ECS, but also with the Reservoir Saturation Tool (RST), which has the additional benefit of identifying coal zones with the carbon/oxygen measurement.

5.3 IMAGING AND MECHANICAL PROPERTIES

Further detailed coal analysis can be achieved with the Formation Micro Imager Log (FMI) with its high 0.2" (0.5 cm) vertical resolution. This downhole tool makes circumferential microresistivity measurements and responds to the various conductivities of laminations, pore space, and some minerals. FMI offers thin coalbed resolution, identification of fractures, cleat type, faults, drilling-induced fractures, dip and strike orientation, and in situ stress determination that can supplement advanced coal analysis. See Figure 5.5 for an FMI example in coal.

In addition, dipole type sonic tools attain the formation mechanical properties for understanding fracture behavior and for designing optimum

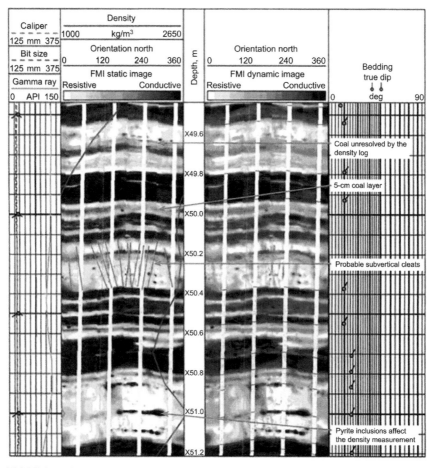

FIGURE 5.5   High resolution coalbed image using the FMI (Formation Micro Imager Log). API, American Petroleum Institute.

fracture jobs. Coal has a higher Poisson's ratio and lower Young's Modulus oftentimes than surrounding beds, so coals tend to transfer overburden stress laterally and maintain higher fracture gradients. The Sonic Scanner (SSCAN) has provided these critical formation property measurements and has shown that coals are typically more stressed than surrounding beds, thereby indicating more fracture growth outside of the coalbed than within. The SSCAN not only determines Young's Modulus, Poisson's Ratio, and Closure Stress Gradient, but also the in situ stress magnitudes and directions to improve hydraulic fracturing results. See Figure 5.6 for a Coal Advisor plus SSCAN-Mechanical Properties Log in coal.

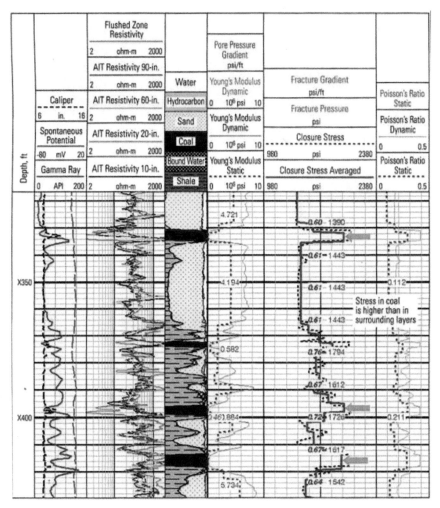

FIGURE 5.6 Coal Advisor example with stress profile using Sonic Scanner log. AIT, Array Induction Tool.

5.4 SUMMARY

Formation evaluation techniques for bulk density, spectral analysis, imaging, and mechanical properties; tested by extensive core study help the industry understand coalbed reservoirs. Log processing methods uncover lithology, coal quality, proximate analysis and permeability data. Local stress information from imaging and dipole sonic tools reveal more detailed variables associated with permeability and fracture behavior. All logging techniques contribute to more precise reserve calculations and more effective completion methods (Anderson et al., 2003).

References

Ahmed, U., Johnston, D., Colson, L. "An advanced and integrated approach to coal formation evaluation", paper SPE 22736, presented at the SPE Annual Tech Conference and Exhibition, Dallas, TX, October 6–9, 1991.

Anderson, J., Basinski, P., Beaton, A., Boyer, C., Bulat, D., Satyaki, R., Ronheimer, D., Schlachter, G., Colson, L., Olsen, T., Zachariah, J., Khan, R., Low, N., Ryan, B., Schoderbek, D., Autumn 2003. Producing natural gas from coal. Oilfield Review 15 (3), 8–31.

Herron, S., Herron, M. "Quantitative lithology: an application for open and cased hole spectroscopy", Transactions of the SPWLA 37th Annual Symposium, New Orleans, LA, June 16–19, 1996, paper E.

6

Vertical Well Construction and Hydraulic Fracturing for CBM Completions

Gary Rodvelt

Global Technical Services, Halliburton Energy Services, Canonsburg, Pennsylvania, USA

6.1 INTRODUCTION

Coalbed methane (CBM) wells are not unlike conventional oil and gas wells in the well construction phase. Freshwater zones must be drilled through and cased off. Production casing is predominately set through all the coal seams with additional rathole provided to pump water from the lowest seam efficiently. An additional casing string may be needed to case off mine voids encountered below surface casing and prior to drilling the active coal seams. Production casing sizes are typically 4½-or 5½ in outer diameter (OD); however, larger sizes might be needed in highly permeable seams like the Powder River to allow larger artificial lift systems. Appalachian, Black Warrior, and Illinois basin wells tend to dewater quickly. Water production subsides to the point that variable speed units are needed to continuously pump the wells or timers are used to cycle pumps on and off.

Most Appalachian wells are air drilled to limit damage to the coal seams. In overpressured areas such as the San Juan and Arkoma basins, it can be necessary to drill with fluid systems to provide hydrostatic control. Cementing needs for surface and intermediates are generally dictated by state regulations using standard cements with accelerators to provide quick early compressive strength. Production string blends can vary depending on desired cement tops, coal seam permeabilities, and economics.

While the recent trend is to drill horizontally in areas with thicker coals, there still exist a number of areas where multiple coal seams are stacked over 800–1400 ft of vertical depth that can only be economically accessed with multistage hydraulic fracturing. The stimulation engineer must work with the reservoir engineer to develop the most cost-effective completion program. Problems unique to coal include fines, excessive fluid leakoff, fluid damage, and high treating pressures. Additionally, as operators look at deeper coals, permeability trends down, reducing the effectiveness of conventional hydraulic fracturing. The stimulation engineer often must rely on practices learned in treating tight gas and shales to bring commercial gas production to the wellhead.

This discussion describes the various drilling techniques, drilling fluids, cementing solutions, and casing designs used by operators in CBM vertical wells. Current best practices for hydraulic fracturing of coals include formation stress measurements, proppant considerations, fluid systems, and specialty additives. In addition, multiple staging techniques and recommendations for treating low-permeability coals will be detailed in the completions section.

6.2 WELL CONSTRUCTION

6.2.1 Drilling

Vertical well drilling in Appalachia is done with small footprint air rigs that can access locations in the steep terrain. Small cuttings pits are necessary to capture returned solids and formation fluids carried back by the air stream. If fluid flows are encountered, a mist or foam system will be implemented using a surfactant and stabilizer to help clean the hole and maintain circulation. When drilling through strip-mined areas (prevalent in the Illinois Basin), the rubble pile, or "spoils", might require air drilling as lost circulation will be encountered with a fluid system. In overpressured areas, a fluid system may be required to maintain control of the well. Bentonite mud systems mixed with freshwater provide fluid loss and hydrostatic control. If soft, unconsolidated formations are encountered, it is preferable to use a mud system for hole stability; it can be as simple as circulating freshwater and native mud picked up during drilling. Most shallow shales are susceptible to sloughing when in contact with freshwater for more than 1 or 2 days; plans to inhibit them should be in place for extended drilling operations such as core holes or mine void areas.

As in situ permeability drops below 2 mD and/or the coal thickness is greater than 2 ft, horizontal drilling often becomes a preferred method to access production. Horizontal drilling processes are covered in a later section.

6.2.1.1 Drill Bits and Hole Size

Hole size should be determined by the amount of fluid that will be produced. High fluid volume wells such as in the Powder River Basin require the ability to move 500–2500 barrels of fluid per day to reduce pressure and begin gas desorption. Hole sizes will be large enough to allow the different casing strings to be landed and cemented in place. These holes would be drilled using a rotary rig that uses air and air hammer bits or fluid and the typical tricone rotary bit. Coal is soft; therefore any bit will penetrate it. The associated boundary layers of shale or sand are the design criteria for endurance and the appropriate fluid being used. Drill rates of 100–200 ft/h are not uncommon in coal while circulating water to clean the hole.

6.2.1.2 Drilling Fluids

Air is the preferred drilling fluid for CBM. It is low cost and environmentally safe. Air allows the driller to maintain an underbalanced hydrostatic head on the coal formation, which minimizes damage to the cleat system and problems associated with lost circulation. If permeability is great enough, the coals may "self-clean" any cuttings damage by flowing water and gas into the wellbore during drilling operations. If too much fluid entry is encountered, a surfactant can be added to the air system to create a mist or foam that will combine with the water and carry itself out of the well. Additional chemical additives for clay control, scaling, corrosion, or bacteria may also be added to the fluid system. Please note that caution is needed when using chemicals in the fluid system as the coal has a high surface area and will adsorb the chemicals, especially those with a hydrocarbon base carrier. Minimal use of surfactants, polymers, and solids will prevent both permeability damage and environmental concerns.

In areas that are overpressured, the air may be replaced with a compatible water-based system. It is still preferred to drill with a slightly underbalanced fluid system to avoid damage to the cleat system. Due to gas influx, annular control will be required as methane volumes are increased at surface conditions. With liquid drilling fluids, containment pits will need to be larger for capturing the increased fluids involved.

6.2.2 Casing Considerations

Casing designs should start by considering fluid volumes of the expected production and work back out of the hole. High-volume water producers like the Powder River basin wells require 7 in production casing whereas Appalachian wells "dry up" in a few months' time allowing the operator to choose 4½-or 5½ in production strings. If drilling in a mining area, an intermediate string will be required to case off this hole section. Unconsolidated surface areas can require a conductor to prevent washout from under the rig.

6.2.2.1 Conductor and Surface Casing

Conductor casing in Appalachia is typically 13⅜ in (with some 12¾ in) followed by 9⅝ in surface casing. For cases where 4½ in will be the production string, an 11 in conductor can be run followed by 8⅝ in surface casing. The conductor is the first string that isolates and protects surface freshwater zones from downhole fluids and prevents washout from under the rig in unconsolidated soil. These strings are cemented in place using standard cement with accelerator to provide a set that allows drillout in 8 h. Cementing regulations vary from one state to another so each state's rules must be consulted before completing the drilling prognosis.

6.2.2.2 Intermediate or "Mine" Casing

Intermediate casing is typically run in Appalachia to provide coverage over an open mine void section. Seven inch casing will be run approximately 50 ft below the mine void with cement baskets (Figure 6.1) above and below the mine void. A small amount of gypsum cement (thixotropic) volume is pumped and a plug dropped behind it to bring the slurry up to the mine void sealing the space below. One inch tubing is then run down the annulus between the 9⅝ and 7 in to fill the annulus from the cement basket above the mine void back to surface. The top out cement is typically standard with accelerator.

6.2.2.3 Production Casing

Production casing must be designed to withstand fracturing stimulation treatment pressures and large enough to allow production of water and gas. For cavitation completions or high-volume water producers, it will be necessary to run 7 in casing to allow room for cleanout and high-volume pumps. In Appalachian, Black Warrior, and Illinois basin wells, the operator has a choice of 4½- or 5½ in casing because the wells produce lesser volumes of fluid. The openhole should be drilled deep enough to allow setting casing 50–100 ft below the lowest producing coal seam. Artificial lift systems are not 100% efficient, and they need some (±10 ft) fluid above the pump to avoid gas locking during production. For 4½ in casing, it is recommended that the production pump be placed 75–100 ft below the lowest perforation to allow gas separation from the fluid. For 5½ and 7 in strings, 30–50 ft below is recommended. Production string cement should be raised at least 200 ft back into the intermediate or surface casing (if not back to ground level). This will ensure additional protection of the groundwater formations.

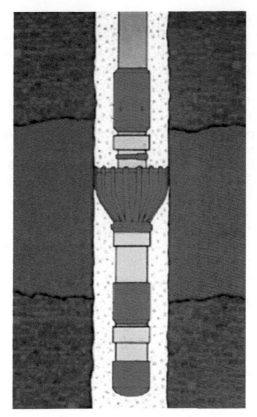

FIGURE 6.1   Illustration of using cement baskets for casing through mined zone.

6.2.3 Cementing

Cementing the casing provides pipe support, zonal isolation, and well control. CBM wells are similar to conventional wells in the process of cementing; however, the cement blends used must control fluid loss and maintain circulation to prevent damage to the cleat system. A slightly overbalanced cement system will maintain control over gas entry. The slurry weight must be balanced with additives to prevent lost circulation yet still be economical. Best practices for the cementing operation include borehole conditioning at optimum flow rates for good mud and cuttings removal (7–10 min of contact time with spacer), pipe centralization, and pipe movement to ensure complete isolation of the target zones.

In air-drilled holes, circulating the hole with water or gel sweeps to remove cuttings and prewet the wellbore will reduce the risk of cement dehydration before final placement. A reactive spacer can be used ahead of severe lost circulation cases to improve cement placement.

6.2.3.1 Lightweight Additives

Lightweight cementing additives include pozzolan material, bentonite, coal or asphalt particles, glass microspheres, fibrous materials, and even nitrogen gas, which is used to generate foamed cement. Standard cement is the base fluid, and some percentage or amount per sack of cement weight is used to form the blend. It is preferred to cement with slurry that is slightly overbalanced to prevent gas entry into the wellbore after placement, but the slurry must not be so heavy that it fractures the coal or begins to enter the cleat system. Standard cement is slurried neat at 15.6 lb/gal; most coals are at or below freshwater gradient of 8.33 lb/gal. Typical blends will be reduced in weight to the 11–13 lb/gal range to successfully place the slurry without breaking down the coals. Each basin will have specific needs that can be addressed by tailoring the cement blend to satisfy requirements. It is best to contact the local service company representative when beginning work in a new area to learn what blends have been successful in providing good zonal isolation.

6.2.3.2 Weighting Additives

When the coals are at an overpressured condition, or the formations encountered are unstable, it may be necessary to drill in an overbalanced condition to provide borehole stability and pressure control. Weighting additives will then be needed in the cement slurry to maintain well control and provide good zonal isolation. A weighted spacer can be required ahead of the cement to displace the drilling fluids and provide a clean formation face and pipe for the slurry to bond against. The additives will usually have a low water requirement, not significantly reduce the strength of the set cement, have minimal effect on the pump time, be chemically inert, compatible with other additives, and not interfere with well logging. Additives that are used to provide additional weight are barite, salt, hematite, and quartz sand; and, by using a dispersant, the neat blend can be densified to a heavier weight than it is normally mixed. For most cases, cementing overpressured coals will only require neat cement blends or densified (using salt and dispersant) as the degree of overpressure can be overbalanced with these solutions. It is best to contact the local service company representative when beginning work in a new area to learn what blends have been successful in providing good zonal isolation.

6.2.3.3 Specialty Additives

In addition to controlling the slurry weight, different classes of additives are available to provide fluid loss control to the formation, lost circulation control, and provide other slurry benefits required in cementing. Additives exist to control air entrainment, retard cement thickening time, accelerate cement thickening time, and to provide "markers" for cases when plugging holes to alert operators of borehole conditions below the cement. For example, red dye can be used in the slurry to denote a radioactive source that has been lost in the hole below. Expansion additives are available to increase expansion of the set cement over time during cement curing and provide a way to eliminate microannulus effects. Fibrous additives can provide resistance to shattering of set cement when added to the slurry. There are numerous additives available for use in any situation that might be encountered. Choosing the right combination should be done with advice from the local service company representative.

6.2.3.3.1 FLUID LOSS ADDITIVES

Filtration controllers minimize dehydration of the cement slurry into porous zones, protect water-sensitive formations like shales or fire clays, and improve squeeze cementing operations. Using a low-fluid loss additive (FLA) provides a thin filter cake of cement across porous zones allowing placement of the cement slurry without premature pressure buildup. A low FLA also provides hydrostatic pressure in the annulus after placement, which prevents gas cutting during set time. Low FLAs include organic polymers and latex. Latex provides the additional benefit of hydrogen sulfide (H_2S) and acid resistance. These properties become more important in areas that produce H_2S or carbon dioxide.

6.2.3.3.2 LOST CIRCULATION ADDITIVES

Lost circulation occurs when the formations encountered while drilling will not support the hydrostatic pressure of the drilling or completion fluid. It can be as simple as a porous sand thieving fluid to as drastic as all the mud or cement being lost into a cavern or void space. Lost circulation additives are then added to the system to slow or stop the rate of loss so that drilling (or cementing) can continue or the formation can be effectively isolated. If lost circulation occurs during drilling, additives should be added to the drilling fluid to "cure" the loss and restore circulation. It is not advisable to continue drilling without some means of cuttings removal as the drill string will become stuck. Logging operations also require that the hole is stable and the well under control.

Pipe-running operations can cause lost circulation because surge pressures may breakdown a zone. Care should be taken to run pipe at a rate that reduces or eliminates surging, and the use of auto-fill float equipment is recommended to aid in minimizing surge pressures. Once pipe is landed on bottom, circulation should be reestablished and "bottoms up" as a minimum volume circulated to ensure system stability.

Cement additives for lost circulation control include granular particles such as coal or asphaltic materials, quick or flash setting additives, and fibrous materials such as polypropylene and cellophane particles. Care must be taken when using any of these materials as they can plug float equipment, reduce thickening time, or plug the annulus if mixed inconsistently or at too high a loading. The local service company representative can assist in determining the proper additive and loading for the situation.

6.2.3.3.3 THIXOTROPIC ADDITIVES

Thixotropic additives provide the cement blend properties to be formulated for a low viscosity during mixing and displacing, but then have a rapid increase in viscosity when the slurry is static. Thixotropic slurries gain viscosity even during pumping when an accelerator is added, and as the circulation rate slows it will begin to build gel strength that resists flow. They should be used in lost circulation cases and across thief zones that might want to take fluid after placement such as fractured or vugular formations. Thixotropic cements improve fill-up because of resistance to loss after placement; they also inhibit gas entry because of the quick gel strength development. Finally, the gypsum additive typically used will provide some expansion during the hydration process, improving pipe bond and formation seal.

6.2.3.4 *Foamed Cement*

One lightweight additive mentioned earlier, nitrogen gas, can be used to form a unique cement blend that appears more as a foam than a fluid slurry. Weights (densities) can be varied based on the amount of gas added and the pressure acting to contain the gas. For this reason, a closed system must be in place before cementing operations to contain the foam within the well. Flowback through a choke manifold is required to maintain backpressure on the well during the cementing process and the annulus must be secured when finished. Foamed cement provides a secure, ductile, and long-lasting cement sheath for many types of wells, but especially CBM wells. The hydrostatic column can be tailored to just exceed the formation pressure without breaking down or fracturing of the coals.

Foamed cement has been tested in the lab and found to be very ductile. It is able to withstand hundreds of pressure expansion and contraction cycles without degradation of the sheath or bond to casing. It expands to fill every void space along the wellbore providing excellent bond logs with no fallback problems. Slurry weights of 8 lb/gal are attainable that have very low permeability and fair compressive strength and thus could be used as primary cement. This is the perfect cementing solution for a weak, fragile cleat system that cannot withstand excessive hydrostatic pressure. A local service company representative should be contacted to help in designing a foamed cement treatment.

6.3 COMPLETION PROCESSES

6.3.1 Introduction

Coal seam completions are similar to oil and gas well completions with some procedure changes required to protect the cleat system that occurs naturally in all coals. This system provides the permeability and porosity characteristic of the coal seam where free gas and water are stored. Most productive coals are of bituminous rank, which also has the lowest compressive strength. Coals contain large surface areas; this translates into many adsorption sites for polymers, hydrocarbon carriers, and surfactants that can damage the cleats or inhibit gas desorption. Coal fines are generated by the drilling and completion operations; a plan must be in place to control or inhibit their movement. Many coal basins are made up of multiple coal seams varying in thickness of 0.5 ft to 3–5 ft. Finally, not only shallow depths, but tortuous paths and multiple fractures/geometries lead to high treating pressures (and possible screen outs) during fracturing operations.

6.3.2 Openhole Completions

To minimize damage to the prospective completion interval, the first completions done in coal were openhole. Pipe was set above the zone before drilling with cable tools or a rotary and clear fluids or air. This technique works well for completing an interval that is to be mined, as it does not place any steel across the coal that might cause a fire hazard later during mining operations. The procedure is documented for the Warrior and San Juan basins by Rogers et al. (Rodgers, 2007). If multiple coal seams are exposed, wellbore sand plugs might be required to help direct fracturing treatments into upper seams. A large drawback of this process is the lack of a stable rathole to allow pumping of the openhole seam. Many times, a shale or fireclay is located below the coal seam that is susceptible to sloughing once it is penetrated and frac or production water passes by it. In some cases, an uncemented liner can be placed in the rathole immediately after drilling to provide wellbore stability.

6.3.2.1 Openhole Cavitation Process

In 1986, Meridian reported prolific CBM production after enlarging the openhole section through a technique called cavitation. This process involves pressuring up the wellbore and surrounding region by shut-in or using injection of air or water, then violently returning the fluid as an "open flow to atmosphere" condition through large diameter manifolding at surface into a pit with the intent of dislodging and rubbilizing the coal around the borehole. After supercharge is flowed off, the rubbilized coal is manually cleaned from the well by circulating or bailing. This cycle might be repeated a number of times to extend the radius of the wellbore from 7 to 8 in out to 10–20 ft.

Seam characteristics that will lead to success using cavitation methods are coals at least 10 ft thick, have permeability greater than 20 mD, low density (low ash), at or above water gradient pressure, and preferably in a high in situ stress regime. The extent of the fracturing of the coal will extend beyond the physical enlargement of the hole as stress is relieved. Rogers et al. provide more detail into the San Juan cavity completions in Chapter 7 of Coalbed Methane: Principles and Practices (Rodgers, 2007).

6.3.3 Cased Hole Completions

6.3.3.1 Cased Hole Conditions

The disadvantages of openhole completions—lack of fracture initiation control, limited rathole, and wellbore stability—can all be negated by running casing to total depth (TD) (including the appropriate rathole) and cementing in place. If one of the seams is to be mined, fiberglass or composite casings are available to replace steel components. Borehole sizes may need to be altered as fiberglass couplings have greater ODs than standard steel couplings. This will allow the completion engineer to stage multiple, thin seams with one of the many different access techniques described below.

6.3.3.2 Slotting Access

Once casing has been cemented in the hole, the operator must then gain access to the formation. One method to do this without rubbilizing the coal involves the use of a jetting tool (Figure 6.2) (Rodgers, 2007) run on jointed or coiled tubing. Once spotted at the calculated depth opposite the zone, friction-reduced water and sand are pumped at high pressure through opposing jets to abrasively remove casing and formation. Slots can be cut most efficiently going down by slowly lowering the tool in the hole while pumping. Slot lengths should not exceed 12–14 in; otherwise, casing integrity might be compromised. Casing may be cut off by rotating the tool in place while pumping. Penetration depths vary between 5 and 7 in (penetration graph); greater penetration (up to 14 in) can be accomplished by adding nitrogen to the system to reduce annular backpressure on the jets.

Care must be taken to control the flowback rates for prevention of surface equipment erosion. A safety hazard exists if methane gas or hydrocarbons are circulated back to surface. Strict safety guidelines must be followed in rig-up upwind and flowback away from the wellhead and pumping equipment. Contacting the local service company representative for help in location layout and jetting design is the best course of action for this job application.

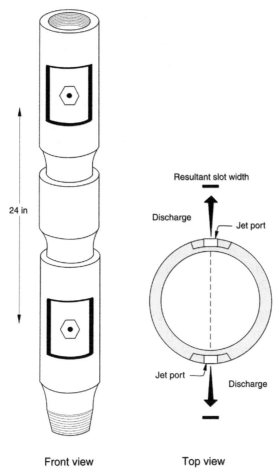

Front view   **Top view**

FIGURE 6.2 Jet slotting tool.

6.3.3.3 Conventional Shape Charge Perforating Access

Perforating with explosive jet charges provides a more conventional approach to gaining access to the coals. Four to six perforations per foot will allow the client entry into the coal seam without heavy rubbilization of the coals, which has been seen at higher density perforating. The chances of placing perforations within a 30° angle of the induced hydraulic fracture direction are increased with the six Jet shots per foot (JSPF), 60° phased perforating, which reduces tortuosity and perforation friction during fracture stimulation treatments. Conventional perforating is low cost and a routine operation by oil field workers. It allows selective opening of target zones and allows the stimulation engineer to design treatments for more complete coverage with the fracturing design. The completion design might need to include a small amount of hydrochloric acid for initial breakdown and cleanup of the perforations or as a spearhead to reduce entry pressures since the acid may remove cementing and perforating damage.

6.3.4 Multiple Seam Completions through a Cased Hole

Multiple seam completions can be staged a number of different ways depending on number of stages, casing size and pressure limits, job volumes (proppant and fluid), and frac rates desired. Sand fill has been used as a temporary plug in both open and cased hole applications. In cased hole, frac baffles, frac plugs, and packer/bridge plug combinations can be used. For larger intervals with many different target zones, coiled tubing fracture stimulation processes begin to make sense both from an economic and a time analysis. Each of these is discussed briefly so that the completion engineer can make the best decision for his/her project. Rodvelt and Oestreich provide greater detail concerning these completions (Rodvelt and Oestreich, 2007).

6.3.4.1 Staging with Sand Plugs

Sand plugs are set at the end of a frac stage by raising the sand concentration to the 10–15 lb/gal range to ensure that the final stage "packs off" the perforations. The "sand plug" volume is typically 250–500 gallons of slurry, depending on casing size and interval separation. Slowing the injection rate as the sand plug gets to the perforations may facilitate a pressure increase, but it can also leave the sand plug top too high and result in a tubing trip to "dress off" the top or bypassing a target interval. This technique is best utilized where at least 500 gallons of space is available between stages. The completion procedure must include the materials and fluid volumes to account for the sand plugs; the cleanout process might require a drill bit and power swivel to circulate the sand plugs out.

6.3.4.2 Staging with Frac Baffles

Casing frac baffles provide an economic way to isolate a portion of the casing once that stage is complete. Baffles of differing inner diameters (IDs) (smallest ID on bottom) (Figure 6.3) (Rodvelt et al., 2001) are run in the casing string during the primary cement job. During the fracturing operations, the respective ball is dropped at the end of the frac stage, usually in (or chased by) a small spearhead acid volume, and pumped to the smallest baffle it will not pass through, bringing an end to pumping. A new zone above the seated ball is perforated and that frac treatment is performed, followed by another ball drop (next larger ball) as before. This process is typically repeated three to five times

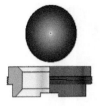

Pipe Size Inches (mm)	Baffle Thread	Baffle I.D. Inches (mm)
2.875 (73.0)	8 Rd. EUE	2.00 (50.8)
	10 Rd. NU	
3½ (88.9)	10 Rd. NU	2.37 (60.2)
	8 Rd. EUE	2.18 (55.4)
4½ (114.3)	8 Rd.	2.37 (60.2)
		2.70 (68.6)
		3.00 (76.2)
		3.31 (84.1)
		3.56 (92.7)

Pipe Size Inches (mm)	Baffle Thread	Baffle I.D. Inches (mm)
5 (127.0)	8 Rd.	3.00 (76.2)
		2.75 (69.9)
		3.06 (77.7)
5½ (139.7)	8 Rd.	3.27 (83.1)
		3.50 (88.9)
		3.87 (98.3)
7 (177.8)	8 Rd.	3.31 (84.1)
		3.5 (89.9)
		3.87 (98.3)
		4.06 (103.1)
		4.5 (144.3)

FIGURE 6.3 Frac ball and baffle with table of sizes. Rd. EUE is Round External Upset, Rd. NU is Round Non-Upset, Rd. is Round.

depending on the number of stages required, with five to six being the practical maximum. Drawbacks to this process include drillout required, metal debris going to bottom, and need for geologist or engineer to determine baffle placement points prior to the primary cement

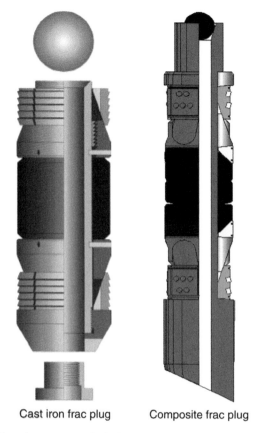

Cast iron frac plug   Composite frac plug

FIGURE 6.4   Frac plug types available for stage isolation.

job. If a premature sand-out occurs during frac, the next planned stage uphole may be covered with sand requiring cleanout or to bypass completion of that pay zone.

6.3.4.3 Staging with Flow-through Frac Plugs

Flow-through frac plugs are an improvement over frac baffles in that they can be run in the hole after the perforations are shot allowing flexibility in spotting the "plug" near the bottom of the new perfs. Once the plugs are spotted, a ball is landed much like the frac baffle technique, and the perfs above can be broken down and fractured. Two kinds of frac plugs are available—cast iron and composite (Figure 6.4) (Reese and Reilly, 1997). As the names depict, the cast iron tools are made of metal and rubber while the composite tools have ceramic, rubber, fiberglass, and epoxy components. The cast iron tools are less expensive to build; however, the composites require much less drillout time and can be

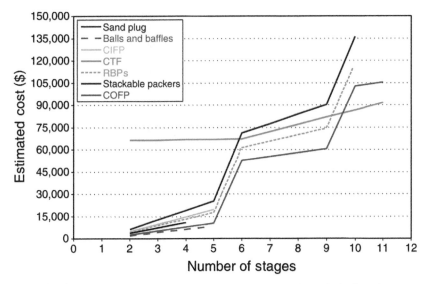

FIGURE 6.5   Cost comparison of different staging options. CTF, coiled tubing fracturing. CIFP is Cast Iron Frac Plug, RBPs is Retrievable Bridge Plugs, COFP is Composite Frac Plug.

the most economical staging method (up through nine stages). Figure 6.5 (Warpinski, 2009) depicts the costs of different staging techniques.

6.3.4.4 Staging with Retrievable Packers/Bridge Plugs

Retrievable tools have been used in cased holes for staging jobs since fracturing began. Bridge plugs (Figure 6.6) (Palmer et al., 2008) can be wireline set at precise points in the casing before or after perforating. Swab-bing operations can be carried out and acid dumped into place or it can be spotted via tubing and circulation. After a frac stage is done, another bridge plug is run by wireline and set (usually from a lubricator that con-tains the well pressure) and the next stage is perforated and fractured. The major drawbacks to this process include tight clearances around tools increasing the risk of sticking them before placement, no flow-through capabilities so treatment fluid is left in the lower zones for an extended period of time, and tubing retrieval with circulating equipment must be available to recover all the bridge plugs. These operations can take several days to a week depending on the number of stages completed. All of the methods described up to now require casing integrity above the target interval in order to treat down casing. In the case of open perfs or a leak, the following method is available.

Stackable packers are variations of using bridge plugs except these tools allow flow through them for treating, and then, with the rotation of pipe, a valve is closed and the tubing disconnected forming a bridge plug. If a second

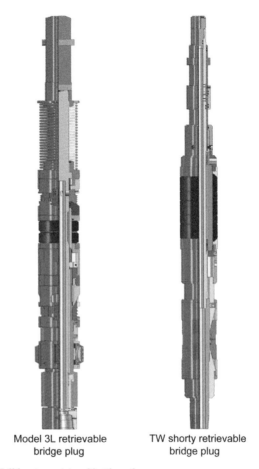

Model 3L retrievable   TW shorty retrievable
bridge plug   bridge plug

FIGURE 6.6   Halliburton retrieval bridge plugs.

tool has been run above, it is pulled into place, set, and the next stage begun; if not, tubing is retrieved, another tool picked up to run into place, and the new stage begun. After all stages are done, these tools have to be retrieved like the retrievable bridge plugs using tubing and a circulation system.

6.3.4.5 Staging with Coiled Tubing

Pin-point completions with coiled tubing offer the completion engineer the ultimate staging process. With large-bore coiled tubing (2⅜ or 2⅞ in OD), treatments can be pumped down the coil into a single seam at lower rates, effectively treating a seam at a time. This method requires perforating before running the coiled tubing treating tools.

Smaller diameter coil strings can be used for higher-rate jobs with the treatment being pumped down the annulus into the target seam. Here, perforations must not be present above the zone being treated. In some of these processes, the perforating will be by hydrajetting using the coiled tubing string. Many of the coiled tubing fracturing (CTF) processes can also be used in horizontal wellbore stimulations.

6.3.4.5.1 COILED TUBING AND PACKER PROCESS

CTF was introduced to the United States operations in coal by Rodvelt et al. in 2001 (Rodvelt et al., 2001). A 2⅜ in OD coil string was used to treat as many as 19 coals over a 2 day period in Buchanan County, Virginia. A proprietary bottomhole assembly (BHA) (Figure 6.7) (Rodvelt et al., 2001) was used to isolate individual seams as close as 7 ft without communication. With this process, all seams to be stimulated are perforated the day before. The coil and BHA are run to bottom to a known depth before pulling up to straddle the first treatment zone. The hole is reverse circulated to clean any fines out of the perfs that may be clear and ensure clean fluid in the tubulars. The frac job is then started down the coil string; flow down the tubing pressurizes the top opposing cup and isolates the annulus above the cups while it directs the treatment through the perforations into the target seam. At the conclusion of the treatment, displacement is pumped, the flow switched to reverse circulation and the tool spotted across the next seam up. Once all the seams have been treated, the BHA can be removed from the well, and the coil tubing runs back in to reverse out any residual sand—there are no baffles or plugs to drillout. Production equipment, rods, and tubing can be installed and the well started pumping. This process does require additional horsepower to pump against the fluid friction inside the coil string, and proppant concentration is limited through the coil. Measured depth is limited to ~7000 ft because of wear on the upper cups.

6.3.4.5.2 COILED TUBING AND JETTING TOOL PROCESS

Because of the limited rate and proppant concentrations in treating through the coil string, a process was developed using the jetting technology discussed earlier and a smaller ID (1¾ in) coil string. In this technique (Figure 6.8), a jetting tool actually replaces the need for conventional perforating. The process is similar to the CTF with packer in that coil with a proprietary jetting tool is run to a known depth. The jet is then spotted at the zone of interest and fluid injection begun down the coil string until the coil volume has been flushed with clean fluid. A ball is dropped to activate the jets and sand is added to a slickwater system to cut through the casing. Once the casing is perforated with the jetting tool, the annulus fluid flow direction is reversed toward the perforations and the frac treatment is commenced as normally designed. With this process, higher injection rates are possible (using the annulus as the injection path) without the friction associated with a small-ID coil string. A small amount of fluid is also injected through the coil during the treatment to ensure that the jets remain

FIGURE 6.7   Bottomhole assembly for coiled tubing fracturing (CobraFrac) process.

clear and the coil does not collapse. At conclusion of the frac treatment for that stage, the jet tool is moved up the hole and a sand plug (as discussed earlier) is spotted to pack off the perforations and provide isolation for the next stage above. The casing is reverse-cleaned to the required depth and

FIGURE 6.8   Jetting tool assembly for coiled-tubing fracturing—CobraMax process.

the jet pulled to start the next stage. This process is repeated to treat all the target seams in the well. At the conclusion of the final stage, the jetting tool is removed and coil run to bottom to cleanout the sand plugs; no baffles or frac plugs need be drilled out. The wellbore is then ready to run production equipment. This process provides maximum conductivity in the wellbore region; and, because of the use of the jetting tool for formation access, eliminates perforation friction and tortuosity problems. There is no depth limit for this process other than the reach of the coiled tubing string.

6.3.4.5.3 COILED TUBING, PACKER, AND JETTING TOOL PROCESS

The final staging technique described is the combination of small-ID coil, advanced jet technology, and a packer included below a proprietary BHA similar to the described above In this process, a packer is run below the jetting tool (Figure 6.9) and this BHA is run to a known depth. The BHA is then pulled to spot the jet tool across the first target zone and the packer set. Circulation is established through the coil string, the activating ball dropped, casing is jetted for access, and then the annular flow reversed to begin the

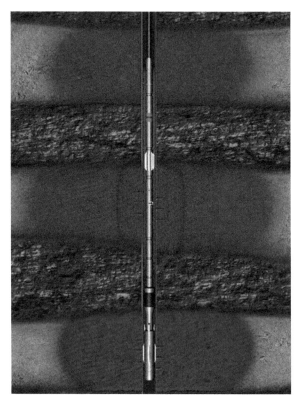

FIGURE 6.9   Bottomhole jet assembly for coiled tubing fracturing—CobraJet process.

fracturing treatment down the annulus. The fracturing design is pumped and at the end of the treatment, flow is stopped and reverse circulation begun while the packer is unseated and the BHA moved to the next zone. Because isolation is provided for each injection stage by the bottom packer, no sand plug is needed to isolate the stages. All target zones can be treated with tailored frac designs. At the conclusion of treating the top interval, the BHA can be pulled from the well and coiled tubing run back to bottom to circulate clean any fill; the operator is then ready to run production equipment. The jetting technology once again eliminates near-wellbore perforation friction and tortuosity enabling lower breakdown and fracture-treating pressures.

6.4 HYDRAULIC FRACTURING

6.4.1 Fracturing Considerations

Early CBM projects started where coal thickness and permeability used cavitation as a stimulation method. As the coals became thinner or permeability dropped, openhole treatments needed to be stimulated with hydraulic fracturing. Hydraulic fracturing, developed in the conventional oil and gas industry, improved the dewatering process allowing quicker methane desorption, and leading to higher production rates and improved project economics. Gains were seen, but need for better control of the treatments pushed operators to cased hole completions and the ability to access a greater number of the thinner, lower perm coals. Hydraulic fracturing of coal required solutions for fines control, fluid compatibilities, and the unique geometries developed while treating.

6.4.1.1 Fines

Fines are a major contributor to fracture initiation difficulties, pressure buildup during the treatment, and production decline early in the life of the well. Some operators (Rodgers, 2007) have even investigated perforating the partings between coal layers or fracture stimulating adjacent sands to reduce the fines generated during frac and subsequent production thereof. If someone picks up a piece of coal, their hands will have black particles adhere to them, especially if it is a well cleated and friable coal. High-volume injections of fluid and proppant at high rates will erode particles of coal that get transported along with the proppant slurry. They can be concentrated at the tip if fluid loss is high, resulting in a pressure buildup during the treatment that may curtail the proppant placement. After the fracturing treatment, these same fines will begin to move back toward the wellbore during production. The operator must recognize this symptom when production declines and alleviate it with a remedial treatment.

A postfracture service has been developed that helps remove wellbore damage and coal fines blockage through the use of a strong back flush. Fines are flushed away from the near-wellbore vicinity and immobilized with a proprietary chemical formulation that makes the surface of the coal particle "tacky", enabling "clots" of particles to stick together in an immobile mass some distance away from the wellbore as depicted in Figure 6.10 (Rodgers, 2007) and restoring conductivity to the wellbore. This proprietary system (marketed by Halliburton as CoalStim® Service) can also be formulated to remove polymer damage from fracturing treatments. Both guar and polyacrylamide polymers have been removed with this treating fluid. Figure 6.11 (Rodgers, 2007) depicts one operator's success in using the process.

Given the propensity for coal to produce fines, the best way to counteract them is with a surface modification agent (SMA) on the proppant grains as proppant is added during the frac job. This will trap the fines before they can travel through the proppant pack maintaining a high well production rate for a longer period of time. This process also reduces proppant flowback allowing the operator to place the production pump

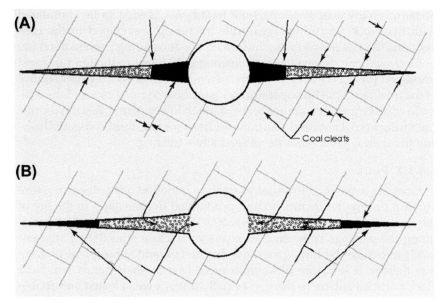

FIGURE 6.10 Removing and holding fines away from the wellbore—CoalStim Service. (A) Fine blockage reduces methane desorption. (B) Blockage pushed away from wellbore and held in place.

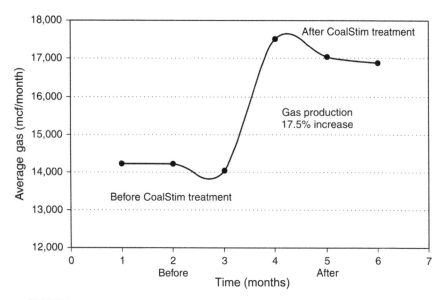

FIGURE 6.11 One operator's success—production increase from controlling fines.

below the lowest coal at a point to operate efficiently without fear of sanding issues. The enhanced conductivity achieved because of reduced proppant settling will improve water recovery and gas flow. One operator in the Fruitland Coal of the San Juan increased production 200 mcf/day after a refrac using SMA on the proppant (Rodgers, 2007).

6.4.1.2 Frac Fluids

Fracturing fluid selection should be determined based on compatibility with the formation and consideration of the in situ permeability of the specific formation that is being stimulated. Moderate- and high-permeability wells need only to bypass damage from drilling operations to connect with the wellbore. Lower-permeability wells may benefit most from foamed treatments that limit fluid load. Some "dry" gas coals should only be treated with 100% nitrogen to prevent relative-permeability damage. Ultra-low permeability coals need technology borrowed from nano-Darcy shale completions where horizontal wells are drilled and complex fracture networks are generated. The measurement of reservoir properties (gas content, adsorption isotherm, permeability, and reservoir pressure) along with fluid compatibility testing on cores early in the life of a project can be the difference between optimum production and a dismal failure.

6.4.1.2.1 WATER

Coal seams that appear water saturated may best be stimulated with a water-based system that contains no additives other than salts (potassium chloride or sodium chloride), scale inhibitor, and bactericide. For exploratory wells, this is especially true as the operator strives to determine gas productive potential based on initial core work. Water and sand fracs may not provide the optimum stimulation or sand placement, but they are usually the least expensive way to provide some stimulation of the coal seam without introducing other "damage" factors such as gels or chemicals. As the operator moves from exploration to development, different types of jobs should be evaluated to determine the optimum.

6.4.1.2.2 SLICKWATER

Reducing pipe-friction losses to deliver more fracturing rate into the formation can be done by adding friction reducing (FR) polymer (typically polyacrylamide) to water to make it "slick". Normal concentrations are 0.5–1 gal/Mgal depending on the amount of friction reduction required. This will reduce the hydraulic horsepower requirements on location, minimizing the surface equipment footprint. The fluid viscosity will still be low so sand transport is poor. This type of fluid is used in shale fracs to generate complex fracture geometry. In addition to scale inhibitor and bactericide, breaker for the polymer is added to aid in removal.

6.4.1.2.3 GELLED WATER

Increasing viscosity with guar, hydroxypropyl guar, or hydroxyethyl cellulose polymers becomes a trade-off between damaging the coal and generating the fracture width to allow proppant placement. The polymer will form a filter cake, which will suppress fluid leakoff and extend the dimensions of the fracture. Friction reduction occurs when these polymers are added to water so horsepower requirements are lowered. Sand transport improves, although suspending it over thick intervals is an issue. Limiting the polymer concentration to 10–20 lb/Mgal has improved the cleanup/regain permeability. Enzyme breakers are very efficient at destroying the polymer at low temperatures. Oxidizer breakers require a catalyst to be effective in the temperature range of most coals (<120 °F).

This system is good for fair permeability coals and where height growth might be an issue with more viscous fluids. In some instances, small amounts of gel (3–10 lb/Mgal) have been used to provide slickwater fluids similar to the FR polymers.

6.4.1.2.4 CROSS-LINKED GELLED WATER

As permeability and/or thickness increase, more viscous fluids are needed to control fluid loss, generate height, transport greater concentrations of proppant, and leave it suspended across the entire thickness. Without adding any additional polymer to the base fluid, a cross-linking agent can be added to bind the polymer strands together and give an apparent viscosity 200–300 times higher than the base fluid. Borate-cross-link systems are the preferred cross-link fluid as they are pH reversible—when the pH drops below 8, they "unlink" back to base-gel viscosity, which is reduced by enzyme breaker to water. They will "reheal" after shear thinning in small tubulars or through perforations and recover their viscosity profile. Low-gel borate (LGB) systems have been used in all parts of the world to aid in job placement, and should be considered the fluid of choice in high-permeability wells. When coupled with the SMA coating on the proppant, an operator in the San Juan Basin created $720,000 of economic value per year (Rodgers, 2007).

6.4.1.2.5 HYBRID CROSS-LINKED GELLED WATER

Pressurizing a coal seam with water (or slickwater) before pumping the main stages of the frac design have shown to be beneficial in frac fluid cleanup (Rodgers, 2007). The thin water prepad can condition the cleat system with treatment fluid containing scale inhibitor, breaker, iron control, and bactericide. The LGB system that follows develops width and places and suspends the proppant. When the job is done and the cross-link is broken, the nongelled prepad fluid provides energy to flush gel residue out of the cleats and back to surface. One operator in the Northern Appalachian Basin improved production 20–30 mcf/day after switching to a hybrid LGB (Warpinski, 2009) completion design vs the thin fluid designs previously used. One key point to remember is that the supercharge from the prepad will dissipate over time, so it is imperative to return the treatment fluids after a short (2–4h) break time, and the well should be put on pump within 2–3 days to reduce fluid lost to the cleat system containing the residuals. This requires the operator to have lead lines laid, pump jack set, and tubing and rods ready to run in hole after a TD check so that the well can begin pumping on schedule. Leaving the well shut-in an extended period of time can require a remedial treatment to remove residue.

In the case of low permeability reservoirs, the prepad stage can be used to generate complex fracturing geometry before the LGB stages. The addition of 80/100-mesh sand may be used to prop this geometry before generating a planar frac with the LGB fluid and SMA-coated sand. Research continues on the best way to control fines in low permeability coals.

6.4.1.2.6 NITROGEN FOAMS

Nitrogen foams are an excellent way to remove 50–80% of the treatment fluid from the process. A gelled water system is pumped that contains a foaming surfactant. When nitrogen is added at quantities above 60% and below 95%, stable foam can be generated that have viscosities equivalent to cross-linked fluids such that proppant can be placed and suspended. Fluid loss is excellent for low to moderate permeabilities as the nitrogen bubbles act to impede fluid loss to the cleat system. After a period of time, the foam dissipates and the base fluid will be carried back out of the well in the nitrogen stream. Foams improve relative permeability to gas since they are 60–90% gas themselves. Less fluid is required to do the job and dispose of later and 70% quality nitrogen foams are the predominant treating fluid for the Central Appalachian CBM completions.

6.4.1.2.7 NITROGEN GAS

A few CBM reservoirs such as the Horseshoe Canyon in Canada contain "dry gas" meaning the cleats do not contain any water to produce before gas desorption can take place. In this type of reservoir, nitrogen gas is the preferred treating fluid. In some instances, a small amount of carbon dioxide may be injected with it to remove any residual liquids. Systems have been developed that are made up of 95%–98% nitrogen that can still carry proppant to stimulate these "exception" reservoirs. Fracturing with straight-nitrogen systems in water-saturated coals has not provided the benefit of foams or other fluid systems.

6.4.1.3 Chemical Aids

Other chemical aids have been used to enhance fluid recovery and provide protection downhole from scaling waters or bacterial action. A typical treatment would contain the following:

- Bactericide to prevent sulfate-reducing bacterial fouling.
- Scale inhibitor, which is in a liquid form for pad and prepad, and a solid form for inclusion with the proppant pack.
- Surfactant that lowers the surface tension of water for more thorough recovery.

The latest additions are the microemulsion agents:

- Iron control to prevent scale nuclei and corrosion.
- Foaming surfactant for foam fracs and jobs that wish to help lighten return fluids by entraining gas. The latest additive is a foamer that is stable above $pH = 9.0$; below that the foam dissipates, which can be advantageous during production.

6.4.1.4 Proppants

As most coal seams completed thus far have been above 4000 ft depth, closure pressure on the proppant has not required anything stronger than natural quartz frac sand. Primary mesh ranges are 20/40, 16/30, and 12/20 API grade. The shallower, more permeable coals should be treated with larger sands to provide the necessary conductivity for a production increase. The use of some 80/100-mesh sand is advocated in very low permeability coals to prop complex fracture geometry and provide a transition to a single, planar frac (using LGB fluid and larger sand) near the end of the job for a highly conductive path at the wellbore. This portion of the job should be treated with SMA to trap fines and enhance the conductivity.

6.4.1.5 Rock Properties

For fracture modeling, it is beneficial to know the rock properties of the coal and the surrounding strata to get an idea of fracture growth during the frac job. Rock properties are best obtained from cores that are analyzed in laboratories. Rock properties can be obtained from electric log information using a dipole sonic log along with resistivity and porosity measurements. Geographic models provide insight into stress levels, gas contents, and water-producing formations when the complete set of logs are available and can be calibrated with whole core data. Since all wells are usually logged, data sets can be used to interpolate between core wells and develop the geomechanics model for future well positioning.

6.4.1.6 Stress Measurements

A direct measurement of the horizontal stresses can be determined from the diagnostic fracture injection test as well as the permeability and reservoir pressure. Pressure dependent leakoff (PDL) character is a good indication of natural fractures or fissures that will contribute to the production. Knowing the value of PDL enables the frac design engineer to provide the most efficient design that will account for this leakoff and propagate the treatment to completion. While log calculations can provide some insight into the stress trend, they will not account for external tectonic stresses that occur in areas where mountains apply significant horizontal stress such as the Cedar Cove Field (Figure 6.12) (Rodgers, 2007). The boundary layers may also be tested to confirm log calculations with small, injection-falloff tests to measure closure stress values. Diagnostic testing performed early in the life of a project is repaid many times over in future optimized designs.

6.4.1.7 Fracture Geometry

Researchers have spent 60 years attempting to understand the geometry of fractures and how they propagate. Fracture geometry and the disposition of the proppant injected influence the production characteristics of a hydraulic fracture. Models exist from the conventional oil and gas industry for use in coals with some success. However, confirmation of the fracture dimensions has been elusive to most operators except those with

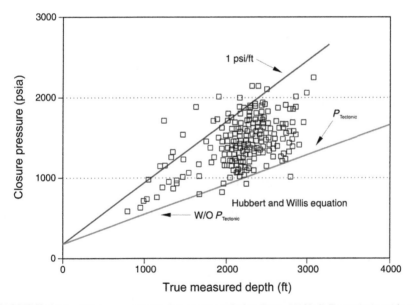

FIGURE 6.12 Minimum principal stresses at Cedar Cove. W/O P Tectonic is without tectonic stress pressure adjustment.

opportunities in mine areas. When a zone can be mined after a fracture has been performed, dimensions can be visually mapped and the geometry better understood. Some operators have performed postfrac injection/falloff testing and determined fracture conductivity and length. An alternative to mine-backs is the use of passive microseismic monitoring to determine fracture dimensions.

6.4.1.7.1 FRACTURE MODEL PREDICTIONS

Commercial software exists to attempt modeling of fractures in coal. Predicting geometries such as a 2 in width, a tee (T)/I-shape, or multiple fractures in three dimensions (horizontal and vertical) are not currently possible. What can be done is to predict a length and height and width assuming some homogenous character, observe production, then go back and recalibrate the model. Any type of additional observation—mine back, postfrac testing or microseismic—will add understanding to the changes needed for model improvement. As discrete fracture network software becomes available, a better model can be built; however, coal is heterogenic so there will always be ambiguity.

6.4.1.7.2 MINE-BACK OBSERVATIONS

A picture is worth a thousand words. Figure 6.13 (Rodgers, 2007) shows a T-shaped frac with a half inch width and most of the proppant deposited in the shale/coal interface at the top. Mine-back observations provide the basis for understanding the complexity of fractures in coal. Stress laws are still obeyed; the major axis of the hydraulic fracture is always in the direction of the maximum stress and the fracture plane is always orthogonal to the least stress, multiple fractures abound, especially with the use of slickwater and thin gel treatments. Coal has many cleats per foot so the opportunity exists for creating many parallel fractures. Here are personal recollections by Dr Pramod Thakur, Consol Energy, who has been in the mine when a hydraulic fracture cuts through a corridor: "First you hear the sound as the fracture approaches, and then you see the crack appear in the coal. Shortly after that the frac fluid appears and copious amounts of fines are expelled followed by the frac sand".

A long fracture was documented by Steidl (Figure 6.14) (Rodgers, 2007) to extend 525 ft away from the wellbore with proppant deposited 352 ft from the well. Maximum fracture width was 0.3 in. Reese and Reilly (Reese and Reilly, 1997) (Figures 6.15–6.18) observed fractures in a Pennsylvania mine back—276 ft in length with widths of 3–12 in—completely filled with frac sand. These observations confirm the complexity of fractures in coal, and corroborate the difficulty in modeling fracture behavior.

6.4.1.7.3 POSTFRAC INJECTION/FALLOFF TESTING

Determining what fracture characteristics to use in forecasting production may best be served by measuring fracture conductivity and length via a

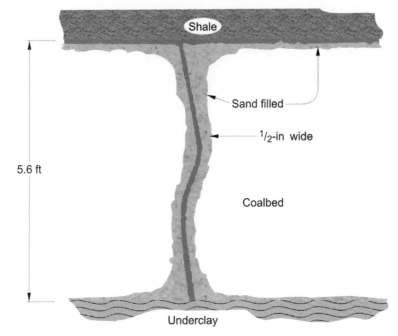

FIGURE 6.13   Mine-through observation of T-shape fracture.

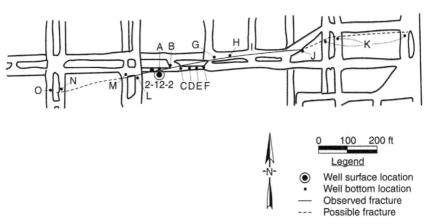

FIGURE 6.14   Mine-through observation of long fracture wings. A through O is points of observation taken during the mineback.

postfracture injection/falloff test. This entails injection below fracturing pressures at rates that will build a pressure wave during the injection cycle, and then observation of the falloff pressure decline that is then analyzed with well test and/or simulation software. A fracture model can help with starting conditions for a simulator as it will give an expected prop length and conductivity.

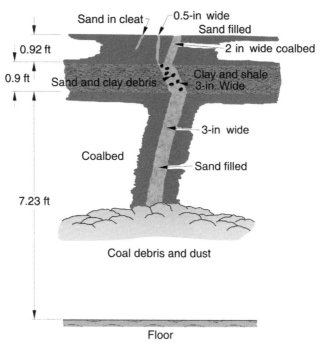

FIGURE 6.15   Mine-through observation in a Pennsylvania mine showing 3 in width.

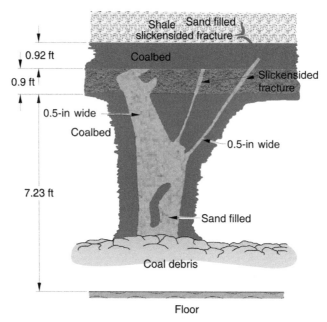

FIGURE 6.16   Mine-through observation showing multiple branches.

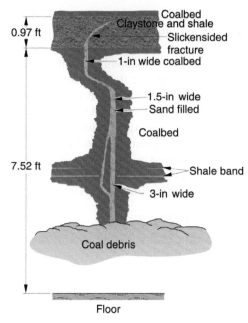

FIGURE 6.17   Mine-through observation showing 1½–3 in fractures.

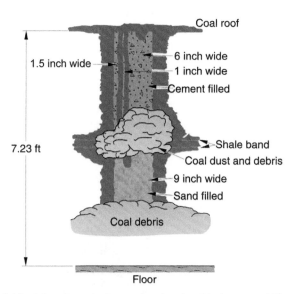

FIGURE 6.18   Mine-through observation showing 9 in fracture width, sand filled.

It is helpful to have a prefrac permeability measurement; otherwise, this becomes a variable to be determined. The injection cycle will be 24–48h with a shut-in period of 7–10days. Bottomhole memory gauges should be used to capture the pressure decline under a downhole shut-in valve. Surface data are then merged with the memory gauge data to provide the data set for analysis. This analysis outcome can then be used in simulation software to predict water and gas production rates for a specified backpressure profile. If the testing solution does not match the fracture model predictions, the model can be "tweaked" to give an improved prediction and used for the next well's frac design. Adjustments to the design might include higher sand concentrations to improve conductivity, larger volumes to create a larger fracture geometry, or lower pumping rate to optimize height growth.

6.4.1.7.4 MICROSEISMIC MONITORING

Microseismic monitoring is a valuable tool in understanding what a hydraulic fracture is doing real time. Microseisms, small earthquakes, are induced by the hydraulic fracturing fluid pressurizing the rock. A set of geophones is run in an offset monitoring well some distance away from the injection well to listen for microseisms. The harder and more uniform the rock, the farther sound travels and the more microseisms can be picked up with downhole telemetry (Figure 6.19) (Warpinski, 2009). This presents some "hearing" problems in coal as it is a softer rock than sandstone or limestone and for CBM wells the maximum listening distance is usually less than 750ft. Adjacent sandstone in the staging package may allow greater distances between the monitoring well and the injection site. Real-time data are plotted to determine height growth, length of the events

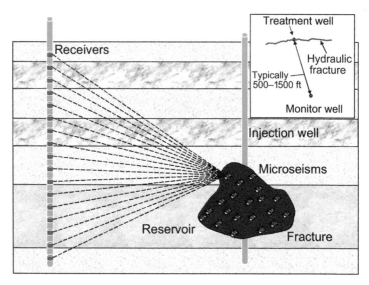

FIGURE 6.19   Microseismic downhole geophone array receiving seismic events.

from the wellbore, and the field width of the events. On-the-fly changes to the fracturing schedule may then become apparent to influence fracture growth. Final analysis after treatment provides a fracture map showing the length, width, and height growth of the fracture (Figure 6.20). This aids the design engineer in sizing frac designs for future treatments. It also allows the reservoir engineer to optimize placement of wells to effectively drain production. In the case of low permeability coals, it can be used to validate a complex fracture network development that would enhance permeability and increase gas production.

6.4.1.8 Branch Fracturing

Low permeability and ultra-low permeability coals require attention to the fracturing design such that it enhances the effective permeability by creating complex fracture geometry. This technology is currently employed by the shale community to create large, surface contact volumes,

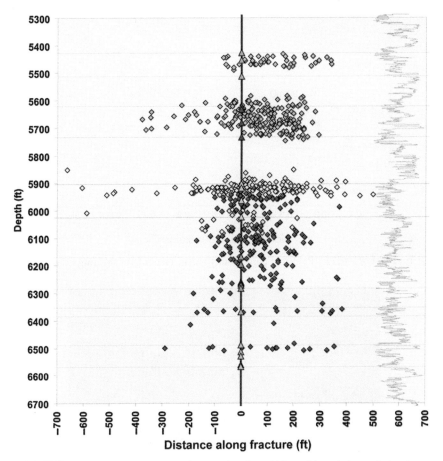

FIGURE 6.20 Fracture mapping showing height, width, and length for optimization.

especially in horizontal wellbores, that unlock the gas in nano-Darcy permeability. Palmer et al. (Palmer et al., 2008) investigated the permeability enhancement that could take place by increasing the shear stress in a coal and found 10–100-fold increases possible (Figure 6.21). Furthermore, if the formation has many natural fractures (cleats), PDL becomes high with increased pressure. This can be used to design fracturing treatments that increase PDL, open up microfissures, prop them, and the end result is enhanced permeability.

A novel approach to increasing pressure in the fracture system has been proposed using sand slugs to screenout an established fracture and cause diversion down a different pathway. This "branch" fracturing method (Figure 6.22) would require precise proppant control near the

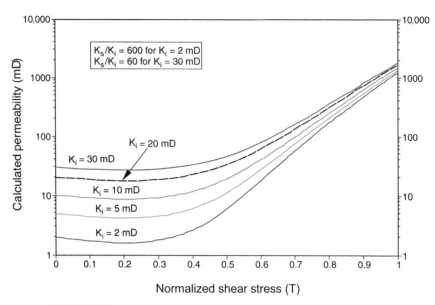

FIGURE 6.21   Predicted permeability enhancement from shearing tests.

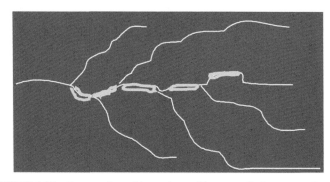

FIGURE 6.22   Fracture branching enhanced by controlled proppant loading.

perforations, which can be accomplished with coiled tubing systems. More work is needed to validate this process and the improved production it may provide.

References

Palmer, I., Cameron, J., Moschovidis, Z., Ponce, J., 2008. Role of natural fractures in shear stimulation: a new paradigm. Submitted to the International Coalbed Methane and Shale Gas Symposium (ICMSS) in Tuscaloosa, May.

Reese, R., Reilly, J., 1997. Case Study: Observations of a Coal Bed Methane Extraction Pilot Program via Well Bores in Greene County, Pennsylvania. Paper SPE 39227 presented at the SPE Eastern Regional Meeting, Lexington, Kentucky, 22–24 October. http://dx.doi.org/10.2118/39227-MS.

Rodgers, R., 2007. Coalbed Methane: Principles and Practices, second ed. (MS).

Rodvelt, G.D., Oestreich, R.G., 2007. Composite-Fracturing Plug Reduces Cycle-Time in a Coalbed Methane Project. Paper SPE 111008 presented at the Eastern Regional Meeting, Lexington, Kentucky, USA, 17–19 October. http://dx.doi.org/10.2118/111008-MS.

Rodvelt, D.G., Toothman, R., Willis, S., Mullins, D., 2001. Multiseam Coal Stimulation Using Coiled-Tubing Fracturing and a Unique Bottomhole Packer Assembly. Paper SPE 72380 presented at the SPE Eastern Regional Meeting, Canton, Ohio, 17–19 October. http://dx.doi.org/10.2118/72380-MS.

Warpinski, N., November 2009. Microseismic monitoring: inside and out. J. Pet. Technol. (Distinguished Author Series) 61 (11), 80.

Horizontal Coalbed Methane Wells Drilled from Surface

Stephen Kravits, Gary DuBois

Target Drilling Inc., Smithton, PA, USA

7.1 INTRODUCTION

In-mine horizontal methane drainage boreholes have been proven very effective in reducing methane emissions to safe levels in advance of mining (Kravits et al., 1999; DuBois et al., 2006; Thakur and Poundstone, 1980). However, the underground access required to drill in-mine horizontal boreholes limits their ability to degasify beyond the perimeter of the current mine workings. In-mine boreholes are primarily drilled to shield methane emissions during longwall development, or to degasify the longwall panel. Their productive life is short, usually less than 2 years before mine through. Hence they cannot be used for commercial gas production. Lastly, drilling in-mine horizontal boreholes and maintaining their wellheads and the underground gas transmission pipelines connected to vertical boreholes used to safely transmit the gas produced to the surface, require routine maintenance and integration with already difficult mining operations.

The in-mine horizontal boreholes are inherently drilled underbalanced because the pressure of the drilling water exiting the bit flushing the coal cuttings out of the borehole after hydraulically powering the downhole motor does not exceed the relatively low, underpressured, in situ reservoir pressure of the coal. This was the critical technical obstacle of drilling horizontal laterals targeting the coal initiated from the surface, hundreds of feet below the surface.

Earlier in the oil and gas industries, and later the coalbed methane (CBM) and coal mine methane industries, surface horizontal or "sideways", directional drilling was recognized as a way of combining the best elements of vertical well and horizontal in-mine drilling. Drilling from the surface is safer than from in-mine, does not hinder mining operations and can be carried out years in advance of mining to maximize degasification effectiveness and economic viability. A long horizontal borehole intersects a much greater volume of the coalbed than a vertical borehole, negating the need for hydraulic fracturing in most cases and the borehole trajectory can be controlled to take advantage of coalbed directional cleat permeability. In addition, a large area 2.6 km^2 (640 acres), can be drained from a single surface site. Compared to vertical wells, 16 vertical wells drilled on 0.16 km^2 (40 acres) spacing greatly reduces the environmental impact of the methane drainage project and results in drilling, infrastructure, and maintenance cost savings.

Understanding the advantages and disadvantages of in-mine horizontal boreholes, and the benefits of degasification years in advance of mining if drilled from the surface, the U.S. Bureau of Mines drilled an experimental CBM horizontal well from the surface in 1978 in southwestern PA to degasify a coal area to be mined in the future (Diamond and Oyler, 1980) (Figures 7.1 and 7.2). The Bureau's CBM horizontal well, often called the

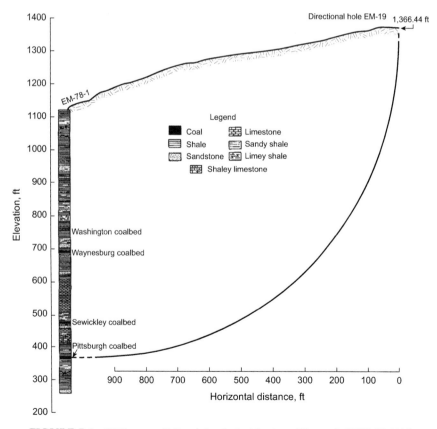

FIGURE 7.1   US Bureau of Mines' slanthole side view. *(Diamond, SPE/DOE 8968)*

"slanthole" proved that the coal could be accessed through a directionally drilled and cased curve intercepting the Pittsburgh coal. This surface CBM well included drilling three laterals in the coal. However, as pioneering as this surface CBM well was at the time, attempts to dewater the coal by drilling several vertical wells near the landing of the well's curve in the coal failed to adequately dewater the coal laterals reducing the well's gas production and degasification effectiveness. Furthermore, during the surface drilling of the coal laterals, drilling fluid pressure induced into the coal exceeded the coalbed's gas reservoir pressure because it was at or near hydrostatic pressure, which exceeded the coal's reservoir pressure, meaning the coal was drilled overbalanced.

Nearly 20 years after the U.S. Bureau of Mines' slanthole, several underbalanced dual well systems (and single well systems) were designed and implemented to drill surface horizontal wells. These designs benefited from directional drilling technology innovations including downhole motors, electromagnetic measurement while drilling (EM MWD) survey systems, and advances in well completion technology. Several companies

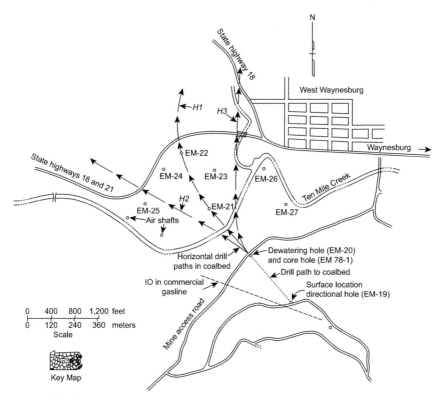

FIGURE 7.2 US Bureau of Mines' slanthole plan view. *(Diamond, SPE/DOE 8968)*

have proven the effectiveness of their surface drilled horizontal CBM dual well systems in numerous coalbeds, including, but not limited to, CDX Gas, LLC (now owned by Vitruvian Exploration); CNX Gas, LLC; and Target Drilling Inc. (TDI). The dual well systems discussed in this chapter have been patented or licensed for use and implemented successfully in the Pocahontas #3, Pittsburgh, Sewickley, Lower Kittanning, Freeport, and Hartshorne coalbeds, to name a few. These dual well methods can utilize air foam (a mixture of compressed air, water, drilling soap, and chemicals), or fluid (water) to directionally drill the coalbed laterals while injecting compressed air down the vertical production well to aerate the coal effluent. This technique is designed to lift and remove cuttings to the surface through the access well while drilling the coal laterals so that the pressure of the effluent circulating to the surface does not exceed the CBM gas reservoir pressure (underbalanced).

It should also be pointed out that horizontal CBM wells drilled from the surface, although significantly more costly than the cost of vertical wells that are hydraulically fractured, have been proven to produce greater than 20 times the gas production, especially in thin coalbeds. Because horizontal CBM wells contact the coalbed reservoir incrementally every foot drilled creating a relatively large drainage area, they reach peak gas production very soon after initial production and dewater the coalbed much faster than vertical wells while free gas and water trapped in the cleats of the coalbed are produced. Consequently, desorption of gas trapped in the micropores or microstructure of the coalbed occurs as the reservoir pressure reduces resulting in a reduced and eventually close to flatline gas production decline curve.

7.2 CRITICAL FACTORS INFLUENCING SURFACE CBM WELLS

7.2.1 Surface Logistics and Strategic Placement of the Coal Laterals

Surface CBM wells, whether drilled in advance of mining, or where mining is not planned or the coal is unmineable, must be permitted as gas wells, and surface right-of-way obtained for the drilling pad, prior to starting or spudding the well. Typically, the drill pad and access road to the pad are designed by an engineering firm and built adhering to standard civil engineering practices for sediment and erosion control. If the CBM well is drilled to degasify the coal in advance of mining, the curve and laterals can be strategically oriented based on various factors. These variables include, but are not limited to: the direction of coal dip; face and butt cleat orientation; future mine plans—to maximize degasification time before mining; appropriate drill hole designs, staying within coal lease boundaries; surface rights and access and surface topography. Well owners have different opinions on the sensitivity of factors mentioned that determine optimum coal lateral orientation and placement, and typically must choose to only satisfy a few of the critical factors on a well basis, especially in Northern and Central Appalachia due to the topography and numerous surface property owners when the surface owner does not own the CBM rights.

In the modeling of multilateral drainage patterns, Maricic et al. (2005) concluded that the optimum well configuration can be determined by considering the total horizontal length, the spacing between laterals and the number of laterals. Longer horizontal length increases the contact with the coal seam and increases yields for more gas recovery, but at the same time increases drilling costs and drilling risks (Figure 7.3). Balancing these factors led operators to more commonly drill a three to four lateral pattern per horizontal CBM well drilled from surface (Maricic et al., 2005).

7.2.2 Underpressured Coalbeds

Compared to deeper, conventional oil and gas reservoirs, coalbeds are usually underpressured whereby the in situ gas reservoir pressure in the coalbed is less than the hydrostatic pressure. Consequently, if the coal is not drilled underbalanced defined as maintaining the drilling fluid pressure exiting the bit in the horizontal wellbore or lateral to less pressure than the in situ gas reservoir pressure in the coalbed, the cleat permeability of the coal will be reduced hindering or preventing the coalbed methane to produce through the wellbore because the drilling fluid will enter and plug off the cleat permeability. The most common method of underbalanced drilling is to drill with air or air foam, whether it is a dual

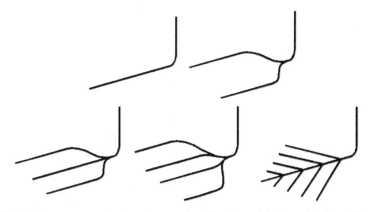

FIGURE 7.3   Various coalbed methane well coalbed lateral designs. *(Maricic, SPE 96018)*

well or single well system. However, because coalbeds are typically water saturated, directional drilling using air or air foam to power the downhole motor used in a single CBM well system has not proven successful to remove coal cuttings and produced coalbed water to maintain underbalanced unless a parasite casing string is installed in the directional access curve in a single well system, injecting compressed air on the outside of the parasite casing to lift coalbed drilling effluent and produced water to the surface through the annular space of the outside of the drill string and inside wall of the parasite casing. In a dual well system while drilling the coalbed laterals, compressed air is injected down the cased vertical production well to aerate the coalbed drilling effluent and produced water assisting its removal via the directional access well to keep the pressure of the drilling fluid underbalanced.

7.2.3 Staying in the Coalbed

Some coalbeds are surrounded above and below, by competent rocks that if directionally drilled with water or air foam, will not deteriorate or slough even after the drill bit traverses in and out of the coalbed and into the impervious rock and back into the coal, while attempting to stay within the coalbed. Although the roof rock or floor rock, respectively, are out of the pay zone and might not add to the gas production, if these rock intervals are stable and do not collapse, or have a history of not collapsing, sometimes drilling will continue in the rock for short distances before the lateral reenters the coal.

Conversely, if the roof and or floor rock are weak, soft, weathered, sedimentary rock, and either is penetrated by the directional string using water or air foam drilling fluid, it can collapse or slough behind the bit around the drill string. This can result in sticking or losing the directional drill string in the lateral that meanders in and out of the coalbed through the unstable rock. To complicate this problem, the weak, sedimentary rock is not much stronger than the coalbed at times, and can be difficult for the driller to know that the lateral has exited the coalbed and is in the rock. Using sophisticated borehole EM MWD guidance systems equipped with natural gamma sensors that measure natural gamma radiation of the formation, but generally are positioned about 6 m (20 ft) behind the bit, and the experience of the "coalbed" directional driller and drill rig operator, have reduced the quantity of rock drilled. One company has developed a downhole motor equipped with a gamma sensor about 1.5 m (5 ft) behind the bit to assist the directional driller to stay within coalbed. Generally speaking, the coalbed has a lower American Petroleum Index (API) natural gamma count compared to a higher gamma count in roof and or floor rock. If intervals of the coalbed lateral are drilled in the roof or floor rock and then steered back into the coal, the roof and floor rock interval could collapse, squeeze or bridge, resulting in drag on the drill string. The drill string could get stuck or broken and lost in the lateral. Longwall mine operators do not appreciate dealing with a directional drill string left in their longwall panel no matter what orientation the coalbed lateral was drilled.

7.2.4 Removal and Disposal of Produced Water from the Horizontal CBM Well

After the horizontal CBM well's coalbed laterals have been drilled, the gas and water production phase begins. As stated earlier, the horizontal CBM well usually produces gas and water very soon after put on production. Various types of downhole pumps are used in the vertical well including, but not limited to, sucker rod downhole pumps; electric submersible and positive displacement progressive cavity pumps.

A detailed explanation of dealing with produced water from horizontal CBM wells drilled from the surface is discussed in another chapter. The inherent nature of a horizontal well with multiple coalbed laterals has the potential of dewatering the coalbed quickly. Some coalbeds in northern Appalachia contain significant quantities of water and are considered aquifers. Due to the potential adverse chemistry of water produced from coalbeds including high volumes of total dissolved solids, a plan to dispose of the CBM water must be developed and approved by pertinent state and federal regulatory agencies, and strictly adhered to. Lastly, the cost of disposing of CBM produced water must be considered in the CBM well operating budget. One positive note has been the use of produced water from surface drilled horizontal CBM wells as a potentially acceptable source of water for hydraulically fracturing Marcellus Shale horizontal and vertical wells reducing the water disposal costs for the CBM wells and the fracking costs of Marcellus Shale wells.

7.3 DIRECTIONAL DRILLING TECHNOLOGY DEVELOPMENT

7.3.1 CDX Gas, LLC (limited liability company) Horizontal Dual CBM Wells

After U.S. Steel Mining had drilled numerous surface horizontal CBM dual wells in Wyoming Co., WV targeting the Pocahontas #3 coalbed in 1997, CDX Gas, LLC began an extensive program drilling dual CBM wells featuring their patented Z-Pinnate horizontal drilling and completion system in 1998 (Figure 7.4). CDX's dual CBM well includes drilling a vertical production well and an access well, several hundred feet from the production well as shown in Figure 7.5. An 8 foot diameter cavity is

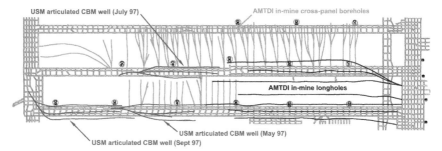

FIGURE 7.4   U.S. Steel Mining articulated (vertical to horizontal) CBM wells drilled at Pinnacle Mine in conjunction with in-mine boreholes drilled by Target Drilling Inc.

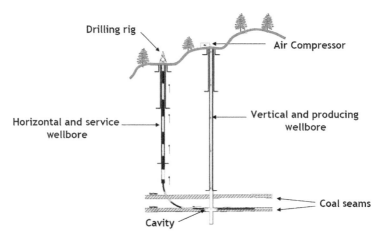

FIGURE 7.5   CDX dual well. *(CDX Gas, LLC, 2005.)*

created where the vertical and producing wellbore intersects the coalbed. After the access well is drilled and cased to a predetermined depth to protect the water table, the vertical section is drilled, and cased to the designed kick-off point, and then the curve is directionally drilled or navigated to intercept the targeted coalbed, landing in the coalbed horizontally. The horizontal wellbore is then directionally drilled to intercept the cavity, and continued to drill various patterns of laterals. During the directional drilling of the coalbed laterals, compressed air is injected down the vertical wellbore to aerate the drilling effluent to assist its removal out of the uncased curve and vertical section of the horizontal wellbore to surface. The coalbed laterals are drilled so that produced water from the laterals drains to the vertical production well for pumping to the surface when the well goes on production. CDX Gas has used this dual well technique to drill laterals in a "pinnate" drainage

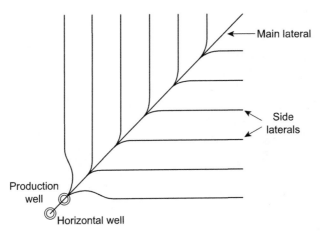

FIGURE 7.6 CDX pinnate. *(EPA, 2009.)*

pattern (Figure 7.6). Four sets of main laterals with associated side laterals can be drilled, forming a 360° drainage pattern that can drain 1280 acres and replace 16 standard 80 acre locations (Figure 7.7). Successful pinnate configurations, drilled in suitable geologic environments in the Appalachian, San Juan, and Arkoma basins in the U.S., have large initial production figures and have dewatered the coal very quickly, resulting in drainage of 80–90% of in situ methane within 2–3 years (EPA, 2009). CDX engineers believed by reducing the number of wells needed to deplete a project area, the Z-Pinnate Horizontal Drilling and Completion System reduces the surface disturbance caused by locations, gathering systems, and production facilities. Furthermore, this technique also reduces project development costs, improves project economics, and minimizes environmental impact.

CDX Gas report costs of $2.2 million for wells targeting coals 275–395 m (900–1000 ft) deep in their Hillman Field in West Virginia. Laterals are drilled in a pinnate pattern for a total drilled length over 6100 m (20,000 ft) and drain 2.4 km² (600 acres). Wells have initial production of over 14 Mcmd (500 Mcfd). In 2008, 21 wells were producing in the Hillman field at a rate of 595 Mcmd (21 MMcfd). The average estimated ultimate recovery (EUR) per well is about 28 MMcm (1 Bcf) per well (Oil and Gas Investor Magazine, 2008). CDX Gas has used their pinnate drilling pattern to drain coal seams at the Pinnacle Mine in West Virginia reporting that 60–65% of all in situ gas was recovered in a 2–3 year period. In 2006, the Pinnacle Mine recovered and sold approximately 130 Mcmd (4.6 MMcfd) of gas from its premine drainage wells (USEPA, 2008). CDX has drilled over 250 pinnate patterns totaling over 5 million feet as of the spring of 2008 in the Appalachian and Arkoma coal basins (Lusk and Jones, 2008).

FIGURE 7.7   CDX Gas Pinnate multilateral network. *(CDX Gas, 2005.)*

7.3.2 CNX Gas, LLC Horizontal Dual CBM Wells

In 2001, CNX Gas began experimenting with producing gas through horizontal CBM wells in southwestern Pennsylvania and northern West Virginia in a play the company designates as Mountaineer CBM field targeting the Pittsburgh and Freeport coalbeds. The Pittsburgh coalbed is between 183 and 395 m (600–1000 ft) deep and about 1.8 m (6 ft) thick, while the Freeport coalbed is 180–240 m (600–800 ft). The Pittsburgh coalbed CBM horizontal wells are drilled to degasify in advance of longwall mining (McLaughlin, 2010).

The CNX Gas Mountaineer well design is a dual design with horizontal access including vertical section, directional curve, and laterals; and vertical production well (Figure 7.8). The vertical section of the wells and in some cases the curve of the access well, are cased or treated with a chemical under pressure prior to intercepting the vertical production well to enhance stability of the wall of the curve borehole and prevent any potential water ingress from shallow water bearing rock. The main laterals can be lined with slotted pipe to prevent borehole collapse after completion. While directionally drilling the horizontal laterals using air foam to power

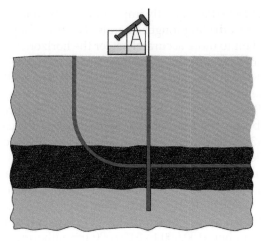

FIGURE 7.8   CNX Gas Mountaineer coalbed methane well design. *(Coal News, March 2010)*

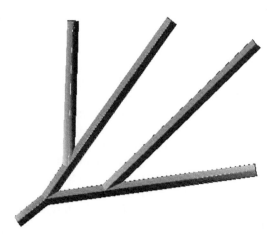

FIGURE 7.9   CNX coalbed methane well asymmetric quad pattern. *(Coal News, March 2010)*

the downhole motor, if the lateral exits the coalbed intersecting the roof or floor of the coalbed, a sidetrack is conducted to bring the well back in coal.

The original well design called for a turkey foot pattern of three laterals each of which was 1219–1524 m (4000–5000 ft) long. The total area to be drained was 640 acres per well. The pattern evolved into one called an asymmetric quad, allowing for four laterals of varying length, but totaling 2800 m (9200 ft). The drainage pattern was reduced to 1.9 km^2 (480 acres) per well requiring more wells to be drilled per given area (Figure 7.9). According to CNX, the asymmetric quad system more efficiently drains the reservoir and the results are known with improved drilling efficiencies compared to drilling longer laterals. For example, drilling times were reduced from 21 to 15 days, greatly improving well economics, in addition to reducing the risk drilling longer laterals. The use of a gamma detector close to the drill bit to more accurately steer the horizontal borehole in the coalbed, further reduced drilling times to 10 days on some wells.

In 2007, 2008, CNX Gas drilled 176 horizontal CBM wells at its Mountaineer CBM field in southern Pennsylvania and northern West Virginia, targeting the Pittsburgh and Freeport coalbeds. Peak production rates from the Pittsburgh coalbed wells varied between 7 Mcmd (250 Mcfd) and 10 Mcmd (360 Mcfd), with total production of 14 MMcm (490 MMcf) (McLaughlin, 2010).

One of the first Freeport coalbed CBM wells produced a peak production of 25 Mcmd (900 Mcfd) (Oil and Gas Investor Magazine, 2008). CNX Gas produced about 0.3 Bcm (12 Bcf) from over 200 Mountaineer CBM wells in 2008. Total cost for a horizontal well ranges between $800,000 and $900,000, with another $100,000 to $150,000 for gathering and processing (Oil and Gas Investor Magazine, 2008).

7.3.3 TDI Horizontal Dual CBM Wells

TDI began drilling horizontal CBM wells from the surface in 2002 after directionally drilling over 150 in-mine degasification boreholes greater than 1219 m (4000 ft) from 1995 to 2002, shielding longwall gateroad development and perpendicular to longwall gateroad development (Figures 7.10 and 7.11). Primarily drilled to degas future longwall mining areas, TDI's dual CBM well patented technique includes reaming the access curve and cementing casing from the surface to the coal (Figure 7.12). Advantages of casing the curve include: (1) reducing torque and drag in the curve when drilling the coalbed laterals maximizing bit weight to drill longer laterals; (2) permits only produced water from the targeted coalbed to be produced, not unwanted water from formations above the targeted coalbed; and (3) provides a guaranteed access to plug the coalbed laterals prior to mining, years after the well was drilled, unlike uncased curves (American Longwalls Magazine, 2007; Coal USA Magazine, 2008).

After the casing has been cemented from the coalbed to the surface, the coalbed is directionally drilled to intercept the vertical production well, and stopped. The main coal laterals are initiated by sidetracking from the connector lateral and are steered left or right to avoid intercepting, and drilling through the vertical production well unlike other CBM dual well systems.

TDI directionally drills main coalbed laterals by steering to purposely intercept the roof rock at drilling intervals of 100–150 m (300–500 ft) even though an EM MWD guidance system is used. Once the roof rock has been intercepted, the bit and directional string are pulled back into the coalbed at a suitable horizontal depth and a sidetrack is conducted to eliminate the rock drilled from the coalbed lateral. If roof or floor rock is intercepted

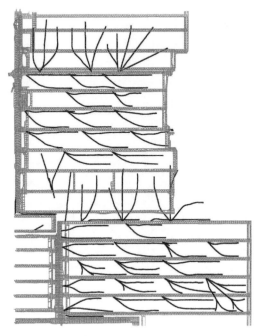

FIGURE 7.10   Target Drilling Inc. drilled in-mine degasification boreholes.

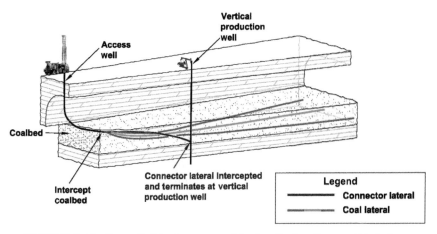

FIGURE 7.11   Target Drilling Inc. patented dual coalbed methane well technique.

unintentionally, a sidetrack is conducted in the coalbed to eliminate the rock
drilled. By sidetracking to eliminate rock intervals drilled, the coalbed lateral,
or parent borehole of the lateral, remains entirely in the coalbed. Keeping the
coalbed laterals entirely in the coalbed by sidetracking when rock is intercepted has resulted in laterals drilled to 1765 m (5789 ft) distance in the coalbed

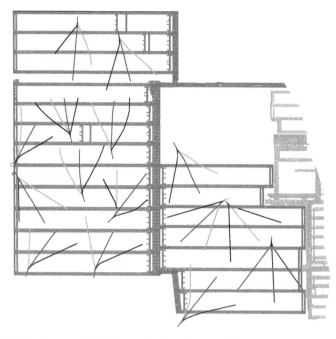

FIGURE 7.12   Target Drilling Inc. coalbed methane wells degasify future longwall panels.

(Figures 7.13 and 7.14). These distances drilled in the coalbed have been achieved by providing pull down force from the surface drill rig onto the drill string to get weight on the bit without using drill string jars or string agitator.

In blocky, relatively high cleat permeability coalbeds like the Pittsburgh coalbed and other North American coalbeds, the longer the coalbed laterals are drilled, the more cleat system or natural fractures are incrementally intercepted and connected to the lateral, increasing gas production (Maricic et al., 2005). Lastly, if the horizontal CBM well drilled from surface is going to be mined into, keeping the coalbed lateral in the coalbed can enhance the effectiveness of plugging it prior to mine-through.

TDI has successfully completed horizontal CBM wells applying its dual well method in the Pittsburgh, Sewickley, Lower Kittanning, and Freeport coalbeds with coalbed laterals with distance drilled in the coalbed exceeding 1764 m (5789 ft). Applying a turkey foot, three lateral pattern, TDI's average total footage drilled in a Pittsburgh coalbed CBM well is 6,033 m (19,789 ft), with 4939 m (16,200 ft) for three laterals and 1094 m (3589 ft) of sidetrack footage with similar coalbed footage in the Sewickley coalbed CBM wells (Figures 7.15 and 7.16).

The initial 30 day average production from TDI's Pittsburgh coalbed dual wells is 19 Mcmd (679 Mcfd) with several CBM wells peaking at greater than 25 Mcmd (900 Mcfd) and 28 Mcmd (1 MMcfd). Four years

FIGURE 7.13   Target Drilling Inc. Pittsburgh Coalbed's coalbed methane access well site. Note black coalbed cuttings in effluent pit.

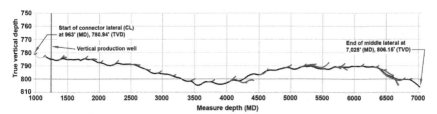

FIGURE 7.14   Elevation side view of Target Drilling Inc. coalbed methane well lateral keeping "parent borehole" of lateral in coalbed. TVD, true vertical depth; MD, measure depth; CL, connector lateral.

after drilled, wells are producing 7 Mcmd (250 Mcfd). The average drilling cost of a TDI CBM three lateral well with approximately 6100 m (20,000 ft) of coalbed lateral including sidetrack footage is $1,100,000 including consumables (casing, cement, etc.).

7.4 CONCLUSIONS

Surface drilled horizontal CBM dual wells have been proven effective in coalbed methane degasification in advance of mining, and gas recovery in coalbed areas where mining is not planned in the near future, or the

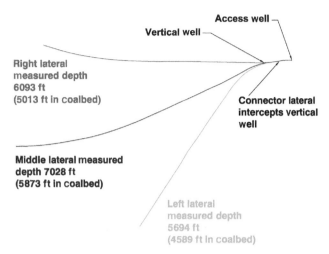

FIGURE 7.15   Plan view of Target Drilling Inc. coalbed methane well in the Pittsburgh coalbed.

FIGURE 7.16   Target Drilling Inc. Sewickley Coalbed's coalbed methane access well site.

coal is unmineable. Several companies including, but not limited to, CDX Gas (now Vitruvian Exploration), CNX Gas (now Consol Energy), and TDI have proven their CBM dual well techniques utilized to drill the coalbed laterals underbalanced have been effective in recovering acceptable volumes of coalbed methane compared to their EUR's. The volume of gas produced per well and the gas price are the two factors with the highest sensitivity determining if CBM dual well's are commercially viable considering the reported drilling costs of a CBM dual well ranges from $800,000 to $2.2 MM depending on various drilling design factors. These CBM dual well techniques have been proven in various coalbeds throughout the United States including the Pocahontas #3, Hartshorne, Pittsburgh, Freeport, Lower Kittanning, and Sewickley.

References

American Longwalls Magazine, May 2007. Target Drilling Hits Bullseye. Aspermont Limited.

CDX Gas, 2005. Unconventional Plays: Enhancing performance with new technologies. Summer NAPE Expo 2005, Doug Wight.

CNX, 2008. Investor Presentation December 2008. www.cnx.com.

Coal USA Magazine, May 2008. Targeting CBM, CMM Industries. Aspermont Limited.

Diamond, W.P., Oyler, D.C., 1980. Drilling Long Horizontal Coalbed Methane Drainage Holes From A Directional Surface Borehole, SPE/DOE 8968, Symposium on Unconventional Gas Recovery.

DuBois, G., Kravits, S.J., Reilly, J.M., Mucho, T.P., 2006. Target Drilling's Long Boreholes Maximize Longwall Dimensions, 11th U.S and International Ventilation Conference; Penn State University.

Coal Mine Methane Recovery—A Primer, U.S Environmental Protection Agency, EPA-430-R-09-013, September 2009.

Kravits, S.J., DuBois, G., Reilly, J., April 1999. Reaching greater depths. World Coal Magazine, 44–47.

Lusk, J., Jones, W., 2008. Presentation—Pinpoint Drilling and Directional, LLC, North American Coalbed Forum, April 30.

Maricic, N., Mohaghegh, S.D., Artun, E., 2005. A Parametric Study of on the Benefits of Drilling Horizontal and Multilateral Wells in Coalbed Methane Reservoirs, SPE 96018, Society of Petroleum Engineers Annual Technical Conference and Exhibition.

McLaughlin, S., March 2010. Degassing Coal Seams. Coal News.

Oil and Gas Investor Magazine, April 2008. Appalachian Cooperation. Hart Energy Publishing, Houston, Texas (Article by Jeanie Stell).

Oil and Gas Investor Magazine, January 2008. CNX: A CBM and Shale Gas Play. Hart Energy Publishing, Houston, Texas (Ellen Chang, contributing editor).

Thakur, P.C., Poundstone, W.N., June 1980. Horizontal drilling technology for Advance degasification. Mining Eng., 676–680.

USEPA, 2008. Identifying Opportunities for Methane Recovery at U.S. Coal Mines: Profiles of Selected Gassy Underground Mines 2002–2006. EPA 430-K-04-003.

CHAPTER 8

Coal Seam Degasification

Pramod Thakur

Coal Degas Group, Murray American Energy Inc., Morgantown,
West Virginia, USA

8.1 ORIGIN OF COALBED METHANE AND RESERVOIR PROPERTIES OF COAL SEAMS

8.1.1 Origin of Coalbed Methane

Coal seams form over millions of years by the biochemical decay and metamorphic transformation of plant materials. This coalification process produces large quantities of byproduct gases, such as methane and carbon dioxide. The amount of these byproducts increases with the rank of coal and is the highest for anthracite, where for every ton of coal nearly 1900 pounds of water, 2410 pounds (20,000 ft^3) of carbon dioxide, and 1186 pounds (27,000 ft^3) of methane are produced (Hargraves, 1973). Most of these gases escape to the atmosphere during the coalification process, but a small fraction is retained in the coal. The amount of gas retained in the coal depends on a number of factors, such as the rank of coal, the depth of burial, the type of rock in the immediate roof and floor, local geologic anomalies, and the tectonic pressures and temperatures prevalent at that time. The gases are contained under pressure mainly adsorbed on the surface of the coal matrix, but a small fraction of gases is also present in the fracture network of the coal. Methane is the major component of gases in coal, comprising 80–90% or more of the total gas volume. The balance is made up of ethane, propane, butane, carbon dioxide, hydrogen, oxygen, and argon.

Methane is released into each mine airway from the coal seam as mining proceeds. Large volumes of air, sometimes as much as 20 ton of air for each ton of coal mined, are circulated constantly to dilute and carry methane away from coal mines. Methane is a colorless, odorless, combustible gas

Copyright © 2014 Elsevier Inc. All rights reserved.

that forms an explosive mixture with mine air in the concentration range of 5–15% by volume. The maximum concentration of methane in mine air is restricted by law to 1–1.25% in all major coal-producing countries, but methane-air explosions are quite common even today. Table 8.1 shows a list of mine explosions since 1970 in the USA. In these 14 explosions, 196 lives were lost, inspite of coal seam degasification taking place in some mines.

Coal has been mined throughout the world for hundreds of years, and the history of coal mining is repleted with mine explosions and consequent loss of lives. Even today, 70 countries around the world mine about 8000 million ton of coal annually with more than 10,000 fatalities per year. Prior to 1950, when coal seam degasification was generally unknown and ventilation was the only method of methane control, mine explosions in the US were much more disastrous with very high fatalities. To mitigate this problem, in many instances mine ventilation can be supplemented by coal seam degasification prior to mining and even after mining.

8.1.2 Reservoir Properties of Coal Seams

Coal seam degasification techniques to be used in a mine depend on the reservoir properties of the coal seams being mined. Good methane

TABLE 8.1   Coal Mine Methane Explosions in the USA Since 1970

Year	Mine	Deaths
2010	UBB Mine, Whitesville, WV	29
2006	Sago Mine, Tallmanville, WV	12
2001	Blue Creek #5 Mine, Brookwood, AL	13
1992	No. 3 Mine, Norton, VA	8
1989	William Station #9 Mine, Wheatcroft, KY	10
1983	McClure #1 Mine, McClure, VA	7
1982	No. 1 Mine, Craynor, KY	7
1981	No. 21 Mine, Whitwell, TN	13
1981	No. 11 Mine, Kite, KY	8
1981	Dutch Creek No. 1 Mine, Redstone, CO	15
1980	Ferrel #17 Mine, Uneeda, WV	5
1976	Scotia Mine, Oven Fork, KY	26
1972	Itman #3 Mine, Itman, WV	5
1970	No. 15 and 16 Mines, Hyden, KY	38

control planning depends on accurate information on the reservoir properties of the coal seam and the total gas emission space created by the mining process. Reservoir properties governing the emission of methane from coal seams can be divided into two groups: (1) properties that determine the capacity of the seam for total gas production, e.g., adsorbed gas and porosity, and (2) properties that determine the rate of gas flow, e.g., permeability, reservoir pressure, and diffusivity of coal. The reservoir properties are highly dependent on the depth and the rank of the coal seam. The most important of these properties is the seam gas content.

8.1.3 Seam Gas Content

Based on their gas contents, coal seams can be classified as mildly gassy, moderately gassy, and highly gassy, as shown in Table 8.2.

By definition, seam gas content is the amount of gas contained in a ton of coal and includes both adsorbed gases and gases in the fracture matrix. Formerly, gas content of a coal seam or the gassiness of a coal seam was measured by the specific emission of methane from the mine, expressed as the volume of methane emitted from the mine per ton of coal produced. Although a rough correlation exists between specific emission and actual gas content of coal, it is not very reliable nor can it be used effectively for forecasting. Today, gas content of a coal seam is best measured directly. If the reservoir pressure is known, an indirect estimate of gas content can also be obtained by Langmuir's equation (Langmuir, 1918) for mono-layer adsorption:

$$V = V_m BP / (1 + BP)$$

where V is the estimated gas content of coal, V_m is the volume of gas for full saturation of coal, B is a characteristic constant of the coal seam, and P is the reservoir pressure. For US coalbeds, the reservoir pressure is roughly correlated with the depth of the coal seam (Thakur and Davis,1977) and is estimated at 0.303 psi/ft or roughly 70% of the hydrostatic head.

TABLE 8.2 Gassiness of Coal Seams

Category	Depth (ft)	Gas Content of Coal (ft³/ton)
Mildly gassy	Less than 600	Less than 100
Moderately gassy	600–1200	100–300
Highly gassy	1200–3000	300–700

Since coal seams and gas in coal are formed together, it is a misnomer to call a coal seam nongassy. All coal seams are gassy by definition but they vary in their degree of gassiness—i.e., gas content per ton of coal. The depth of a coal seam and its rank are good indicators of its gassiness, but direct measurement of gas content is highly recommended.

Figure 8.1 shows the gas content of coal vs gas pressure for some US coals (Kissell et al., 1973). Gas content, in general, increases with depth and higher rank.

8.2 A THRESHOLD FOR COAL SEAM DEGASIFICATION

Generally, it is economically feasible to handle specific methane emissions[a] from a mine up to $1000\,ft^3$/ton with a well-designed ventilation system. At higher specific emission rates, a stage is reached where ventilation

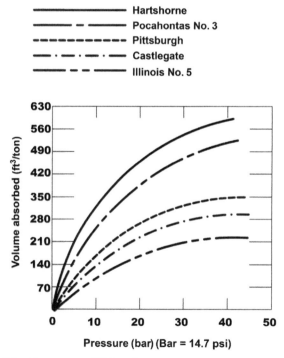

FIGURE 8.1   Gas content of U.S. coals vs reservoir pressure (Kissell et al., 1973).

[a]Specific methane emission is the total amount of gas released from the mine divided by the total amount of coal mined.

cost becomes excessive or it becomes impossible to stay within statutory methane limits with mine ventilation alone. However, with a well-planned methane drainage system and a well-designed ventilation system, even highly gassy mines with specific methane emissions in excess of 4000 ft³/ ton can be safely operated.

In some mines, there is often a choice regarding how much methane should be drained and how much should be handled by mine ventilation air. Figure 8.2 illustrates a generalized optimum point; the actual optimum point depends on a number of factors, including the rate of mining, size of longwall panel, specific methane emission, and the cost of ventilation and methane drainage.

Advantages of coal seam degasification can be summarized as:

1. Reduced methane concentrations in the mine air, leading to improved safety.
2. Reduced air requirements and corresponding savings in ventilation costs.
3. Faster advance of development headings and economy in the number of airways.
4. Improved coal productivity.
5. Additional revenue from the sale of coal mine methane.
6. Additional uses of degasification boreholes, such as water infusion to control respirable dust.
7. Advance exploration of coal seams to locate geological anomalies in the longwall panel.

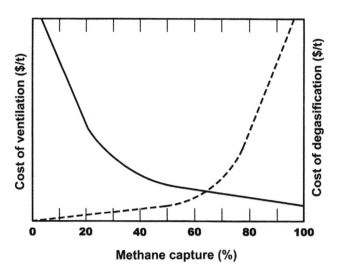

FIGURE 8.2   Generalized optimum degasification vs ventilation.

8.3 METHANE EMISSIONS IN MINES

Underground mining is done in two phases: (1) development and (2) pillar extraction. Development work involves the driveage of a network of tunnels (entries) into the coal seam to create a large number of pillars or longwall panels to be mined later. This driveage is usually done with a continuous mining machine. This machine cuts and loads coal into a shuttle car which in turn hauls and dumps the coal onto a moving belt. The coal travels out of mines on a series of belts and is finally brought to the surface via a slope or shaft. Figure 8.3 shows a simplified illustration of a typical longwall panel layout in a US coal mine.

All methane produced during the development phase of mining is from the coal seam being mined. Methane is emitted at the working face as well as in the previously developed areas. All emitted methane is mixed with ventilation air, diluted to safe levels, and discharged on the surface. Methane drainage during or prior to development becomes necessary if the development headings experience a high rate of methane emissions. This is called premining methane drainage. Horizontal drilling of longwall panels prior to mining also falls into this category.

The second phase of underground mining involves complete or partial extraction of the coal pillars. Smaller pillars are extracted by continuous mining machines by splitting them into even smaller pillars, but larger panels of coal (up to 1500 ft × 15,000 ft) are extracted by the longwall method of

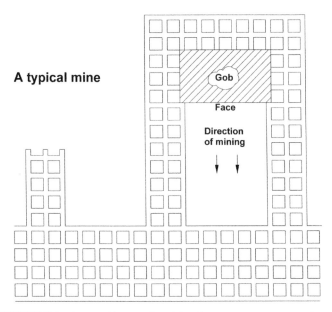

FIGURE 8.3   A typical longwall and development section in a coal mine.

mining. In either case, the mined coal produces methane; in addition, the extraction of these pillars or longwall panels causes the overlying strata to subside[b] and the underlying strata to heave. The ventilated mine workings constitute a natural pressure sink, into which methane flows from the entire disturbed area or what is known as the "gas emission space". Figure 8.4 shows the limits of the gas emission space (Lidin, 1961; Thakur, 1981; Winter, 1975; Gunther and Belin, 1967). The gas emission space may extend 270 ft below the coal seam being mined and approximately 1000 ft above it. It depends on the width of the panel.

The gob methane emission rate mainly depends on the rate of longwall advance, the geology, the size of the longwall panel, and the gas content and thickness of coal seams in the gas emission space.

Various methane drainage techniques are used to capture the gas from the gob so that the mine ventilation air does not have to handle all of it.

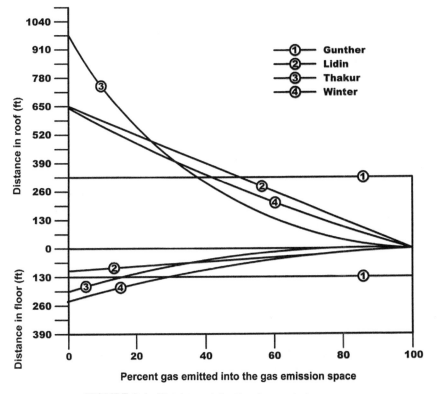

FIGURE 8.4   Heights and depths of gas emission space.

[b] In the US, the subsided region is called a "gob".

Depending on the magnitude of the problem, methane drainage can be performed prior to mining, known as premining methane drainage. Methane drainage can also be performed during mining and after the area is completely mined out and sealed. These two stages are generally grouped together as postmining methane drainage.

8.4 METHANE DRAINAGE TECHNIQUES

8.4.1 Premining Methane Drainage

Techniques for premining drainage can be broadly classified into four categories:

1. Horizontal inseam boreholes.
2. In-mine vertical or inclined (cross-measure) boreholes in the roof and floor.
3. Vertical wells that have been hydraulically fractured (so-called "frac wells").
4. Short-radius horizontal boreholes drilled from surface.

8.4.1.1 Horizontal Inseam Boreholes

Early work in premining methane drainage was done with short horizontal inseam boreholes (Spindler and Poundstone, 1960). Figure 8.5 shows the two most commonly used variations of degasification with inseam horizontal boreholes. Success of the technique is predicated on good coalbed permeability (≥ 5 mD). The horizontal drilling technique and its application to degas coal seams are well-documented in published literature (Thakur and Davis, 1977; Thakur and Poundstone, 1980; Thakur et al., 1988). In highly permeable coal seams, e.g., the Pittsburgh seam of the Appalachian Basin, nearly 50% of the in situ gas can be removed by

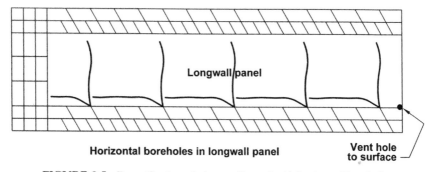

Horizontal boreholes in longwall panel   Vent hole to surface

FIGURE 8.5   Degasification of a longwall panel with horizontal boreholes.

this technique prior to mining. The major drawback of this technique is that only about six months to a year—the time between development and longwall extraction—is available for degasification.

8.4.1.2 In-mine Inclined or Vertical Boreholes

Short vertical or long inclined boreholes have been drilled from an existing mine (or roadways expressively driven for this purpose) to intersect other coal seams in the gas emission space, allowing for the seams to be degassed prior to mining. Again, success depends on high permeability. A far better way to degas these coal seams lying in close proximity to each other is to use vertical frac wells.

8.4.1.3 Vertical Frac Wells

Vertical frac wells are ideally suited to highly gassy, deep, low-permeability coal seams where it takes several years prior to mining to adequately degas the coal. These wells are drilled from the surface on a grid pattern over the entire property or only on longwall panels to intersect the coal seam to be mined in the future.

Vertical wells drilled into the coal seam seldom produce measurable amounts of gas without hydraulic stimulation. High-pressure water (or other fluids) with sand are pumped into the coal seam to create fractures (Figure 8.6). The fluid (water) is then pumped out, but the sand remains, keeping the fractures open for gas to escape to the well bore. Under ideal conditions, if the vertical frac wells are drilled more than 5–10 years in advance of mining, 60–70% of the methane in the coal seam can be removed prior to mining.

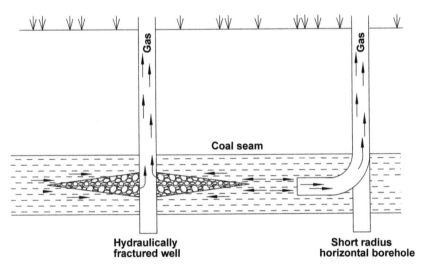

FIGURE 8.6 Vertical frac wells and short radius horizontal boreholes.

Vertical frac wells have been very successful in the Appalachian and San Juan Coal Basins of the US. They have also been attempted in the UK, Germany, Poland, China, and Australia, but met with only limited success. Major reasons for the lack of success overseas are cost and lack of sufficient permeability, demonstrated as follows:

- The cost of drilling and hydrofracing a well in Europe and Australia is typically three times the cost in the US. The cost of permitting and site preparation is also higher. In many countries, the drilling and hydrofracing equipment are not conveniently available.
- Lower permeability (<1 mD) of many European, Asian, and Australian coal seams contributes to the limited success of frac wells. Even well-designed and well-executed frac jobs in Bowen Basin, Australia, were ineffective. A solution for this problem may lie in "gas flooding", i.e., injection of an inert gas such as nitrogen or carbon dioxide to drive methane out (Puri and Yee, 1990). Increased methane production is, however, obtained with an increase in the inert gas content of the produced gas. This may affect the marketing of produced gas adversely.

8.4.1.4 Short-Radius Horizontal Boreholes

In coal seams with high permeability, methane drainage can be performed with vertical boreholes drilled vertically from the surface and then turned through a short radius to intersect the coal seam horizontally. The horizontal extension can be up to 3000 ft. Methane then flows from the coal seam under its own pressure as shown in Figure 8.6. The technique is well-proven in oil fields, but it has found a limited application in coal mines for two reasons:

- *Cost*: A short-radius borehole drilled vertically to a depth of 1000 ft and horizontally extended to 3000 ft may cost up to 0.5 million dollars (US).
- *Water accumulation in the horizontal borehole*: As can be seen in Figure 8.6, any water accumulation in the horizontal leg of the borehole will seriously inhibit gas production. A solution may lie in deepening the vertical leg below the coal seam being drilled and installing a dewatering pump in it as is commonly done for vertical frac wells.

Of the above four techniques, vertical frac wells have been the most effective option for premining degasification of most coal seams. Vertical frac wells also allow access to all coal seams in the gas emission space for predrainage. Such access becomes necessary in highly gassy mines in order to achieve high productivity. The only possible exception is for shallower, very permeable coal seams where in-mine drilling is sufficient and more economical. In shallow formations, the fracture system created by hydrofracing is like a horizontal pancake and is not very productive because the fracture system does not extend far enough from the bore-hole[c]. Strong roof and floor are also necessary to contain the fracture system within the coal seam.

Recently, many short-radius horizontal boreholes drilled from the surface have been used to recover methane from permeable coalbeds. In the future, CO_2 flooding may be used.

8.4.2 Postmining Methane Drainage

Techniques for postmining drainage can be broadly classified into four categories.

1. The packed cavity method and its variants.
2. The cross-measure borehole method.
3. The superjacent method.
4. The vertical gob well method.

8.4.2.1 The Packed Cavity Method and Its Variants

This is a technique primarily used in Russian coal mines. Early methods of methane control consisted of simply isolating the worked out area in the mine using pack walls, partial or complete stowing, and plastic sheets or massive stoppings. A network of pipeline which passed through these isolation barriers was laid in the gob and methane was drained using vacuum pumps. Lidin (1961) has reviewed several variants of this technique. Figures 8.7 and 8.8 show typical layouts for caving and partially stowed longwall gobs.

Methane capture ratios achieved in practice are shown in Table 8.3. The ratios generally seem to improve in going from caving (20–40%) to fully stowed longwall gobs (60–80%). In Figure 8.7, the gate roads are protected by a pack wall against the gob. Pipelines are laid through the pack wall to reach nearly the center line of the gob, then manifolded to a larger diameter pipe in the gate road. In Figure 8.8, the partially stowed longwall gob, cavities are purposely left between alternate packs. The overlying strata in the cavity area crack and provide a channel for gas to flow into these packed cavities. Pipelines are laid to connect the cavity with methane drainage mains. Methane extraction is usually done under suction.

8.4.2.2 Cross-Measure Borehole Method

This is by far the most popular method of methane control on European longwall faces. Figure 8.9 shows a typical layout for a retreating longwall

---

[c] The ideal fracture is vertical, entirely within the coal, less than an inch wide, and extends upwards of a 1000 ft from the borehole, bilaterally.

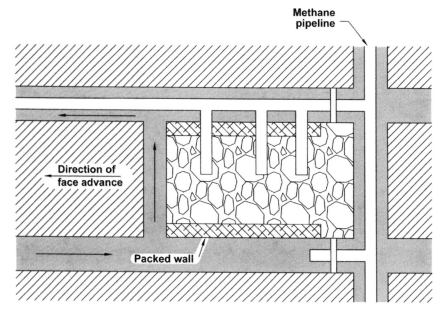

FIGURE 8.7   Methane drainage by packed cavity method.

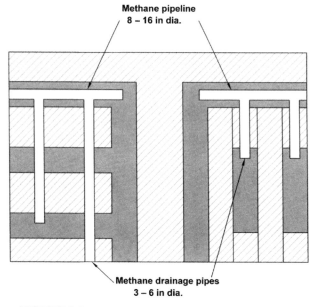

FIGURE 8.8   Methane drainage from partially filled gobs.

TABLE 8.3   Methane Capture Ratios for Postmining Methane Drainage Techniques

Methane Drainage Technique	Methane Capture Ratios (5%)	Remarks
Packed cavity method After Lidin (1961)	20–40 30–50 60–80	Caving longwalls Partially stowed longwalls Fully stowed gobs
Cross-measure boreholes After Kimmins (1971)	59–70	Highly gassy mines with specific emissions 3000–6000 ft^3/ton
Superjacent method	50	For multiple coal seams in the gas emissions space
Vertical gob wells	30–80	The methane capture efficiency depends on the number of gob wells per longwall panel and production techniques

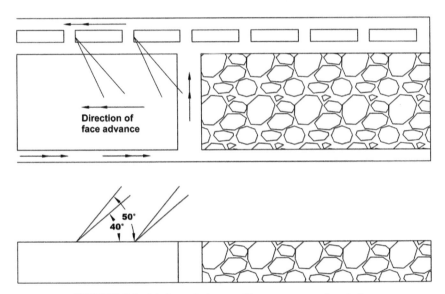

FIGURE 8.9   Methane drainage with cross-measure boreholes.

face. Boreholes 2–4 in in diameter and about 80 ft apart are drilled from the tail gate to a depth of 60–500 ft.

The angle of these boreholes with respect to horizon varies from 20° to 50°, while the axis of the borehole is inclined to the longwall axis at 15°–30°. At least one hole in the roof is drilled at each site, but several boreholes in roof and floor can be drilled at varying inclinations depending on the degree of gassiness. These holes are then manifolded to a larger pipeline system and gas is withdrawn using a vacuum pump. Vacuum pressures vary from 4 to 120 in of water gauge.

The amount of methane captured by the drainage system, expressed as a percentage of total methane emission in the section, varies from 30% to 70%. Some typical data for British and U.S. mines are given by Kimmins (1971) and Thakur et al. (1983), respectively, and are shown in Table 8.3.

The cross-measure borehole method is generally more successful for advancing longwall panels than it is for retreat faces. The flow from individual boreholes is typically 20 ft^3/min, but occasionally it can reach 100 ft^3/min for deeper holes. Sealing of the casing at the collar of the borehole is very important and is usually done with quick-setting cement. Sometimes, a perforated liner (a pipe of smaller diameter than the borehole) is inserted in the borehole and sealed at the collar to preserve the production from the borehole even when it is sheared by rock movements.

8.4.2.3 *The Superjacent Method*

This method was mainly used for retreating longwall faces in highly gassy seams in French mines. Figure 8.10 shows a typical layout. A roadway is driven 70–120 ft above the longwall face, preferably in an unworkable coal seam. The roadway is sealed and vacuum pressures up to 120 in of water gauge are applied. To improve the flow of gas, inclined boreholes

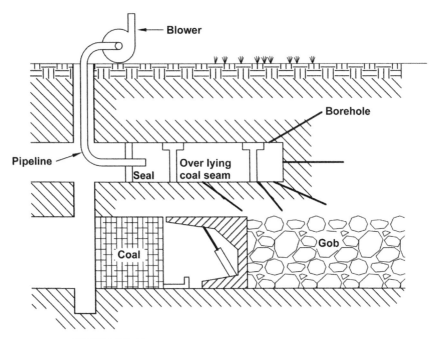

FIGURE 8.10   Methane drainage by the superjacent method.

in the roof and floor are drilled to intersect with other gassy coalbeds. If the mining scheme proceeds from the top to the bottom seams in a basin, the entries in a working mine can be used to drain coal seams at lower levels. Methane flow from these entries is high, averaging 700–1000 ft^3/min for highly gassy seams. Nearly 50% of total emissions have been captured using the superjacent method.

8.4.2.4 Vertical Gob Well Method

This technique, most commonly used in longwall mining in the United States, is relatively new and it differs from European systems in several ways. US coal seams are generally thin, shallow, and relatively more permeable. Typically, only one seam is mined in a given area and retreat longwall mining is the only method being practiced at present. Methane emission rates from gobs in various coal basins vary depending on the geological conditions, but deep-seated longwall gobs (e.g., those in Pocahontas No. 3 seam in Virginia and in the Mary Lee seam in Alabama) produce methane in the range of 1800–18,000 ft^3/min. Multiple entries (typically four) are driven to develop longwall panels so that necessary air quantities can be delivered to the longwall faces via the mine ventilation system. In many cases, however, some sort of additional methane control becomes necessary.

The most popular method of methane control is to drill vertical boreholes above the longwall prior to mining as shown in Figure 8.11. Depending on the length of the longwall panel (typically 10,000 ft) and the rate of mining, 3–30 vertical gob degas boreholes are needed. The first hole is usually within 150–500 ft of the start line of the longwall face. The borehole is drilled to within 30–90 ft from the top of the coal. The casing is cemented through the fresh water zones near the surface, and a slotted liner is provided over

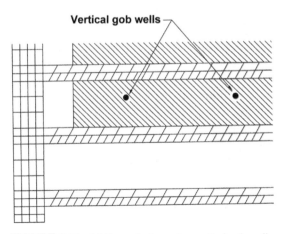

FIGURE 8.11   Methane drainage by vertical gob wells.

the lower open section to prevent closing of the hole by caving. These boreholes are completed prior to mining. Usually no measurable methane production is realized until the longwall face mines past the borehole.

Early experiences with this method of gob degasification have been described by Moore et al. (1976) for the Lower Kittanning seam, by Moore and Zabetakis (1972) for the Pocahontas seam, and by Davis and Krickovic (1973) and Mazza and Mlinar (1977) for the Pittsburgh seam. Many gob degasification boreholes produce naturally when the longwall face intersects them, but vacuum pumps are often added to further improve the flow and in some cases to prevent the reversal of flow. The capture ratios vary from 30% to 80% depending on the number and size of gob wells per panel and the size of vacuum pumps.

A comparison of methane capture ratios for various postmining methane drainage techniques is shown in Table 8.3. Although each technique offers high capture efficiency in some cases, it is the author's experience that the vertical gob wells, if properly designed, offer the most universal application with consistently high capture ratios. In addition, this technique is a natural outgrowth of the premining degasification technique using vertical frac wells. These frac wells can be converted easily into postmining gob wells with minimal additional expense.

8.5 HOW TO TRANSPORT GAS IN UNDERGROUND MINES

In-mine horizontal drilling and cross-measure boreholes drilled to degas longwall gobs produce large volumes of gas. This gas must be conducted out of the mine without being allowed to mix with the mine ventilation air. The US coal industry, working with the US Mine Safety and Health Administration (MSHA), has developed general guidelines for installing and operating these pipelines, as follows:

1. Underground methane pipeline will be made of high molecular weight polyethylene plastic or steel as detailed in Figure 8.12.
   a. All underground steel pipelines will be 3.5 to 8 in O.D. schedule 40 pipes joined together with threaded couplings. These pipes will be made up tightly using a good grade of thread lubricant. Mill collars will be broken out, doped, and remade. A flange connection will be used every 10 joints (approximately 210 ft apart) so that a section of the pipeline can be removed without cutting the line if one or more joints need to be replaced later.
   b. All underground plastic line will be 3–8 in high-density polyethylene pipe. Plastic flange adapters will be fusion bonded to the pipe ends in fresh air. Steel flanges back-up rings installed prior to fusion bonding will be used to connect plastic to plastic and plastic to steel.

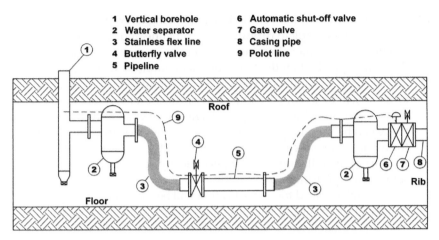

FIGURE 8.12 Underground gas pipeline for methane drainage.

2. The entire length of pipeline between the bottom of the vent hole and the well head will be pressure tested to 1.25 times the shut-in pressure of the borehole or 90 psi, whichever is greater.
3. Pipeline will be generally laid in the return airway and will not be buried. Whenever the pipeline must cross a fresh air entry, it will be conducted through a steel or plastic conduit.
4. No hoses will be used in the system except while a hole is being drilled. Stress relieving, flexible tubing will be used at critical points such as the well head to pipeline connection. This will be stainless steel tubing with a triple wire braid cover.
5. The steel pipeline will be firmly supported with no unsupported span greater than two feet.
6. A gas water separator will be installed at the bottom of the vertical vent hole to remove condensation that falls back down the casing. Other separators will be installed on the holes or on the pipeline if water production from coal warrants. All separators will be preferably be commercially made. Water drains will be provided on the line wherever necessary.
7. If steel pipeline is used, a potential survey will be made and cathodic protection provided where needed.
8. Automatic shut-in valves will be installed at each well head. These will be held open by nitrogen or air under pressure contained in a fragile plastic pilot line running parallel to and secured on top of the pipeline. Any roof fall or fires serious enough to damage the pipeline will damage the pilot line first and close the boreholes immediately.
9. The pipeline system will be inspected weekly by a competent person familiar with system operation.

10. If the quantity of mine ventilation air flowing over the pipeline is such that a complete rupture of the pipeline and consequent discharge of methane in mine air will raise its concentration above the limits specified by the law, a methane monitor will be used.

11. At the surface installation (Figure 8.13), a commercially made flame arrestor will be installed within 10 ft of the top of the vent stack. A check valve shall be used to guard against reversal of flow. The check valve can be manually defeated if it is desired to purge the pipeline for repairs. Flow measurements should be taken with a pilot tube. All surface installations will be periodically inspected to ensure satisfactory performance.

12. All boreholes drilled for degasification will be accurately surveyed either during drilling or after drilling is completed using commercially available borehole surveying tools. These boreholes will be accurately plotted on mine maps to prevent any inadvertent mining through them.

13. Should an occasion arise in the future when cutting into an abandoned, unplugged borehole will be necessary, a detailed mining plan will be submitted to MSHA.

14. A compressor will be required at the surface if beneficial use is made of the gas at a future date. Plans for the installation will be discussed with MSHA at the appropriate time.

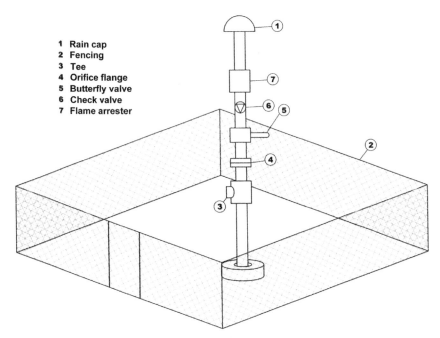

1 Rain cap
2 Fencing
3 Tee
4 Orifice flange
5 Butterfly valve
6 Check valve
7 Flame arrester

FIGURE 8.13   Surface installation for methane gas disposal.

8.6 ECONOMICS OF COAL SEAM DEGASIFICATION

In general, unless the specific methane emission from the mine (ft^3 of methane per ton of mined coal) is high (above $3000\,ft^3$/ton), it may not be profitable to process the gas for marketing. The cost of compressing and processing the coal mine methane and a complete economic analysis to reflect rates of return on the investment is beyond the scope of this chapter. A rough estimate of costs associated with coal seam degasification can be derived, however, as shown here.

For all underground longwall mining, a generalized scheme of degasification depending on the gassiness of the coal seam has been proposed by Thakur and Zachwieja (2001). The following assumptions were made:

1. The longwall panel is 1000 ft wide and 10,000 ft long.
2. The coal seam has an average thickness of 6 ft.
3. The coal block to be degassed is 1300 ft by 10,000 ft, assuming that the width of chain pillars is 300 ft.
4. The cost of contract drilling for the in-mine horizontal drilling is $50/ft.
5. The cost of a gob well is $50,000 to $200,000, depending on the depth of the mine and the size of the borehole.
6. The cost of hydrofracing a well is $250,000.

If the total cost of in-mine drilling, including all of the underground pipeline costs, all vertical frac wells, and all other gob wells is added and then divided by the tons of coal in the longwall block, the result is the cost of coal seam degasification per ton of coal.

8.6.1 Estimated Cost for Mildly Gassy Coal Seams Less Than $100\,ft^3$/ton (See Table 8.2)

Premining degasification: For coal seams with gas contents less than $100\,ft^3$/ton, there is generally no need for premining degasification.

Postmining degasification: Two gob wells are recommended for the longwall panel. The first gob well should be installed within 1000 ft of the setup entry, and the second one installed in the middle of the panel.

The total cost is $100,000 or $0.03/ton.

8.6.2 Estimated Cost for Moderately Gassy Coal Seams $100–300\,ft^3$/ton (See Table 8.2)

Premining degasification: The longwall panel should be drilled horizontally at 1000 ft intervals and development boreholes should be drilled to degas development sections. Total in-mine drilling footage for a typical panel may total 25,000 ft.

Postmining degasification: In moderately gassy coal seams, a proposed longwall panel may need 5–6 gob wells. The diameter and size of exhaust fans will depend on local conditions.

The total cost is approximately $1.55 million, or $0.50/ton.

8.6.3 Estimated Cost for Highly Gassy Coal Seams over 300 ft³/ton (See Table 8.2)

Premining degasification: Highly gassy coal seams must be drained several years ahead of mining with vertical frac wells (wells that have been hydraulically fractured). These frac wells can be placed at about a 20-acre spacing. Frac wells drilled about 5 years ahead of mining can drain nearly 50% of the in situ gas prior to mining, but this may not be sufficient. Additional degasification with in-mine horizontal drilling can raise the gas drained to nearly 70%. Horizontal boreholes are drilled 200–300 ft apart to a depth of 900 ft. Assuming a 200 ft interval, nearly 45,000 ft of horizontal drilling and about 15 vertical frac wells may be needed to properly degas the panel.

Postmining degasification: Because of very high gas emissions from the gob, the first gob well must be installed within 50–100 ft from the setup entry. Subsequent gob wells may be drilled at a 6- to 15-acre spacing, depending on the rate of mining and the gas emission per acre of gob. In the US states of Virginia and Alabama, two states with some highly gassy coal seams, gob wells are generally 9–12 in in diameter. Powerful exhaust fans capable of a suction of 5–10 in of mercury are needed to capture up to 70% of gob gas emissions.

The total cost of degasifying a longwall panel in a highly gassy coal seam is approximately $11 million, or $3.52/ton. Coal seam degasification is needed for mine safety and high productivity, but in highly gassy mines it becomes quite expensive. In these mines, the processing and marketing of coal mine methane becomes necessary to defray the cost.

References

Davis, J.G., Krickovic, S., 1973. Gob Degasification Research – A Case History. U.S. Bureau of Mines. IC8621.

Gunther, J., Belin, J., 1967. Prevision du degasment du grisou en taille pour le grisements en Plateure. In: The 12th International Conference on Safety in Mines Research Institute, Dortmund, Germany, 1967.

Hargraves, A.J., March 1973. Planning and operation of gaseous mines. CIM Bulletin.

Kimmins, E.J., 1971. Firedamp drainage in the northwestern area. Colliery Guardian, Annual Review, 39–45.

Kissell, F.N., McCulloch, C.M., Elder, C.H., 1973. The Direct Method of Determining Methane Content of Coalbeds for Ventilation Design. U.S. Bureau of Mines. RI 7767, 17 pp.

Langmuir, I., 1918. The adsorption of gases on plane surfaces of glass, mica, and platinum. Journal of the American Chemical Society 40 (9), 1361–1403.

Lidin, G.D. (Ed.), 1961. Control of Methane in Coal Mines. English translation by Israel Program for Scientific Translation, Jerusalem, 1964 (original in Russian 1961).

Mazza, R.L., Mlinar, M.P., 1977. Reducing Methane in Coal Mine Gob Areas with Vertical Boreholes. Final report on U.S. Bureau of Mines Contract No. H0322851.

Moore, T.D., Zabetakis, M.C., 1972. Effect of Surface Borehole on Longwall Gob Degasification (Pocahontas No. 3 Coalbed). U.S. Bureau of Mines. RI 7657.

Moore, T.D., et al., 1976. Longwall Gob Degasification with Surface Ventilation Boreholes above the Lower Kittanning Coalbed. U.S. Bureau of Mines. RI 8195.

Puri, R., Yee, D., 1990. Enhanced coalbed methane recovery. In: Proc. 65th Annual Technical Conference and Exhibition of the SPE, New Orleans, LA, September 23–26, 1990.

Spindler, G.R., Poundstone, W.N., 1960. Experimental work in the degasification of the Pittsburgh coal seam by horizontal and vertical drilling. In: AIME Annual Meeting. Preprint 60F106.

Thakur, P.C., Davis, J.G., October 1977. How to plan for methane control in underground coal mines. Mining Engineering, 41–45.

Thakur, P.C., Poundstone, W.N., June 1980. Horizontal drilling technology for advance degasification. Mining Engineering, 676–680.

Thakur, P.C., 1981. Methane control for longwall gobs. In: Ramani, R.V. (Ed.), Longwall–Shortwall Mining, State-of-the-Art. A.I.M.E, pp. 81–86.

Thakur, P.C., Cervik, J., Lauer, S.D., 1983. Methane drainage with cross-measure boreholes on a retreat longwall face. Preprint #83–398. In: SME/AIME Annual Meeting, Salt Lake City, Utah.

Thakur, P.C., Christopher, D.A., Bockhorst, R.W., 1988. Horizontal Drilling Technology for Coal Seam Methane Recovery. In: Gillies, A.D.S. (Ed.), Proceeding of the 4th International Mine Ventilation Congress, Brisbane, Australia, pp. 201–207.

Thakur, P.C., Zachwieja, J., 2001. Methane control and ventilation for 1000-ft wide longwall faces. In: Longwall USA International Exhibition and Conference, June 13–15, pp. 167–180.

Winter, K., 1975. Extent of gas emission zones influenced by extraction. In: International Conference on Coal Mine Safety Research, Washington, DC, pp. V3.1–V3.17.

Gas Outbursts in Coal Seams

Fred N. Kissell[1], Anthony T. Iannacchione[2]

[1]Mine Safety Consulting, PA, USA, [2]Mining Engineering Program, University of Pittsburgh, PA, USA

9.1 INTRODUCTION

A gas outburst is a sudden, violent blowout of coal and gas from a coal mine working face, leaving a cavity ahead or to a side of the face.[1] The coal in gas outbursts is often described as "pulverized" (Campoli et al., 1985) or "of fine size" (Hargraves, 1958) and "flowing". The amount of coal involved can range from a few to hundreds of tons. The gas release can be correspondingly large, and if the gas released is methane, the presence of an ignition source will magnify the destruction.

Thousands of gas outbursts have occurred around the world, resulting in hundreds of fatalities. For the U.S., gas outbursts have been relatively rare, probably due to a combination of shallower coalbeds with less structural stress, lower gas pressures, and more permeable coal that allows coalbed gas adjacent to the face to drain off more quickly.[2]

Campoli et al. (1985) and Lama and Bodziony (1998) have provided comprehensive overviews of gas outbursts worldwide. In their views, the primary factors contributing to gas outbursts are the following:

1. high gas content, corresponding to high gas pressures;
2. low permeability, which ensures a steep pressure gradient close to the mine workings;

[1]Hargraves (1958) defined a gas outburst as: "… the sudden disintegration of coal, and its projection from the seam, without deliberate initiation and accompanied by, and followed by enormous gas emission. The gas has the effect of carrying the broken coal for considerable distances. This projected coal is invariably of fine size. The gas pressures and volumes associated are sometimes sufficient to penetrate the intake roadways for considerable distances and to blow out stoppings."

[2]Another factor may be the low CO_2 content of U.S. coals, not over 10% in actively mined coalbeds.

3. high stress fields in the rock mass found when the overburden is great;
4. structurally weak coal with a well-developed microcrack structure that has the ability to desorb gas rapidly;
5. a mining method accessing unmined portions of the coalbed, such as longwall gateroad development; and
6. geological anomalies, such as fault zones, clay veins, or igneous intrusions.

With so many factors contributing to gas outbursts, it is difficult to identify the most important, but two of the more significant have to be gas pressure (gas content) and coalbed permeability. Lower permeability coals at high pressure will be more outburst-prone because of their steeper pressure gradient close to the face. This combination of high pressure and low permeability is shown in Figure 9.1, a pressure gradient curve from the Pocahontas #3 coalbed where the overburden was 600 m (Kissell, 1972).

The gas pressure deep in this seam was about 4.5 MPa (650 psi), corresponding to a gas content of 20 m³/mt (cubic meters of gas per metric ton of coal). Close to the working face, the gas pressure is still quite high, even 360 h after mining stopped. It is not difficult to imagine the gas pushing the coal into the mine entry, especially in thicker coalbeds.

Recent research on the origin of gas outbursts has suggested that factors related to the coal may also be critical. For example, Cao et al. (2001)

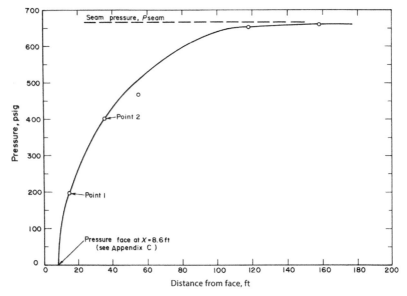

FIGURE 9.1   Gas pressure vs distance measured 360 h after mining stopped.

have associated gas outbursts with faulted and tectonically altered coals. Cao et al. noted that gas outbursts in China are associated with tectonic activity that has produced regional thrust and reverse faulting. Coal near such faults is often crushed and pulverized, resulting in altered desorption and permeability characteristics. In another study, Beamish and Crosdale (1998) suggested that coals with high vitrinite and/or inertodetrinite contents were more likely to retain the large quantities of gas needed to produce a gas outburst.

9.2 GAS OUTBURST PREVENTION

Pressure relief techniques reduce the ability of the immediate face to store the energy necessary for gas outbursts. Fracturing of the immediate face ensures that both high gas pressures and structural stress conditions do not exist for the next cut. The structural stress and higher pressure gas is shifted away from the face and deeper into the coalbed where it is less of a threat.

One relief technique involves the drilling of large diameter boreholes, typically 10 cm in diameter, into the face. Another is to cut slots in the face. Kolmsov and Bolsminskii (1981) have reported success with slots cut into outburst-prone longwall faces, with slot depths ranging from 20 to 100 cm.

In the U.S., the Mid-Continent Resources Dutch Creek Mine near Redstone, CO, used volley firing[3] to reduce the gas pressure gradient in both development and longwall faces (Varley, 2009). The Dutch Creek mines had weak, impermeable coal with high in-seam stresses making gas removal with large diameter drill holes impractical. In addition, the Mid-Continent Resources mines often used probe drilling to evaluate the potential of gas outburst hazards. If a probe hole produced blowing gas, this was taken as evidence that a hazard existed. Volley firing was then used. A successfully distressed area would not contain gas pressure within probe holes drilled to a depth of three seam heights or 10 m of planned advances (Varley, 2009).

In Australia, drainage holes, 9.6 cm in diameter, are drilled in-seam mainly from one set of development headings, across the proposed longwall panel, to a distance of 10–50 m beyond the next adjacent panel. Holes are drilled in fan patterns to minimize rig moves. The target spacing at the end of the holes ranges from 20 to 50 m, although occasionally a remarkably low 6–8 m is required for adequate drainage. Drainage is sometimes

---

[3] In volley firing, explosives are used to fracture the coal face to a certain depth before mining. The method is used prior to face advance or entry development to advance the high stress zone away from the working face. Both volley firing and slot cutting (described in the previous paragraph) are also used to prevent coal bumps.

inhibited because microfractures in the coal are filled with a carbonate cement that reduces the permeability (Wynne, 2003) (Aziz et al., 2005). The cost of gas drainage in Australia is high, $37,000,000 annually for 370,000 m of drainage holes.

The criterion for adequate drainage is a sufficiently lowered gas content. For example, Bulli seam mines in Australia operate under a "Section 63" notice, which prohibits the mining of coal that contains gas at contents above a specified threshold value. Such notices were applied following a fatal gas outburst in 1994. Draining gas to lower the coalbed gas content below the threshold value has successfully prevented gas outbursts. Threshold values fall into the range of roughly 7–10 m^3/mt, depending on whether the seam contains CH_4 or CO_2. Coal seams containing CO_2 are more prone to gas outbursts. To illustrate the effectiveness of drainage, Harvey (2002) showed that in the 4 years prior to the Section 63 notice, there were between 6 and 16 gas outbursts recorded each year in Bulli seam mines, where most Australian gas outbursts have occurred (Black et al., 2009). In the 7 years following, there were a total of nine gas outbursts with no injuries. Nevertheless, Black et al. have contended that there is justification to raise the threshold limit values to 12 m^3/mt for coal containing 100% CH_4 and 8 m^3/mt for coal containing 100% CO_2.

9.3 DIFFERENCES BETWEEN OUTBURSTS, GAS OUTBURSTS, AND BUMPS

As more is learned about the various kinds of bursts, the terminology has changed.

Ulery (2008) has discussed the historical record of outbursts in the U.S. in which the event was described as an outburst because of a release of gas that was then ignited. These included mine explosions in 1915, 1922, 1947, and 1958. However, these could have been categorized as "bumps", since some gas will also be released as bumps occur.[4]

Whyatt (2008) has described both outbursts and bumps as "dynamic failures". He further described outbursts as a brittle failure of the immediate margin of an opening that expels material into the opening. This definition of an outburst is broader than a "gas outburst" given by Hargraves (see Footnote 1) in which the release of high amounts of gas is emphasized.

---

[4] Ignitions of methane gas and dust associated with coal bumps, where full extraction mining was occurring, are somewhat rare, but at least one incident has occurred. In this disaster at the Kenilworth Mine in 1945, a pillar bump caused an electrical short in a loading machine, producing an electrical arc. Gas and coal dust were ignited, severely burning 12 miners, six fatally.

Gas outbursts will only occur where a steep gas pressure gradient, such as that shown in Figure 9.1, exists. Therefore, failure at locations such as pillars and the corners of retreat longwall faces would be better described as bumps or bursts because of their lack of high gas pressure.

Almost all U.S. coal bumps have occurred in two areas. In the east, along a northeast trend from the Harlan coalfield in Kentucky to the Beckley coalfield of West Virginia and, in the west, within the Wasatch Plateau and Book Cliffs in Utah and the Carbondale and Somerset coalfields of Colorado. In a report by Iannacchione and Zelanko (1995), 172 specific bump events were analyzed. This study found that most bumps occurred at overburdens greater than 300 m. They also typically occur in conjunction with full extraction mining, i.e., 35% pillar retreat mining, 26% barrier splitting, and 25% longwall mining. The remaining 14% occurred during development mining. Bumps that occurred in conjunction with full extraction mining were not distinguished by elevated levels of gas emissions. When development bumps occurred, the overburdens where generally very high, greater than 600 m. Mining activity was not always directly associated with a bump since 22% of the events occurred during nonproduction shifts.

Rice (1935) suggested that "a structurally strong coal" not prone to crushing easily would favor bumps. However, Babcock and Bickel (1984) performed laboratory studies on coal samples and concluded that many coals can fail violently given the proper conditions of stress and constraint.

The importance of constraint to elevated levels of stress within coal pillars has been discussed by Iannacchione (1990). If the contact between the coal and the roof or floor rock is smooth and sharp with a low shear resistance, the coal can deform laterally lowering pillar stress levels. Conversely, if this same contact is rough with a high shear resistance, the coal cannot easily deform and excessive stress conditions are possible. Constraint to the coal supplied by the roof and floor contact conditions may be an important condition for both coal bumps and gas outbursts.

As is noted from the discussions above, the dividing line between a coal bump and an outburst is not always sharp. In general, there are several conditions that are shared between bumps and outburst. The most notable are high vertical stress conditions and constraint of the coal, most probably provided by the character of the contact between the coal and the roof and floor strata (Figure 9.2). These conditions allow the coal, and the gas contained within it, to become highly strained with an enhanced potential to release its stored energy suddenly in a dynamic fashion. There are also several conditions that appear to be dissimilar between bumps and gas outbursts. Bumps typically occur during full extraction mining while gas outbursts are more often associated with development mining. In full extraction mining, the coalbed has been sufficiently intersected by mine openings, reducing its in situ gases. In this way, the pressurized gases

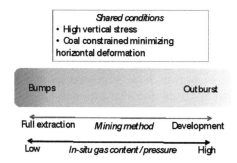

FIGURE 9.2   Conditions shared and dissimilar between bumps and gas outbursts.

escape into the mine atmosphere slowly enough to be well controlled by the ventilation system. Hence, the statement that in situ gas contents/pressures within bumps are low while the opposite is true of gas outbursts.

9.4 DESCRIPTION OF GAS OUTBURSTS FROM ONE U.S. MINING COMPANY

The U.S. mines with the most significant gas outburst problems were operated by Mid-Continent Resources from the late 1950s to the mid-1980s. The Dutch Creek Mine had its first major explosion on December 28, 1965. A combination gas and coal dust explosion fatally injured nine miners (Freeman et al., 1965; Richmond et al., 1983). The U.S. Bureau of Mines accident report by Freeman et al. (1965) revealed that several other gas outbursts had occurred during the years 1957, 1959, and 1963 injuring a total of 14 miners. The 1965 disaster occurred in a development heading of a retreat room and pillar section and was made worse by a gas bleeder in the floor and a ventilation system compromised by stockpiling coal at the face. When the gas outburst occurred, large volumes of very fine coal helped to initiate a secondary dust explosion.

A similar, an equally, deadly event occurred at the Dutch Creek Mine in 1981. In this disaster a massive blowout of gas and coal occurred in a two-entry development section about 2 h after mining through a fault. The gas was ignited by a lighting system on the continuous mining machine that had not been maintained in permissible (explosion proof) condition. Fifteen miners were fatally injured in the resultant mine explosion. The MSHA (Mine Safety and Health Administration) Report of Investigations (Elam et al., 1981) describing this disaster has a valuable description of the situation in this mine prior to the explosion:

> "Due to the amount of overburden and other geological conditions which create extreme stresses, outbursts and bumps of coal from the face and rib frequently occurred during mining, especially in development sections. Pressures exerted on the

ribs outby the face area or in the face area where no coal was discharged were locally referred to as "bumps" or "bounces," and were evidenced by noises and tremors. The outbursts, which were referred to locally as "pushes" or "pushouts" when they occur in the face area, released varied amounts of methane, coal, and fine coal dust into the ventilating air current. According to statements of the miners and mine officials, these outbursts occurred as often as two or three times a shift. When these outbursts occurred the coal and methane was described as "flowing" rapidly from the face out into the entries. In most instances, the outburst discharged approximately 30 to 40 tons of coal, but there were instances of outbursts which expelled sufficient coal from the face to cover the continuous mining machine and shuttle car behind it. The flow of coal was often accompanied by a release of methane, varying considerably in quantity. At times hazardous methane concentrations were found in the intake aircourses 100 ft or more outby the faces, indicating that sufficient methane was released by these outbursts to overcome the intake air current. The area between the line curtain and the rib was also sometimes blocked by the pushouts disrupting ventilation."

9.5 SUMMARY

Gas outbursts are a serious safety issue in coal mining worldwide. In the U.S., they have been relatively rare, largely because of the shallow mining depths and the permeable nature of many U.S. coalbeds. However, as deeper U.S. coalbeds are extracted in the future, there is a potential for this hazard to become more significant. It is therefore important that health and safety research continue to focus on this issue so that appropriate controls can be developed in anticipation of increased gas outburst frequency and intensity.

References

Aziz, N., Sereshki, F., Bruggeman, D., 2005. Status of Outburst Research at the University of Wollongong. In: Coal 2005: Coal Operators Conference. University of Wollongong & the Australasian Institute of Mining and Metallurgy, pp. 283–290.

Babcock, C.O., Bickel, D.L., 1984. Constraint—the missing variable in the coal burst problem. Paper in Rock Mechanics. In: Dowding, C.H., Singh, M.M. (Eds.), Productivity and Protection. Proceedings, 25th Symposium on Rock Mechanics. Soc. Min. Eng, Northwestern Univ., Evanston, IL, June 25–27, pp. 639–647.

Beamish, B.B., Crosdale, P.J., 1998. Instantaneous outbursts in underground coal mines: an overview and association with coal type. International Journal of Coal Geology 35, 27–55.

Black, D., Aziz, N., Jurak, M., Florentin, R., 2009. Outburst Threshold Limits—Are They Appropriate? Coal 2009: Coal Operators' Conference. University of Wollongong & the Australasian Institute of Mining and Metallurgy.

Campoli, A.A., Trevits, M.A., Molinda, G.M., 1985. An Overview of Coal and Gas Outbursts. 2nd US Mine Ventilation Symposium, Reno, NV, 1985.

Cao, Y., He, D., Glick, D.C., 2001. Coal and gas outbursts in footwalls of reverse faults. Int. J. Coal Geol. 48, 47–63.

Elam, R.A., Lester, C.E., O'Rourke, A., Strahin, R.A., Thompson, T.J., Kawenski, E.M., April 15, 1981. Report of Investigations, Underground Coal Mine Explosion, Dutch Creek No.1 Mine—I.D. No. 05–00301. Mid-Continent Resources, Inc., Redstone, Pitkin County, Colorado.

Freeman, J., Dimitroff, A.Z., Moschetti, A.C., 1965. Final Report of Major Mine-explosion Disaster, Dutch Creek Mine, Mid-continent Coal and Coke Company, Redstone, Pitkin, County, Colorado, December 28, 1965. U.S. Bureau of Mines, Lakewood, CO. 23 p.

Hargraves, A.J., June 1958. Instantaneous outbursts of coal and gas. Proceedings of Australian Institute of Mining and Metallurgy 21–72. Paper No. 186.

Harvey, C., 2002. History of outbursts in Australia and current management controls, Proc. 2002 Coal Operators' Conference, Wollongong, February 2002.

Iannacchione, A.T., 1990. The Effects of Roof and Floor Interface Slip on Coal Pillar Behavior, Rock Mech. Contributions and Challenges: Proc. of the 31st Symp. Colorado School of Mines, Golden, CO. June 18–20, pp. 153–160.

Iannacchione, A.T., Zelanko, J.C., 1995. Occurrence and remediation of coal mine bumps: a historical review. In: Mechanics and Mitigation of Violent Failure in Coal and Hard-rock Mines. U.S. Bureau of Mines Special Publication (SP 01-95), pp. 27–67.

Kissell, F.N., 1972. Methane Migration Characteristics of the Pocahontas No. 3 Coalbed. Bureau of Mines Report of Investigations. p. 7649.

Kolmsov, O.A., Bolsminski, M.I., 1981. The development of methods of dealing with sudden outbursts of coal and gas in the Donesk Basin. XIX Int. Conf. on Mine Safety Research, Oct. 6–14, Katowice, Poland.

Lama, R.D., Bodziony, J., 1998. Management of outburst in underground coal mines. International Journal of Coal Geology 35 (1–4), 83–115.

Rice, G.S., 1935. Bumps in coal mines of the Cumberland field, Kentucky and Virginia— Cause and Remedy. USBM RI 3267, 36 pp.

Richmond, J.K., Price, G.C., Sapko, M.J., Kawenski, E.M., 1983. Historical Summary of Coal Mine Explosions in the United States, 1959-81. U.S. Bureau of Mine. IC 8909.

Ulery, J.P., 2008. Explosion Hazards from Methane Emissions Related to Geologic Features in Coal Mines. NIOSH Information Circular. p. 9503.

Varley, F., 2009. Personal Communications.

Whyatt, J.K., 2008 Dynamic failure in deep coal: recent trends and a path forward. Proc. 27th Int. Conf. on Ground Control in Mining, Morgantown, West Virginia, 2008.

Wynne, P., 2003. Problems from a miner's perspective—Tahmoor Colliery. In: Hanes, J. (Ed.), Outburst Needs Workshop, Wollongong, February 12, 2003.

10

Production Engineering Design

Mark V. Leidecker

Jesmar Energy Inc., Holbrook Pennsylvania, USA

10.1 INTRODUCTION

This chapter outlines a simple approach for designing equipment needed to produce coalbed methane. Design for dewatering pumps, well head compressors, and well head dryers are the main focus. An understanding of the coal gas reservoir is helpful when determining the amount of gas and water a well will produce. The engineer has to address potential problems such as corrosion or scale deposits when picking the proper equipment. This chapter concludes by briefly discussing the design of a coalbed methane gas gathering system.

In order to design equipment utilized after a coalbed or coal mine well completion, the engineer has to have a working knowledge of the behavior of the coal gas reservoir. Understanding the flow of gas and water in the coal matrix will help in achieving maximum rates of gas and water. Proper dewatering equipment design, compressor design, and gas gathering design all contribute to this goal. Other factors such as corrosion and scale deposition need to be considered when planning production equipment design.

10.2 COAL GAS RESERVOIR

Figure 10.1 depicts a three-dimensional view illustrating the flow of gas and water within a block of coal. Two fracture systems (face and butt cleats) control the flow of gas and water within the coal seam. The fracture system is initially 100% saturated with water. No gas can flow from the coal until the water saturation is reduced.

185

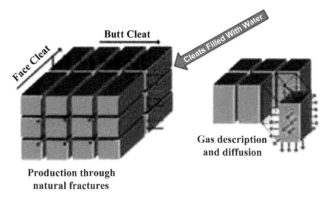

FIGURE 10.1   Methane flow in the coal fracture matrix.

Water flow within the fractures is governed by Darcy's law. The parameters affecting this law are differential pressure between initial reservoir pressure and pressure at the face, fracture permeability, water, and gas viscosity. The amount of gas to be drained from one coalbed methane well is governed by Darcy flow as well as the diffusivity of coal.

After the well produces an amount of water, gas begins to diffuse into the fracture system. In some cases, coalbed methane wells produce only water for many months before gas production is observed.

Decreasing pressure in the fracture system causes gas to desorb from the micropores, and diffuse into fractures. The face and butt cleats are connected to horizontal holes or to induced hydraulic fractures. Desorbed gases then travels through fractures into a hydraulic induced fracture or a horizontal wellbore, then to the surface, where gas is compressed, dried, and sent to a treatment plant.

Production engineers in the oil and gas industry routinely use the productivity index equation to predict gas production. This equation can be applied to predict the gas flow within the butt and face cleats.

$$Q = C * \left(\rho_s^{\,2} - \rho_w^{\,2} \right)$$

where:

Q = production, Mcfd
C = constant
ρ_s^2 = static reservoir pressure, psia
ρ_w^2 = coal face flowing pressure, psia (Craft et al., 1959)

This equation is governed by Darcy. Maximum gas production is realized when the coal face pressure is reduced to zero.

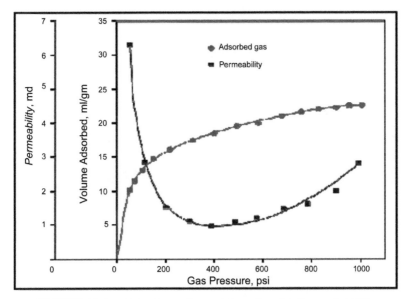

FIGURE 10.2 Klinkenberg effect and deswelling (Mullen et al., 2007).

Even though there are other regimes influencing gas flow in a coal gas reservoir, we can see mathematically how a large differential pressure $(\rho_s^2 - \rho_w^2)$ can increase gas flow in the fracture section of the reservoir.

Cleat permeability can increase over the life of a coalbed methane well. As the reservoir pressure declines, Figure 10.2 clearly illustrates much higher permeability at low gas reservoir pressure. This phenomenon is known as the Klinkenberg effect. It states that gas slippage along the cleats occurs at low pressures increasing the effective gas permeability. The coal matrix shrinks as the gas is desorbed, increasing the fracture permeability. The increasing permeability factor in the coal gas reservoir helps counter act the reduced gas flow caused by declining reservoir pressure. This may explain why coalbed methane reservoirs decline at a much slower rate than conventional gas reservoirs.

Figure 10.3 is a gas production plot of a coal mine well that produced 1.7 Bcf in 20 years of production. A screw gas compressor was installed at the well head. The initial reservoir pressure was 32 psig; the well head pressure was reduced to −6.0 in Hg after 6 years of production. The production decline rate was reduced by 3% per year after the well head pressure went to vacuum.

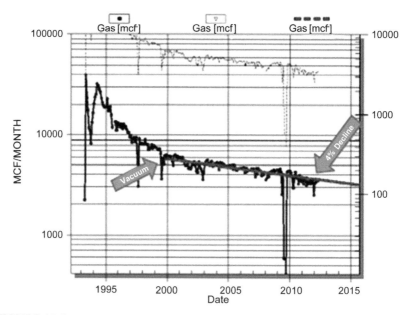

FIGURE 10.3   Production decline plot varner #1 well, Washington County, Pennsylvania. *Molli Computer Services, Inc.*

10.3 COALBED COMPLETIONS

Coalbed methane wells are completed by three methods:

1. vertical well,
2. horizontal openhole, and
3. horizontal cased hole

10.3.1 Vertical Well

Vertical wells are drilled into multiple coal seams. Production casing is cemented through the coal seams. The casing is perforated and the coal seams are hydraulically fractured in various stages. Stages are treated separately by isolating each stage using balls and baffles, frac plugs, or coiled tubing. Baffles and/or frac plugs are drilled out. Production equipment is installed next.

10.3.2 Horizontal Openhole

Horizontal openhole completion consists of drilling long lateral horizontal holes in coal in a predetermined direction. This type of completion was used by the U.S. Bureau of Mines utilizing underground drilling technology to degasify coal seams in advance of mining. Boreholes drilled in

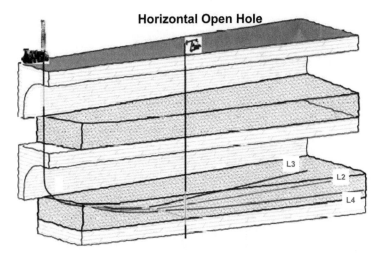

FIGURE 10.4   Coalbed methane well using horizontal bore holes.

a direction that intersects the face cleat perpendicularly produce considerable more gas than holes drilled in other directions.

Figure 10.4 depicts a well plan that drills three 4000 ft lateral wells into
the coal seam from a surface location. These laterals intersect the face cleat
perpendicularly. Water is removed via a vertical well that is in direct communication with the three laterals through a shorter lateral hole. A conventional
oil field pumping unit is installed with 3.5 in tubing and a special downhole piston pump. The vertical well is normally drilled with a 200 ft rathole
(an extention of the borehole into the mine floor). The downhole pump is placed
100 ft below the coal seam intersect to prevent gas locking. The extra 100 ft
of hole located below the pump intake serves as a collection sump for fines.

Maximum production is achieved by reducing the bottomhole producing pressure to zero. This is achieved by reducing well head pressures to a
vacuum and lowering the producing water level below the coal intersect.

Initial gas rates between 1000 and 750 Mcfd are common with this type
of completion in very thick coal seams. Figures 10.5 and 10.6 are example
production plots for openhole completions. Gas production rates are not
available but are expected to better then open boreholes.

10.3.3 Horizontal Cased Hole

Horizontal cased hole completions are primarily used for deep coal seams.
Deeper seams need to be fractured treated due to the low cleat permeability.
Completion procedure is very similar to horizontal shale completions. Fracture treatments are staged usually isolated with frac plugs. The plugs are
drilled out and the well put in production. Normally, the treatments volumes
are considerably less than frac treatment volumes used in shale completions.

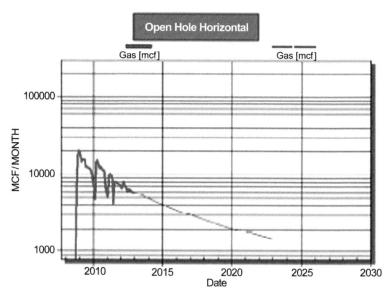

FIGURE 10.5   Example 1-Coalbed methane well in the Pittsburgh coal seam. *Molli Computer Services, Inc.*

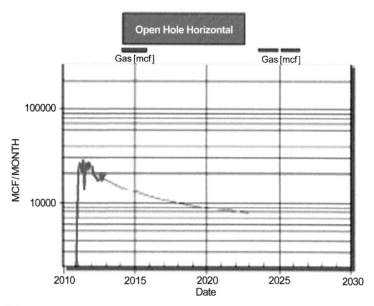

FIGURE 10.6   Example 2-Coalbed methane well in the Pittsburgh coal seam. *Molli Computer Services, Inc.*

10.4 WELL HEAD EQUIPMENT

Well head equipment needs to accomplish three functions:

1. dewater the coal seam,
2. reduce the bottomhole pressure (BHP), and
3. reduce water content in the gas stream.

10.5 DEWATERING OPERATIONS

Removing the water from the coal seam is critical to the flow of gas. The dewatering equipment has to have enough capacity to remove water efficiently. It has to have enough capability to lower the fluid level below the coal seam during the early life of the well. Keeping the fluid level low reduces back pressure on the coal seam. The design engineer must consider the following questions when contemplating what type of dewatering equipment should be utilized.

1. What are the maximum and minimum expected daily water volumes?
2. What is the maximum expected daily gas volume?
3. Are coal fines present?
4. Is corrosion present? What type?
5. Is scale present? What type?

Coalbed methane operators consider all these parameters when choosing dewatering equipment. Normally, three pump designs are utilized, Electrical submersible pumps (ESP), progressive cavity pumps (PCP) and downhole piston pumps. Gas lifts are used sparingly in coalbed methane dewatering. Gas lift or plunger lift equipment requires bottomhole pressure to operate. Since the goal is to reduce to the Bottom Hole Pressure (BHP) to zero, usually gas lift apparatus is not considered.

10.5.1 Electrical Submersible Pump

Figure 10.7 represents a sectional view of an ESP. The pump assembly consists of a downhole electrical motor, downhole pump with a series of impellers, and a downhole gas separator. This assembly is attached to production tubing with armored cable. The motor Horse Power (HP) and number of impellers or stages are determined by the depth and produced water volume. The deeper the well, and the greater the water volume, the higher the horsepower and greater the number of stages. A special well head is installed on the surface to insulate the cable and production tubing from corrosion and electrical shock. Surface transformers supply electric current to the pump motor.

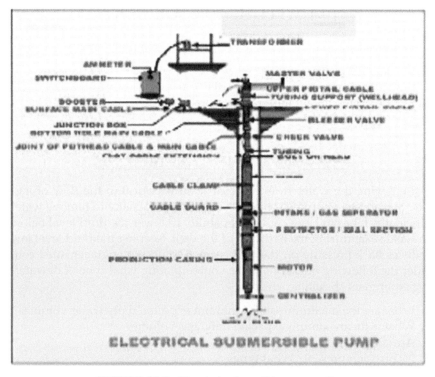

FIGURE 10.7   Electrical submersible pump wells.

ESP pumps perform better at high water volume applications. Larger diameter pumps can pump greater than 5000 bbd (barrels per day). They tend to gas lock, if the pump intake is installed above the completed zone. The motor has to be cooled continuously. It will overheat if the water flow stagnates. Coal fines may settle in and around the impeller section if the pump is cycled. Larger volume requires three-phase current. This could be a problem in areas where 3-phase power is not available. Smaller diameter pumps (2 in or less) can run on single-phase current.

10.5.2 Progressive Cavity Pump

PCPs (Figure 10.8) are driven by gas engines, hydraulic motors, or electric motors. They consist of a rotor and stator assembly driven by a sucker rod string connected to a top drive. They handle fines and solids well. The design should allow for continuous pumping. Stator begins to wear at pump off conditions. Carbon dioxide concentrations greater than 4% can cause corrosion with the rotor and stator. A minimum of 60 ft of head

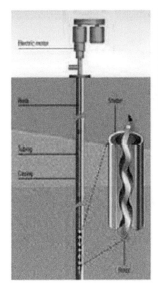

FIGURE 10.8   Progressive cavity pumps.

above the pump intake is required. The pump intake should be set at least 80 ft below the completed interval to avoid gas locking. Polyrod guides should be installed (three per rod) to minimize tubing wear. Rod guides could be a problem in wells that produce a heavy amount of fines. Fines could settle around the guides when pump is idle.

10.5.3 Downhole Piston Pump

Downhole piston pumps generally perform better than the other pumps in removing water from coal seams. It does a reasonable job in pumping coal fines. Gas lock can be eliminated or reduced by setting the pump 60 ft below the completed interval. If the pump has to be set above the completed interval, a downhole gas separator or gas anchor can be effective.

Figure 10.9 is a sectional view depicting the component parts of a downhole pumping system. The downhole pump is seated in a seating nipple located at the base of the production tubing. It is driven by a sucker rod string. At the surface, a pumping unit lifts and drops the rod string creating the movement need to operate the downhole pump. The surface unit can be driven by an electric motor or a gas engine.

Experience has proven that a top hold down pump is easier to unseat than a bottom hold down. When the pump is seated using a bottom hold

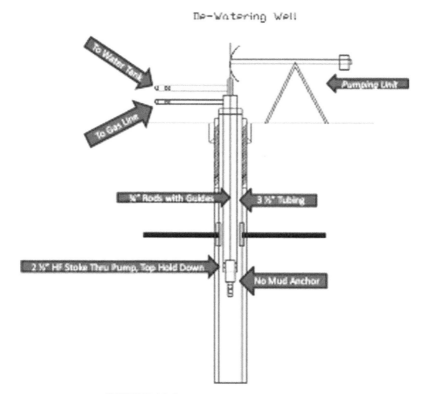

FIGURE 10.9   Downhole pumping stations.

down, fines can accumulate around the pump barrel making it difficult to pull.

A Harbison–Fischer (HE) 2.5 in diameter downhole Harbison–Fischer (HF) stroke through pump is recommend for an openhole horizontal completion. The stroke through feature assists the pump in producing gaseous fluids. The plunger can handle moderate fines flow. It can displace 12.5 Gallons per minute (GPM) of water, pumping at 14 strokes/minute (SPM), utilizing a 46 in stoke length.

Low revolutions per minute (RPM) gas engines are recommended in cases where electric power is not available. These engines generally are more expensive than the higher RPM engines. However their low maintenance costs offset the higher capital cost, especially in situations where the production is long term.

Changing RPMs for the prime mover is essential. It can prevent the well from pumping off. Fluid accumulations that occur during pump off can severely damage the down-hole pump and rod string. Gas engines can change speed by reducing the throttle setting. Electric motors can

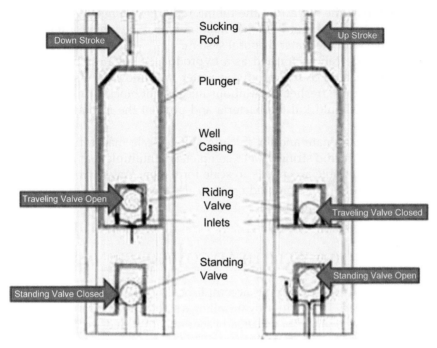

FIGURE 10.10 Sectional view downhole pump operation.

reduce speed by using a variable frequency drive (VFD). A VFD can be installed on a single-phase electric motor when couple with a three-phase converter.

The downhole piston pump is the pump of choice for coalbed methane operations; however it is not perfect. Rod and tubing corrosion, scale formation and gas lock can cause operational problems. As stated earlier gas lock can be minimized by setting the pump intake below the completed interval. The annular space between the tubing and production casing serves as a gas separator. Figure 10.10 is a sectional view of a downhole pump. It depicts the action of the piston and valves during the up stroke and down stroke operations. During gas lock, a volume of gas is trapped between the traveling and standing valves. The peak pressure of the trapped gas on the down stroke is insufficient to overcome the hydrostatic head on the traveling valve. Then, the pressure is not reduced enough on the upstroke to allow the standing valve to open and admit new fluid. Both valves are effectively stuck in the closed position and the pump refuses to pump.

The presence of carbon dioxide can cause severe corrosion in rod and tubing strings. CO_2 corrosion can be reduced by injecting a corrosion inhibitor in the annular space between the tubing and casing. The inhibitor coats the surface of the tubing rods preventing CO_2 corrosion. Inhibitors may be somewhat toxic, make sure that the inhibitor residue is compatible with the water disposal system.

Iron sulfide can be formed as a byproduct of sulfate reducing bacteria. The bacteria live in produced water in shallow seams where the water tends to be fresher. Continuous injection of biocide down the casing annulus should kill the bacteria and prevent the formation of iron sulfide.

Calcium carbonate and/or calcium sulfate scale may form in pump barrels or on the rod strings. Wells completed in multiple seams appear to be more susceptible acceptable to scale formation. Waters from different seams mix in the wellbore and create scale deposit. These scales can be treated with proper inhibitors.

10.6 WELL HEAD COMPRESSION

Zero bottomhole pressure is accomplished by pulling a slight vacuum at the well head. This is done by installing a screw compressor close to the well head. When pulling a well head vacuum, four inch diameter or larger piping should be used on the suction side of the screw. Also, minimize the number of elbows between the compressor suction flange and the well head. Larger diameter pipe and a low number of elbows reduce friction loss when operating in a vacuum. This design is mostly used on horizontal openhole completions. Usually, the flow rate is significant enough to economically justify a compressor for each well head.

Many times operators will link multiple wells together. They use one compressor connected with small diameter gathering lines. Pressures at the well heads usually remain positive. But, high well head back pressure reduces the ability for the coal seam to desorb gas.

10.7 WELL HEAD DRYING

Water content in the gathering system is reduced by installing a desiccant dryer (Figure 10.11) at the compressor discharge. The amount of water vapor is reduced from a saturated condition to 25 lb/MMcf. Water will not fall out at this concentration.

Operating costs can be reduced and down time minimized by investing more capital in well head equipment. Lower labor costs are realized by using automation. Each well head compressor is equipped with communication devices, which give the operator data to perform their duties

FIGURE 10.11   A well head desiccant dryer.

from a remote location. Operating costs associated with gathering line maintenance are reduced due to well head drying. For example, a coalbed methane operation consisting of 35 field compressors and well dryers plus a gas processing plant is operated 24/7 with one supervisor and three support personnel.

10.8 GATHERING SYSTEM

Figure 10.12 is a map of the western leg of a coalbed methane gas gathering system located in South Western Pennsylvania. This system is comprised of horizontal openhole wells and gob wells connected together transporting gas to a processing plant. The solid black lines represent gathering lines ranging from 6 to 10in in diameter. The entire system utilizes SDR-11 grade polyethylene. A maximum working pressure of 160psig was adhered to when laying out the pipeline design. The overall design was accomplished using a software model utilizing nodal analysis. This software simulated pressure drops and flow rates through compressors, regulators, valves, and fittings. The model predicted flow rates and pressures at various points throughout the gathering pipes. The actual pressures and flows were within 3% of what the model predicted.

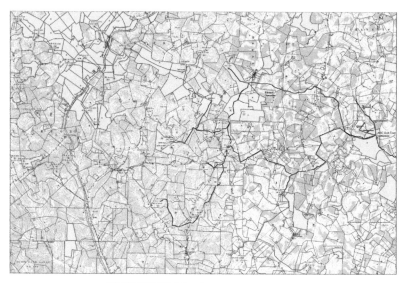

FIGURE 10.12   Gas gathering system.

10.9 CONCLUSIONS

Several factors need to be considered to maximize the gas production potential of the well, after the coalbed methane well is completed. The design engineer should have a general knowledge of the coal gas reservoir in order to design the well head equipment and gathering lines needed to transport gas to market. Bottomhole pressure should be reduced to close to zero shortly after the well is put into production. The type and size of the dewatering equipment, the well head compressor and dryer design all contribute to reducing well head pressure. Gas gathering pipelines need to be designed properly. Poor initial pipeline design can be very costly if modification has to be done in the future.

The design engineer must recognize potential obstacles that may affect the operation of coalbed methane equipment. Scale deposits and corrosion can reduce production and cause considerable down time.

The philosophy of spending capital on well head equipment proved to be successful in reducing operating costs and reducing down time. The capital expenditures include costs for automation, instrumentation, compressors and dryers for each well head.

Gas gathering pipelines need to be designed using nodal software. This software can predict pressures and flow rates with about 3% error. Polyethylene pipe should be used for coalbed methane operations. Installation costs are less than costs associated with steel pipe. Polyethylene resists corrosive methane gases.

Acknowledgments

The author would like to thank Gary Rodvelt and Steve Kravits for their permission to use their slides and information.

References

Craft, B.C., et al., 1959. Applied Petroleum Reservoir Engineering. Prentice-Hall Inc., Englewood Cliffs, NJ.

Mullen, M., et al., 2007. Coalbed Methane: Principles and Practices. Oktibbeha Publishing, LLC, Starkville, MS.

Coalbed and Coal Mine Methane Gas Purification*

Michael Mitariten

Guild Associates Inc., USA

11.1 INTRODUCTION

Natural gas (primarily methane) production from coal deposits is significant and accounts for almost 10% of the U.S. gas production. The gas from conventional reservoirs is generally of high quality requiring only compression and dehydration (typically H_2O, 7 lb/MMcf) before being admitted to the pipeline grid. The removal of water vapor is widely done with high pressure glycol absorption systems.

The gas from coal seams is commonly contaminated with carbon dioxide that can make the gas inadmissible to the transportation pipeline. The carbon dioxide limit for pipeline injection ranges from 1% to 3%. Carbon dioxide content of coalbeds varies with the 2% coal seams. Carbon dioxide levels of 6–10% for Eastern coals are common while the prolific San Juan Basin, for example, has many carbon dioxide removal plants treating feeds in the 20% carbon dioxide concentration range. A variety of technologies are used for carbon dioxide removal, but amine solvent technology is the most commonly used technique.

The gases produced during postmining, whether from GOB gases or abandoned mines are contaminated with a high level of nitrogen in addition to carbon dioxide and water vapor. In general, the total inerts must be 4% or less to meet pipeline specifications. Technologies used for nitrogen

---

*This chapter is a high level discussion of the gas processing options for treating coal mine gases and such information is published in many sources. For further details, especially for dehydration and carbon dioxide removal, we recommend that the GPSA Data Book be consulted. This data book can be obtained from the GPSA at http://www.gasprocessors.com/gpsa_book.html.

removal include the Molecular Gate™[1] adsorption technology, cryogenic fractionation, carbon-based adsorption, and membranes. Molecular Gate is most common with 12 Molecular Gate units installed for nitrogen rejection from coal mine gases.

11.2 DEHYDRATION (WATER VAPOR REMOVAL)

The gases produced from coalbeds both from premining and postmining operations, are invariably saturated with water at the temperature and pressure of the source. Since the coal-based wells are at low pressure the water concentration of the gas is high (several percent). After compression and cooling, and bulk separation, removal of water as a condensed liquid is achieved.

Since commonly the gas produced from the coalbeds is of high quality, lean in heavy hydrocarbons and predominately methane the only gas processing required is the dehydration and removal of water. Pipeline requirements for permitted water levels vary within a small window from location to location but in most cases, seven pounds per MMscf (scf stands for standard cubic feet) (about 150 ppm) is the maximum allowable water content and with a tighter water specification of four pounds per MMscf (about 85 ppm) required in some low temperature locations such as the Rocky Mountains and Canada.

After the gas is produced from the mines, it is compressed to a higher pressure for injection and transmission by the sales gas pipelines. As a general rule, these pipelines operate at high pressure, commonly in the 600 psig to 800 psig level, and for such applications, the compression and subsequent cooling of the water containing methane gas reduces the water level in the stream to the range of a few thousand parts per million. Water condenses due to compression and cooling but the cooling temperature is a critical component in determining the residual water level of the stream. The impact of pressure is significant as shown in Figure 11.1 (for a 100°F feed stream). Note that though the water level declines significantly as the pressure is increased, the low levels required by the pipeline cannot be achieved simply by compression and nominal cooling.

The impact of pressure is significant as shown in Figure 11.1 (for a 100°F feed stream) but temperature has a very significant impact as shown in Figure 11.2 (for the relatively low pressure of 100 psig shown). Roughly for each 20°F increase in temperature the amount of water contained in the feed stream doubles.

While a producer may have little control over the required pipeline pressure or ambient temperature, a good design practice would be to

[1] Guild is a licensee of Engelhard's Molecular Gate® Adsorbent Technology and is solely responsible for all representations regarding the technology made herein.

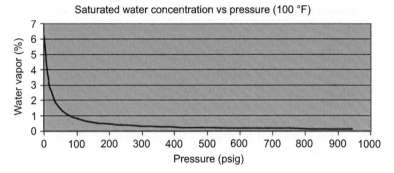

FIGURE 11.1   Water concentration vs pressure at 100 °F.

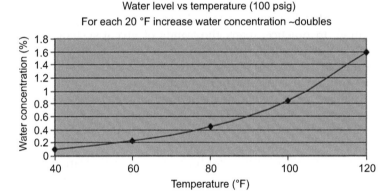

FIGURE 11.2   Water concentration vs temperature at 100 psig.

design the compressor, after-cooler with a view to maximize the water removed by compression/cooling and thus to minimize the load on the downstream dehydration unit.

In high pressure dehydration systems, glycol absorption is the most widely accepted process for water vapor removal. In the glycol process, the compressed and cooled, gas feed stream is contacted in a packed tower against a downward flowing liquid stream of lean glycol. Various glycols are used in this application including diethylene glycol, triethylene glycol (TEG) and tetraethylene glycol with TEG as the most commonly used medium. The downward flowing glycol absorbs water as it flows countercurrently through the contactor against the upwardly flowing methane stream. The glycol removes the water. The dry methane is then suitable for pipeline injection.

The contact of the glycol with the feed stream transfers the water vapor to the glycol and makes the glycol water-rich and this glycol leaves the bottom of the contactor. The water-rich glycol must be regenerated before it can be reused and the regeneration is conducted in two steps. Initially the pressure of the glycol solvent is reduced and this pressure reduction causes a portion of the contained water to flash from the solvent and be removed from the system. Flashing of the solvent can only partially regenerate the solvent and for the required further stripping of the solvent the partially water depleted glycol solvent is heating in a regenerator tower in which the absorbed water is driven off the glycol at low pressure. For most glycol designs heat is input by the combustion of a small portion of the methane from the coal mine gas. Once regenerated, the lean glycol is pumped and cooled and flows back to the contactor to repeat the dehydration process (Figure 11.3).

In design and practice of glycol dehydration systems, the level to which water is removed and thus the pipeline dew point is dependent upon the circulation rate of the glycol fluid, the relative performance of the contactor and, the extent of regeneration of the glycol solvent. In general, the operation of glycol systems is straightforward requiring minor operator attention. In practice, many thousands of glycol systems have been installed for the dehydration of natural gas from a wide variety of feed conditions. Glycol dehydration systems are available from equipment suppliers and can also be leased or rented.

As a rule of thumb, glycol circulation rates range from 2 to 5 gallon of TEG per pound of water absorbed thus the amount of water in the feed has a great impact upon the size of the glycol system. This means that high pressures and low temperatures are preferred.

An alternate to glycol system for high pressure dehydration, especially for lower flow rates, is to utilize a bed of nonregenerable salt tablets, such

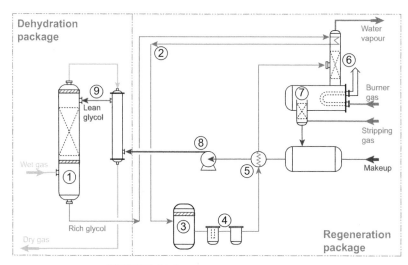

FIGURE 11.3   Typical glycol flow sheet.

as calcium chloride ($CaCl_2$), to dehydrate the gas stream. The dehydration is achieved by the reaction of the calcium chloride forming a calcium chloride brine solution and a dry product stream. Because such fixed bed desiccants have an ongoing cost, an economical evaluation is required to determine if they are cost effective. As a general rule, systems that require less water removal and smaller flow rates can make such a nonregenerable desiccant beds attractive, especially due to the nonmechanical operation of such a system. For such systems, as the desiccant is consumed it is replaced with fresh material. Brine disposal can require some research for an acceptable disposal site. Multiple strength desiccant tablets are available, which have a range of performance. The tablet type is selected as needed to achieve the required water removal. Since higher performance tablets have a higher cost, a lower cost, bulk removal tablet can be followed with a higher cost/higher performance tablet to achieve more stringent pipeline requirements.

Overall, for smaller flows, a nonregenerable desiccant bed can be an alternative to glycol units due to their simplicity of operation. Large flows or low pressure/high temperatures result in higher water removal requirements and make such nonregenerable systems less attractive since the operating cost is proportional to the amount of water to be removed. It should be recognized that the nonregenerable desiccant forms by-product brine, which must be disposed of properly. In turn, the amount of water in the feed is largely dependent upon the operating pressure and temperature.

For both the glycol system and the nonregenerable salt tablets, low pressures lead to higher quantities of water to remove and make these systems more expensive and thus less attractive as compared to high pressure systems. In cases where the pipeline pressure is low, coalbed methane sales may permit lower pressure operation and in such an application, it can be desirable to operate the dehydration system at low pressure. Thus, in this application, regenerable solid beds of a desiccant media (alumina, silica gel, molecular sieves) have been used to remove the relatively large quantity of water vapor from the feed stream. Typically for pipeline requirements, the water level of the product permits the use of activated alumina or silica gels, although molecular sieves can also be used (while generally more commonly used where feed streams are to be dehydrated to water levels lower than that required by the pipeline such as for conventional natural gas NGL (natural gas liquids) recovery facilities).

In a typical solid desiccant dehydration unit, two absorber vessels are used where one tower is on stream adsorbing water from the gas while dry sales gas is produced at the outlet of the adsorbent bed. Beds are typically sized to operate on the adsorption step for as short as 2h to as long as 12h or more. While the first tower is adsorbing and removing the moisture from the methane stream, the second tower is regenerated by heating and cooling. Regeneration is typically by passing a heated slip stream of the product gas in the opposite direction to that of the feed stream. This heating step is typically conducted at temperatures above 300 °F and depending on the amount of water removed typically requires a flow rate of 10% or so of the raw gas flow. The gas leaving the tower contains the previously adsorbed moisture along with the methane and other components present in this regeneration stream.

Because the regeneration requirements for a tower are in the range of 10% or more, the application suffers from a loss of methane if an open loop regeneration system were used. For this reason the methane in the regeneration can be recovered by recirculating the regeneration stream back to the feed compressor suction. This recycle allows the methane to be recycled back through the tower on adsorption while the water leaves the system as a liquid after compressing and cooling and there is no loss of methane from the regeneration stream.

After heating the regenerating tower has been stripped of moisture, however, the bed itself is near the regeneration temperature and must be cooled before it can be placed once more in adsorption service. Thus, the regenerated bed is cooled by removing the heat input and the slip stream of the product at essentially feed temperature passes through the bed being regenerated to cool the adsorbent and ready it for the subsequent adsorption step.

Since the adsorption capacity of the adsorbent is affected by pressure as well as temperature, the bed on regeneration is depressurized to low pressure before heating and cooling. After regeneration is complete, the bed is repressurized to near feed pressure before switching places with the adsorbent bed that was previously removing moisture. Thus, the use of two towers permits one vessel to be on the adsorption step removing moisture at all times while the second bed is depressurized, heated, cooled, and repressurized. A simplified schematic of a two tower cycle is shown in Figure 11.4.

The choice between the adsorbents used in solid desiccant dehydration is a function of the required pipeline level of moisture. As a general rule, alumina and silica gels require less heat for regeneration than the molecular sieves and for that reason, would be preferred for simple pipeline dehydration purposes.

11.3 CARBON DIOXIDE REMOVAL

Though many coal-based gases require only dehydration, carbon dioxide is commonly found in both the gases from virgin coalbeds and that from coalmining operations. Pipeline requirements specify that carbon dioxide be 1–3%, with 2% fairly typical. Carbon dioxide presents the possibility of corrosion in pipeline systems and, hence, pipeline

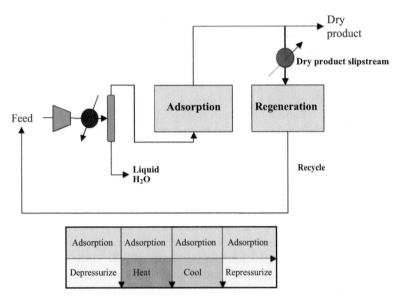

FIGURE 11.4 Typical two bed temperature swing adsorption (TSA) schematic.

companies pay close attention to carbon dioxide levels in any injected natural gas.

The most commonly applied technology for carbon dioxide removal is a chemical reaction process using aqueous amine solvents. Similar to the glycol process, the amine process is favored by operating at high pressures (and thus higher carbon dioxide partial pressures). In the amine process, carbon dioxide reacts with the amine in a reversible manner. The chemistry is multistep and complex and not addressed here but the carbon dioxide is removed by reacting with the downward flowing amine as it contacts an upward flowing carbon dioxide contaminated methane gas streams. The overhead gases from the contactors provided a product stream at near feed pressure with the carbon dioxide removed to the required level. However, because the amine solvent is aqueous based, the overhead product is water saturated and requires downstream dehydration before it can be admitted to the pipeline. Thus the amine system is commonly followed by a glycol unit.

The rich amine solvent, which has reacted with the carbon dioxide, leaves the high pressure contactor and is reduced in pressure, which liberates part of the carbon dioxide. As with the glycol system the pressure reduction is not sufficient to fully regenerate the amine solvent and the low pressure carbon dioxide-rich solvent flows to a stripping (regeneration) column. In the stripping column heat is input through a solution reboiler located in the lower part of the stripping column.

Heating the amine solvent reverses the chemical reaction and the carbon dioxide is liberated from the solvent. The overhead hot carbon dioxide gas and some amount of amine vapor is cooled, which condenses much of the amine vapors and minimizes amine loss while the carbon dioxide is rejected at high purity and low pressure and typically vented to the atmosphere.

The gas from coalbeds is lean in heavy hydrocarbons. There are many amine solvents that can be applied for the separation, conventional diethanol amine (DEA) is the most commonly used. The flow rate of the DEA solution is a function of the amount of carbon dioxide to be removed and the carbon dioxide partial pressure. Process design and optimization addresses items, such as the carbon dioxide absorption by the solvent, heat input requirements to the reboiler, and required carbon dioxide product levels among other process conditions.

As noted, amine systems for carbon dioxide removal are widely used however, field operational attention is required. Within the design of the amine plant, attention must be given to corrosion issues and thus the selected materials of construction (Figure 11.5). The gas from coalbeds is typically low in hydrogen sulfide. Hydrogen sulfide establishes a layer of corrosion protection that is not present where carbon dioxide alone is to be removed. Thus, amine plants treating gases from coal mines can be subject to high corrosion rates. Corrosion is also an issue due to the possibilities of high velocities, formation of carbonic acid due to the presence of

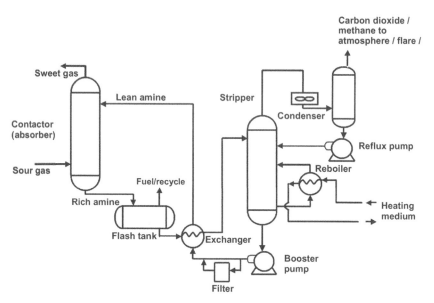

FIGURE 11.5   Typical carbon dioxide removal amine plant schematic.

carbon dioxide, and a wide variety of degradation products that can form as the amine reacts. As an overall concern for corrosion issues, design and operating attention is given to materials of construction at certain points within the plant, operating the stripper column at low, reboiler temperatures, minimizing solid and degradation products within the system, which may require the use of activated carbon beds for solvent cleanup and the addition of reclaimer systems on the solvent.

Importantly, the low pressure gas from the coalbed wells will often have a level of oxygen in the system and oxygen leads to chemical reactions and the formation of highly corrosive components within the system. It is difficult to keep a coalbed or coal mine methane treating system entirely free of oxygen and, thus, corrosion due to the presence of oxygen may be unavoidable and at the minimum should be given careful consideration in the design and operation of the amine plant.

Foaming of the amine solvent also requires attention. Foaming reduces the solvents ability to remove the carbon dioxide and can require a reduction of the capacity of the plant. There are many reasons foaming can occur, including degradation of products, impurities in the solvent, suspended solids, and corrosion inhibitors in the plant.

Despite the challenges of operating an amine plant for carbon dioxide removal, they are widely applied and practiced and are the most common means of removing carbon dioxide. Amine plants can be purchased as new units; however, there is also a ready market for used amine plants as well as companies that will treat the gases for carbon dioxide removal under an operating contract, which eliminates the need for the gas producer to take a hands-on role in gas cleanup. It should be recognized that although the amine plant does not have a tail gas stream resulting in a high loss of hydrocarbons, the fuel input for the reboiler is required and depending on amine gas circulation rates and the amount of carbon dioxide to be removed, the fuel requirements can consume several percent of the hydrocarbons available in the feed stream.

Given the operating issues with amine plants, the use of membrane technology for carbon dioxide removal has become well established (Figure 11.6). As with amine plants, membrane units favor high pressure feed streams and, thus, like the amine plant, would be placed on the coalbed gas to be treated after compression to high pressure pipeline requirements. There are several membrane suppliers offering membrane units for the removal of carbon dioxide and though there are physical differences in the structure and materials in the membrane, there is none clearly superior to all others.

In the membrane operation, the compressed high pressure coalbed methane is passed over the high pressure side of a polymer fiber while a low pressure (permeate side) is maintained on the opposite side of the membrane fiber. The carbon dioxide partial pressure conditions, the

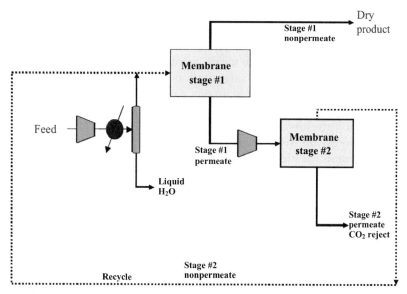

FIGURE 11.6   Typical two-stage carbon dioxide removal membrane schematic.

polymer used and the pressure differential across the membrane fiber are optimized and carbon dioxide permeates from the high pressure side to the low pressure side of the membrane. Because the selectivity in the permeation between the carbon dioxide and the methane from the wells is not absolute, some methane permeates across the membrane with the carbon dioxide. The high pressure, residual gas reduced in carbon dioxide concentration to pipeline requirements is directed to the pipeline. The permeate gas contains a level of methane, which typically represents an unacceptable high loss of hydrocarbons and is often compressed and treated in a second stage membrane. This second stage membrane then produces a methane-rich recycle stream and a reject low pressure permeate stream that is enriched in carbon dioxide and rejected either to fuel or atmosphere/flare.

Despite the loss of methane in the membrane unit, which is typically several percent or more in the use of a two-stage unit, the membrane offers advantages in many cases due to fairly low cost and ease of operation. Though amine treating is more widely applied in upgrading coalbed and coal mine gases, membrane units have been successfully applied and for any carbon dioxide removal requirement they should be given consideration where high pressure sales gas is a requirement. In the optimization of such a membrane unit, the system can become attractive where the low pressure reject from the second stage membrane can be used as fuel gas to compressors or other fuel needs within the gas processing facility. One advantage as compared to the amine plant is that since water permeates from high pressure to low pressure across the membrane unit, the membrane system produces a sales gas stream that meets the pipeline water removal requirements without the use of a separate downstream dehydration unit.

One item of great importance in the selection of a membrane design for carbon dioxide removal is to account for the fact that the membrane can be damaged for a number of reasons including the carryover of liquids and heavy hydrocarbons along with solids and particulate materials and items added to the wells such as corrosion inhibitors. Because the membrane system can be physically damaged, careful attention to membrane feed pretreatment is required. Most membrane vendors advise of pretreatment requirements while recognizing that in a competitive marketplace, such pretreatment adds cost. Consideration should be given to using inlet coalescent filters, nonregenerable adsorbent guard beds, particulate filters, and elevating the temperature of the feed stream with a heater to optimize the operation of the membrane and protect from condensation of water or heavy hydrocarbons.

An alternative to the high pressure amine treating or membrane system for carbon dioxide removal is the use of a Molecular Gate system for carbon dioxide removal. The Molecular Gate technology in carbon dioxide removal consists of typically three, four, or six adsorber vessels in which a feed stream containing water and carbon dioxide is introduced into a bed of adsorbent where the water and carbon dioxide are adsorbed and removed while methane flows through the bed and is available at near feed pressure as sales gas. Subsequent to the adsorption and removal of the carbon dioxide, the adsorbent bed is depressurized and purged (typically under vacuum) and producing a reject tail gas stream. Thus, in a single step the Molecular Gate technology splits the inlet gas into a sales gas stream at near feed pressure and a low pressure reject stream which as with the membrane's reject stream can be used as fuel or vented/flared (Figure 11.7).

The Molecular Gate technology is most appropriate for low pressure feed streams and typically operates at 100 psig. For carbon dioxide removal, feed streams lower than 100 psig and as high as several hundred psig can be treated with 100 psig as near optimum. Thus, a Molecular Gate Carbon Dioxide Removal System can be attractive where the coalbed gases are compressed, typically in an oil flooded screw compressor to 100 psig and where a further gathering system or sales gas line at low pressure can be utilized.

Figure 11.8 depicts the overall flow process for a typical Molecular Gate Adsorption System applied to upgrading coalbed methane for carbon dioxide removal. Feed gas from the wells is compressed from near atmospheric pressure to typically 100 psig where it is introduced into the Molecular Gate Adsorption System. For coalbed and coal mine methane, a screw compressor to 100 psig is typically applied.

FIGURE 11.7   Tidelands oil production company carbon dioxide removal unit.

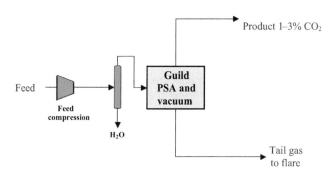

FIGURE 11.8   Typical Molecular Gate carbon dioxide removal schematic.

To maximize the working capacity of the adsorbent to remove the carbon dioxide and water a single stage of vacuum is used to enhance the regeneration. The swing between the high adsorption pressure and regeneration at low pressure is completed in rapid cycles, on the order of a few minutes, to minimize the adsorbent inventory.

Typical methane recovery percentage of 93–98% is achieved in the process. The rejected tail gas, containing the lost hydrocarbons, may be suitable as fuel to gas engines driving gensets or compressors and making use of the otherwise lost methane is a key part of the process optimization.

As with the membrane system, the Molecular Gate unit removes water when carbon dioxide is removed and is a relatively simple system to operate.

11.4 HYDROGEN SULFIDE REMOVAL

The gas from coal mines is typically low in H_2S at 0 to about 10 ppm. At this low level nonregenerable sulfur scavengers can be applied and used to polish the H_2S from the gas where H_2S must be polished to low levels (typically 4 ppm) before a pipeline. A number of technologies can be applied led by the Sulfatreat technology, which for low levels of H_2S is simple and cost effective. In such a technology, a bed of nonregenerable adsorbent media is installed on the feed stream to be treated and H_2S reacts and is removed onto the bed of adsorbent. When exhausted, the bed is disposed of and replaced. A number of configurations can be used, which typically involves a choice between a single vessel and two-vessel, lead–lag, system.

In the evaluation for H_2S polishing, consideration can also be given to iron oxide beds and injected liquid chemicals for scavenging low levels of H_2S from the feed stream. Each of the technologies has its merits and should be evaluated on a case-by-case basis.

In the gas treating process, the amine, membrane and Molecular Gate technology will typically remove such low levels of H_2S to a level of 4 ppm or less. Thus where carbon dioxide removal is used, separate removal of H_2S is not a requirement though the H_2S is present in the reject stream. In the nitrogen rejection Molecular Gate process H_2S is also removed.

11.5 NITROGEN REJECTION

Nitrogen is found in the gases produced during postmining, from GOB (mined out areas) gases or from abandoned coal mines. Occasionally nitrogen is found on the virgin coals but this is a lesser issue than the gases produced during postmining. Nitrogen reduces the heating value of natural gas and pipelines typically limit the amount of nitrogen to 4%. Almost invariably, the gases produced during postmining also contain nitrogen, a smaller amount of oxygen, and carbon dioxide. Processes used in coalmining for the removal of nitrogen include the Molecular Gate process and cryogenic processing.

The Molecular Gate process for nitrogen rejection from coalbed and coal mine methane takes advantage of a unique molecular sieve that has the ability to adjust pore size openings within an accuracy of 0.1 Å. The pore size is precisely adjusted in the manufacturing process and allows the production of a molecular sieve with a pore size tailored to size-selective separations.

Molecular Gate—pore contracted to 3.7 Å

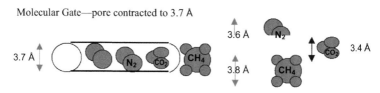

FIGURE 11.9   Schematic view of the Molecular Gate pore size and relative molecule size.

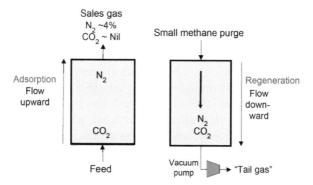

FIGURE 11.10   Schematic of the Molecular Gate adsorption and regeneration steps.

Nitrogen and methane molecular diameters are approximately 3.6 and 3.8 Å, respectively. In the Molecular Gate Adsorption System for upgrading nitrogen-contaminated methane, a pore size of 3.7 Å is used. This adsorbent permits the nitrogen (and carbon dioxide) to enter the pore and be adsorbed while excluding the methane, which passes through the fixed bed of adsorbent at essentially the same pressure as the feed. This size separation is schematically illustrated in Figure 11.9.

Carbon dioxide is an even smaller molecule than nitrogen at 3.3 Å and is easier to remove than nitrogen. One major advantage of the Molecular Gate process in upgrading nitrogen-contaminated feeds is that such feeds almost always have a level of carbon dioxide contamination and, in the process carbon dioxide is completely removed in a single step with the nitrogen removed to pipeline specifications.

The Molecular Gate adsorbent is applied in a pressure swing adsorption system (PSA), which operates by "swinging" the pressure from a high pressure feed step that adsorbs the impurity to a low pressure regeneration step to remove the previously adsorbed impurity. Since methane does not fit within the pore of the adsorbent, it passes through the bed at the feed pressure (Figure 11.10).

PSA is widely used in light gas separations with thousands of units in operation in the oil refining, petrochemical, and air separation industries. The system is characterized by automatic and simple operation with high reliability.

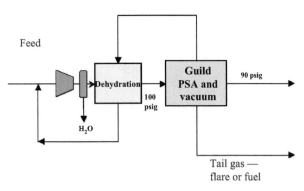

FIGURE 11.11 Molecular Gate process flow schematic. PSA, pressure swing adsorption system.

Figure 11.11 depicts the overall flow process for a typical Molecular Gate Adsorption System applied to upgrading coalbed or coal mine methane. Feed gas from the wells is compressed from near atmospheric pressure to typically 100 psig where it is introduced into the Molecular Gate Adsorption System. For coalbed and coal mine methane, a screw compressor to 100 psig is typically used. Since this gas is water saturated it is first dehydrated in a two-bed thermal swing adsorption system.

The process generates a low pressure recycle stream that is rich in methane and recirculated back to the suction of the feed compressor. By incorporating this recycle stream, the methane recovered as product sales gas is increased without adding additional compressors. The recycle rate is typically 10–20% of the raw feed rate. Further since this stream is dry it is used to regenerate the dryer without the need for an on purpose regeneration stream.

Typical methane recovery rates of 88–93% are achieved in the process. The rejected tail gas, containing the lost hydrocarbons, may be suitable as fuel to gas engines driving gensets or compressors and making use of the otherwise lost methane is a key part of the process optimization.

The feed gas composition from the mines can change over time and in this consideration, the ability to remove both carbon dioxide and nitrogen along with flexibility for changes in composition is critical for continual production of pipeline quality gas. The Molecular Gate Adsorption System is flexible to such changes and can always produce pipeline quality gas regardless of the feed level of carbon dioxide or nitrogen.

Figure 11.12 shows a Molecular Gate unit that was installed in the Illinois Basin and has operated since 2002. Since many wells feed this unit and the sources vary, the system sees a range of feed compositions. In general, the feed contains up to 6% carbon dioxide and up to 15% nitrogen with a pipeline specification of 4% inerts. As with most units in low pressure applications, the feed gas was compressed to 100 psig for treating.

FIGURE 11.12   2.5 MMscfd Molecular Gate unit processing coal mine methane (CMM).

After treating, the product gas after the Molecular Gate Unit was further compressed from 90 psig to pipeline pressure.

The basic equipment required for the removal of the impurities consists of multiple adsorber vessels filled with adsorbent and a valve and piping skid that is placed alongside the adsorber vessels and serves the purpose of switching flows between adsorber vessels as they cycle between the process steps of adsorption, depressurization, regeneration, and repressurization. The overall control system, often including feed and product compression, is controlled by an integrated control system that provides for push-button start-up. The control system operator interface can be supplied with the ability to remotely monitor the system via a modem connection.

For nitrogen removal from GOB or abandoned coal mines the economics can be favorable even at small flows of 0.5 MMscfd (scfd stands for standard cubic feet/day) where the nitrogen is less than 30%. In addressing the nitrogen in the feed, it is also important to recognize that the system will remove carbon dioxide and a portion of any oxygen in the feed without separate processing.

Cryogenic processing has also been used to upgrade the gases postcoal mining, though it is typically evaluated for large flows. The process offers high methane recovery rates but requires multiple step processing for carbon dioxide and water dehydration before the nitrogen rejection system since these can freeze at the temperatures of the process. Oxygen removal is also required to avoid an explosion hazard.

The cryogenic process operates by compressing the coal mine gases to high pressure of typically 600 psig and subsequently processing the gas for removing the oxygen, carbon dioxide, and water impurities. The gas is then cooled to cryogenic temperatures by expanding the remaining methane and nitrogen to the point where refrigeration to condense the gases as a liquid occurs. This refrigeration from expansion is captured in a series of heat exchangers and the temperatures of the design are low enough to liquefy the methane/nitrogen mixture. The liquid is then distilled and the gases largely separated into their individual constituents. The nitrogen leaves the system as a vapor at a midrange pressure and the methane exits at relatively low pressure for recompression and with the actual pressure dependent on the cryogenic system design.

11.6 OXYGEN REMOVAL

Since coalbed methane is produced at low pressure (vacuum is sometimes pulled to increase production) the likelihood of the gas containing oxygen exists. In GOB gases and gases from abandoned coal mines oxygen is generally present at levels that can be in the 1–5% range. The primary issue with oxygen in a methane feed is safety and this requires review for any project.

While pipeline are fairly consistent in their specification for permitted nitrogen, carbon dioxide, sulfur, and water vapor the accepted levels of oxygen range widely from a low of 10 to 20 ppm.

Oxygen can be removed with a variety of technologies. For low parts per million requirements by the pipeline a catalytic system is required and in such a design the feed gas is heated to several hundred degrees Fahrenheit and reacted over a catalyst bed. In such a design the methane and oxygen react to form water and carbon dioxide. These added impurities will require that the stream be dehydrated and the formed carbon dioxide may require removal as well. Such a catalytic system uses precious metal catalyst, which requires a guard bed to remove sulfur to very low levels to avoid catalyst deactivation though sulfur tolerant catalysts have been used in this application.

The oxygen consuming reaction is exothermic and gives off heat. Thus depending on the amount of oxygen in the feed, the heating can be limited to that needed to initiate the reaction. Heavy hydrocarbons also react with the oxygen more readily and thus if present in the feed can allow lower initiation temperature or can be injected into the gas stream before the catalyst bed (recognizing that sulfur can be present).

For pipelines that permit a higher level of oxygen and where a membrane is used for carbon dioxide removal, the membrane will permeate a portion of the oxygen with the carbon dioxide. This one-step removal of both components can be attractive for such carbon dioxide removal plants.

Oxygen is also a small molecule at about 3.5 Å and fits within the pore of the Molecular Gate adsorbent and in doing so is also partially removed in nitrogen rejection systems. Since carbon dioxide drives the oxygen off the adsorbent bed, Molecular Gate units targeted at carbon dioxide removal do not remove oxygen.

11.7 SUMMARY

In addition to the growth of interest in coalbed methane, gas production from GOB gas and coal mine methane is growing. The need for the removal of nitrogen, carbon dioxide, oxygen, and water is thus on the rise and a variety of technologies to remove the impurities are available.

Current and Emerging Practices for Managing Coalbed Methane Produced Water in the United States

Richard Hammack

Office of Research and Development, National Energy Technology
Laboratory, Pittsburgh, USA

12.1 INTRODUCTION

An EPA survey of U.S. coalbed methane (CBM) producers (United States Environmental Protection Agency (U.S. EPA), 2009) found that CBM wells produced 2,001,602,899 Mcf of natural gas in 2008, about 8.8% of the total gas production in the United States for that year. CBM production came from 56,049 wells in 14 coal basins (Figure 12.1 and Table 12.1) with the largest production coming from the San Juan Basin in New Mexico and Colorado and the Powder River Basin (PRB) in Wyoming and Montana.

12.1.1 CBM Produced Water

Initially, large volumes of water must be pumped from CBM wells to lower the hydrostatic pressure in the coalbed and allow methane to desorb from the coal. Then gas production can start. After methane production commences, water production declines to a lower, sustained water yield that persists for the production lifetime of the well (Figure 12.2). CBM produced water exhibits a broad spectrum of water quality, ranging from drinking water quality to hypersaline water that cannot be reused without prohibitively expensive treatment. Produced water quality varies by basin and within basins. For example, produced water from CBM wells located

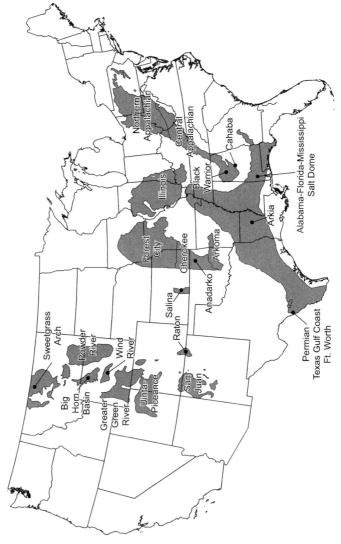

FIGURE 12.1   Coal basins of the United States. *U.S. EPA (2009).*

TABLE 12.1 CBM Basins Listed in Order of Total Gas Production

CBM Basin	CBM Wells, #	Total Gas Production, Mcf
San Juan	7048	755,273,032
Powder River, Wyoming	20,692	591,989,139
Appalachian and Illinois	6190	146,335,694
Raton	3725	128,690,594
Black Warrior	5153	104,412,592
Cherokee/Forest City	5278	85,669,776
Uinta/Piceance	1117	69,544,369
Arkoma	2321	65,972,085
Anadarko	2846	18,612,851
Wind River/Green River	337	15,979,115
Powder River, Montana	898	14,244,971
AL/FL/Salt dome/Cahaba	387	3,756,571
Arkla	42	850,223
Fort Worth/Permian/Texas Gulf Coast	15	271,887
Total	56,049	2,001,602,899

CBM, coalbed methane.
Data are results of a 2008 EPA screener questionnaire that was sent to CBM operators; data were presented by Carey Johnson at the 2009 International Petroleum and Biofuels Environmental Conference in Houston, TX on November 5, 2009.

near the margins of coal basins (where there is significant groundwater recharge) often is of high quality and may not require treatment to be used for a multitude of purposes including drinking water (ALL Consulting, 2003). However, most CBM produced water quality ranges from brackish to saline and requires some form of treatment before it can be reused or discharged to surface waterways. Less commonly, CBM produced water can be hypersaline, having a total dissolved solids (TDS) concentration from 35,000 to 180,000 ppm TDS. The cost of treatment for such water prevents reuse; hypersaline water is injected into deep, brine formations.

This chapter discusses current and emerging management practices for CBM produced water in U.S. coal basins. It is not the intent of this chapter to provide a comprehensive discussion of technologies appropriate for the treatment of CBM produced water. For that, the reader is invited to consult the following excellent reviews that are available online. The Produced Water Management Information System (PWMIS), developed by Argonne National Laboratory is an online resource of technical and regulatory

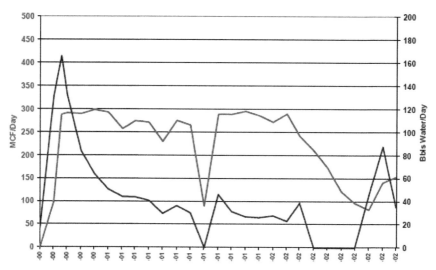

FIGURE 12.2   Monthly history of gas and water production from a coalbed methane well. *ALL Consulting (2003).*

information for managing produced water, including current practices, state and federal regulations, and guidelines for optimal management. PWMIS also includes an interactive decision tree to guide operators to appropriate produced water treatment technologies. The most comprehensive catalog of produced water treatment technologies (IOGCC and ALL Consulting, 2006) was compiled by the Interstate Oil and Gas Compact Commission (IOGCC) and ALL Consulting in 2006. More recently, the Colorado School of Mines (CSM) prepared an updated catalog of produced water treatment technologies (Colorado School of Mines, 2009) that includes a cursory evaluation of each technology. An online presentation by Cath (Cath, 2009) discusses emerging treatment technologies that are appropriate for CBM produced water.

12.2 CBM PRODUCED WATER MANAGEMENT— CURRENT PRACTICES

12.2.1 Underground Injection

The 2009 EPA screener questionnaire determined that most CBM produced water was injected into deep saline aquifers (U.S. EPA, 2009). In fact, all CBM produced water from the Anadarko, Arkoma, San Juan, and Cherokee–Forest City coal basins is deep injected. About 80% of CBM produced water from the Green River and Piceance basins is deep injected; 68% of CBM produced water from the Appalachian Basin is deep injected. Surface discharge (with or without treatment) predominates over underground injection only in the PRB and the Black Warrior Basin (32% deep injection and 5% deep injection, respectively).

12.2.1.1 Deep Injection

The two factors responsible for the predominance of deep injection in some basins are (1) local availability of formations favorable for injection and (2) poor CBM produced water quality that makes the water unsuitable for reuse or surface discharge without prohibitively expensive treatment. Because deep injection of CBM produced water is occurring in arid basins where agriculture, industrial development, and population growth are water limited, produced water may become a future source of water in these areas, especially if lower cost treatment technologies can be developed. However, legal questions regarding the ownership of produced water must be decided before the water can be put to profitable use.

The injection capacity in some coal basins (i.e., PRB, Appalachian Basin) is limited and cannot accommodate the large volume of water initially produced by a developing CBM field. In these basins, CBM produced water must compete with liquid industrial wastes and produced water/ flow back water from other oil and gas sources for the limited injection capacity that exists. Further, the same injection formations are being targeted as reservoirs for natural gas storage and future CO_2 storage. When the capacity of existing injection wells is limited and already committed, CBM operators either must: (1) transport produced water to injection wells with unused capacity; (2) increase the injection capacity of existing injection wells; (3) develop new injection wells; or (4) treat the water for surface discharge or beneficial use.

The economics of transporting CBM produced water to existing injection wells are site specific and heavily dependent on: (1) distance to injection well; (2) volume of produced water; and (3) duration of water production. If the duration of water production is expected to be short, transporting water to existing injection wells may be less expensive than alternatives that require large capital investment in infrastructure and/or may require a lengthy permit review process.

The injection capacity of existing wells sometimes can be increased by perforating the well above the current injection zone to access overlying permeable zones or by extending the well to access deeper formations. The addition of horizontals can maximize contact with the injection formation and greatly increase the injection capacity of formations with low permeability. Hydraulic fracturing or other forms of stimulation can increase injectivity and reservoir capacity; periodic hydraulic fracturing can extend the lifetime of declining injection wells. Further, the EPA Underground Injection Control (UIC) Program, on a case by case basis, will permit injection wells to operate at pressures exceeding the fracture pressure of the receiving formation if it can be demonstrated that the integrity of the confining layer(s) is not compromised (PC, 2009). Possible means to demonstrate confining layer integrity include preinjection fracture modeling and microseismic monitoring during injection to assess the progression of hydraulic fracturing. Contemporaneous monitoring for changes in the chemistry of overlying aquifers is also an option.

New injection wells are needed for CBM production areas where there are no existing injection wells with surplus capacity or where the transportation cost to existing injection wells is prohibitive. Operators can sometimes avoid the high cost of drilling new injection wells by converting depleted or nonproductive oil and gas wells into injection wells. New injection wells or converted oil and gas wells can take advantage of horizontal drilling and hydraulic fracturing technologies to improve formation contact and injectivity.

One gas producer in southwestern Pennsylvania has been injecting CBM produced water into a flooded coal mine. This EPA UIC permitted injection well is shallow compared to other injection wells and has almost limitless injectivity because it is injecting into a flooded void left after coalmining. Recently, the well's operator has voluntarily suspended injections until the source(s) of high TDS detected in a nearby stream can be identified. The future use of coal mine void injection wells is uncertain.

12.2.1.2 Shallow Injection

Devices have been developed that separate gas and produced water downhole (BCEW) thereby enabling produced water to be discharged into overlying aquifers without having to be pumped to the surface (Figure 12.3). Obtaining permits for this practice is usually simpler than obtaining an NPDES permit for a surface discharge of CBM produced water. The operator must show that the quality of the CBM produced water is equal to or better than the water quality of the receiving aquifer. From a practical standpoint, the receiving aquifer must be hydrologically isolated from coalbeds where methane is being produced. Otherwise, pumping would have diminished effect on reducing the hydrostatic pressure in CBM producing coalbed (because pumped water would flow from the receiving aquifer back to the producing coalbed). Downhole separation and reinjection saves the CBM operator, the cost of building and operating surface facilities for the storage, transport, and treatment of produced water. This practice conserves produced water as a groundwater resource and, in some cases, may noticeably improve the quality of shallow aquifers.

In areas where methane will be produced from multiple coalbeds, water production can be minimized by combining a downhole gas/water separation with sequencing the order of gas production from individual coalbeds. Hypothetically, the uppermost coalbed will be exploited first and the water produced from that coalbed will be managed on the surface or injected into an overlying aquifer. When gas production from the

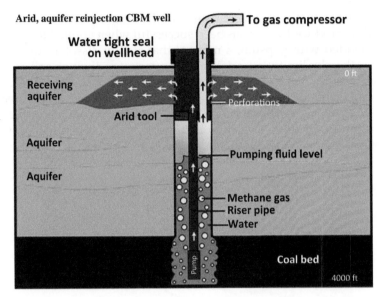

FIGURE 12.3   Cartoon of a downhole gas/water separation device Aquifer Recharge Injection System (ARID) manufactured by Big Cat Energy Corporation. *PC (2009)*.

uppermost coalbed has declined to subeconomic rates, water will then be pumped from a deeper coalbed and injected into the low-pressure pore space created in the upper coalbed by the production of gas and water. This process is repeated with increasingly deeper coalbeds until the gas from all coalbeds has been produced. This procedure would help restore coalbed aquifers previously depleted by gas production. It would also offer cost savings to operators by avoiding the cost of storing, transporting, and treating produced water on the surface.

12.2.1.3 Near-Surface Injection

Subsurface drip irrigation (SDI) and infiltration impoundments are two means of managing CBM produced water that fall within the broad category of a Class V injection well as defined by EPA's UIC program. Class V is a catchall category that includes wells, which are "usually shallow and simply constructed devices" to introduce water into the shallow subsurface for a wide variety of purposes (U.S. EPA).

12.2.1.3.1 SUBSURFACE DRIP IRRIGATION

Irrigation with CBM produced water can sometimes promote crop growth in areas where agriculture is water limited. However, the high sodium content of CBM produced waters can cause an impermeable crust to form in soil that reduces water and air infiltration. Crop yield will decline. Until the soil is mechanically ripped to break up the surface crust, soil productivity will remain low. Traditional surface irrigation with CBM produced water has been successful when applied to soil that was amended with gypsum, a mineral that slowly dissolves to release calcium, thereby limiting the deleterious effect of sodium on soil permeability. Recently, SDI has been promoted as a better means of using CBM produced water for irrigation. SDI applies CBM produced water directly at the base of the root zone where the soil naturally contains sufficient concentrations of soluble calcium and magnesium minerals to prevent loss of soil permeability (Figure 12.4). The salts that form with continued

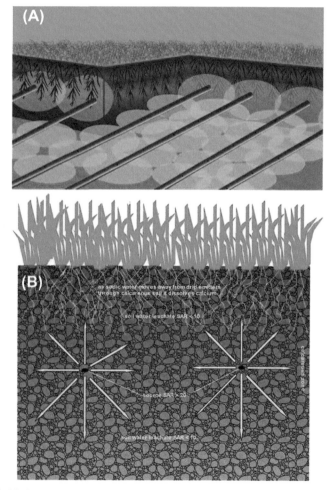

FIGURE 12.4 Cartoon of subsurface drip irrigation using coalbed methane produced water. SAR, sodium adsorption ratio. (A) Cutaway of soil showing the areas irrigated by subsurface drip irrigation. (B) Vertical soil section showing position of drip tubes relative to the root zone.

application of CBM produced water are expected to be "parked" below root depth in the vadose zone where they will not affect crop growth or contaminate shallow groundwater. The National Energy Technology Laboratory (NETL) and the U.S. Geological Survey are evaluating the long-term effect that SDI with CBM produced water has on soil properties, groundwater quality, and crop yield. The study is taking place at a 200-acre SDI area along the Powder River that was previously dry farmed. A dense network of monitoring wells, piezometer nests, surface water sampling stations, and soil moisture sensor arrays has been installed at the site to monitor the effect that SDI with CBM produced water has on the chemistry of surface water, groundwater, and soil water. Semiannual electromagnetic surveys also are performed to identify vertical and lateral migrations of soluble salts (Figure 12.5). After 1 year of CBM produced water application, no adverse impacts to soil structure or groundwater chemistry have been noted. Because the alfalfa crop was planted just after SDI installation and is still becoming established, it is too early to determine the improvement in crop productivity that can be attributed to SDI.

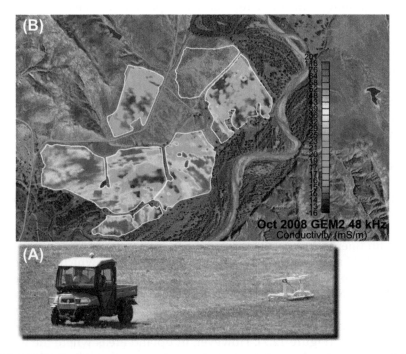

FIGURE 12.5 (A) Apparatus used to conduct ground electromagnetic surveys and (B) conductivity map of a potential subsurface drip irrigation area. Conductive areas are red; resistive areas are blue.

12.2.1.3.2 UNLINED IMPOUNDMENTS

Unlined impoundments (also called infiltration impoundments, ponds) are another means of introducing CBM produced water into the shallow subsurface. In 2006, there were approximately 3000 CBM produced water infiltration impoundments in Wyoming that were either in use or in the permit application stage (Wheaton, 2006). The purpose of infiltration impoundments is to hold CBM produced water until it infiltrates or evaporates. The expectation of permitting agencies was that infiltrating water would augment shallow groundwater resources. In areas where CBM produced water quality is better than the quality of native groundwater (often the case beneath alluvial planes along the Powder River), the infiltrating water was expected to improve the quality of shallow groundwater. Unfortunately, two problems were observed with this practice: (1) infiltrating water sometimes encountered layers of soluble salt that dissolve and degrade groundwater quality and (2) infiltrating water often travels laterally (not downward) through the shallow subsurface to reemerge downslope from the impoundment. The Wyoming Pollution Discharge Elimination System (Wheaton, 2006) estimates that water from about 50 impoundments has reemerged below impoundment dams.

In Wyoming, the discharge of CBM produced water to unlined impoundments is not permitted if the underlying groundwater is suitable for domestic or agricultural use (Class I or Class II, respectively). However, discharge to an unlined impoundment is permitted if underlying groundwater is designated Class III (livestock use) or Class IV (industrial use). For permitted impoundments overlying Class III aquifers, quarterly monitoring of the aquifer is required because of the potential for salt dissolution/migration and resulting aquifer degradation. Groundwater monitoring is not required if the groundwater is classified as Class IV (industrial use) or if no groundwater is encountered within the required depth of investigation.

In 2003 and 2004, NETL performed helicopter electromagnetic (HEM) surveys (Figure 12.6) over developing CBM areas in the PRB to: (1) determine optimum locations for unlined impoundments and (2) determine the fate of CBM produced water infiltrating from unlined impoundments. Figure 12.6 contains a location map of HEM surveyed areas. HEM surveys detect conductors in the ground, which in the PRB are likely to indicate the locations of (1) shallow, high TDS groundwater, (2) salt layers, and (3) clay layers. Along the floodplains and alluvial terrace of the Powder River and its major tributaries, areas of high conductivity generally indicate the presence of shallow, high TDS aquifers (Figure 12.7). However, in upland areas where the water table is apt to be deeper, conductive areas generally indicate the presence of partially solvated salt deposits or clay layers. These conductive areas should be avoided when constructing an unlined impoundment

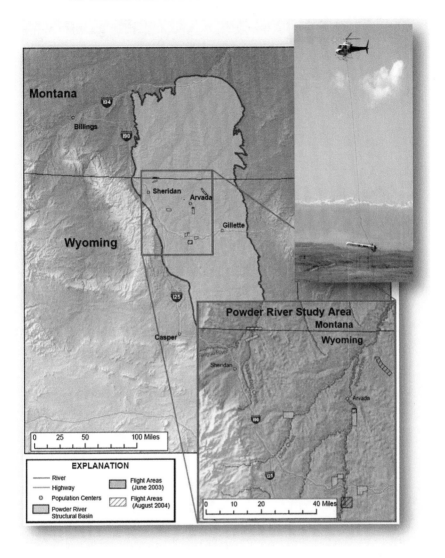

FIGURE 12.6 Map showing the locations of helicopter electromagnetic surveys in the Powder River Basin conducted by the National Energy Technology Laboratory.

because: (1) if the conductivity is from salt layers, the salt will dissolve in the infiltrating water and degrade local aquifers or (2) if the conductivity is from clay layers, the clay may act as an impermeable barrier to the infiltrating water. Figure 12.8 is a HEM near-surface conductivity map of an upland area of the PRB where the high conductivity areas (red) probably denote the presence of salt or clay layers in the shallow subsurface.

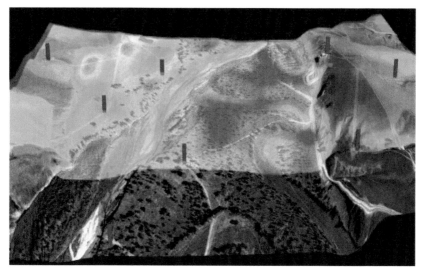

FIGURE 12.7   Conductivity map of the Powder River floodplain. Red areas likely indicate the presence of shallow, high total dissolved solids groundwater.

FIGURE 12.8   Conductivity map of Tongue River Valley showing conductive (red) areas on upland terraces away from the river that likely indicate the presence of salt deposits or clay layers.

12.3   CBM PRODUCED WATER TREATMENT

12.3.1   Current Practices

The quality of produced water from CBM wells is typically better than the quality of water produced from conventional oil and gas activities. In many cases, no treatment or only minimal treatment is needed for CBM produced water to be put to beneficial use or discharged. This is particularly true for produced water from CBM production in the Wyoming.

In December 2008, 54 facilities in Wyoming were treating CBM produced water using technologies that included ion exchange, reverse osmosis (RO), gypsum and acid addition/injection for barium and sodium adsorption ratio (SAR) control, and land application (Table 12.2). These plants were treating 9.55 million gallons of CBM produced water per day, about 10% of the water produced from CBM wells in Wyoming (Fischer, 2009). Ninety percent of the treatment was for lowering TDS and SAR; 10% was for barium removal. In 2008, 13 additional treatment plants were being constructed; including a nanofiltration and a capacitive desalination plant.

12.3.1.1 Ion Exchange

The IOGCC and ALL Consulting (IOGCC and ALL Consulting, 2006) describe ion exchange as "a reversible chemical reaction wherein positively or negatively charged ions present in the water are replaced by similarly charged ions present within the ion exchange resin. The resins immersed in the water are either naturally occurring inorganic zeolites or synthetically produced organic resins. When the replacement ions on the resin are exhausted, the resin is recharged with more replacement ions". Regeneration of the resin results in a concentrated waste solution containing ions that have been removed from the treated water. Evaporation is sometimes used to further concentrate this waste stream before it is injected into disposal wells. Alternatively, the waste stream can be evaporated to dryness with solids disposal at a landfill approved to receive these wastes.

Ion exchange has been used to remove salts, heavy metals, radionuclides, and other undesired elements from produced water. In the PRB, ion exchange is primarily used to remove sodium and TDS. CBM produced

TABLE 12.2 Coalbed Methane Produced Water Treatment Facilities in Wyoming as of December 2008

Treatment Facility	In Operation	Under Construction	Under Review
Ion exchange	15	4	0
Reverse osmosis	2	1	2
Gypsum for Ba/SAR removal	34	0	0
Acid addition	1	5	0
Land application	11	1	0
Solution gypsum injection	1	0	0
Nanofiltration	0	1	0
Capacitive desalination	0	1	0

SAR, sodium adsorption ratio.
Data presented to Powder River Basin Interagency Working Group by Don Fischer, Wyoming Department of Environmental Quality.

water typically requires no pretreatment prior to ion exchange except for the removal of suspended solids or organics, if present. However, some posttreatment of the water with gypsum or limestone may be needed for SAR control because ion exchange preferentially removes divalent cations (Ca^{2+} and Mg^{2+}) over monovalent cations (Na^+).

Ion exchange treatment systems for produced waters fall into two categories: recirculating bed and fixed bed.

12.3.1.1.1 RECIRCULATING BED

The most popular ion exchange treatment system used for CBM produced water treatment is the Higgins Loop, a continuous countercurrent ion exchange method that was invented in 1951 and licensed to EMIT by Severn Trent Services (EMIT, 2010). As of December 2008, there were 12 EMIT Higgins Loop ion exchange treatment plants in operation in Wyoming with three more under construction (Fischer, 2009). In November 2009, Beagle (Beagle, 2009) reported that there were 25 EMIT systems in operation with sites in Wyoming, Montana, and Colorado.

CSM (Colorado School of Mines, 2009) describes the Higgins Loop as "a continuous countercurrent ion exchange contactor for liquid phase separations of ionic components, primarily sodium. The unique aspect of the Higgins Loop is that it performs resin regeneration continuously during the process, with minimal need for system downtime during regeneration". In reality, the treatment is semicontinuous because treatment is shutdown for short periods to allow the resin to be incrementally "pulsed" through the regeneration loop, a vertical cylindrical loop that is separated into four operating zones by butterfly valves. These operating zones (Figure 12.9) include pulsing (A), regeneration (B), adsorption (C), and backwashing zones (D).

A description of the operation of the Higgins Loop based on information from the EMIT Web site (EMIT, 2010) follows.

CBM produced water enters into the adsorption zone of the Higgins Loop (Zone C, Figure 12.9), which contains a strong acid cation resin. Here, cations from the produced water are loaded onto the resin with the concomitant release of hydrogen (H^+) ions. Typically, the resin extracts 95% of the cations present in CBM produced water.

In the regeneration zone of the Higgins Loop (Zone B, Figure 12.9), resin loaded with cations from the produced water is regenerated with hydrochloric acid. Hydrogen ions displace cations from the resin to regenerate the resin and create a waste stream of acidic brine containing the cations removed from produced water. This waste stream, which comprises about 1% of the total water throughput, is usually neutralized by contact with limestone before disposal in deep injection wells. Regenerated resin is rinsed with water to remove acid from its pores prior to reentering the adsorption zone.

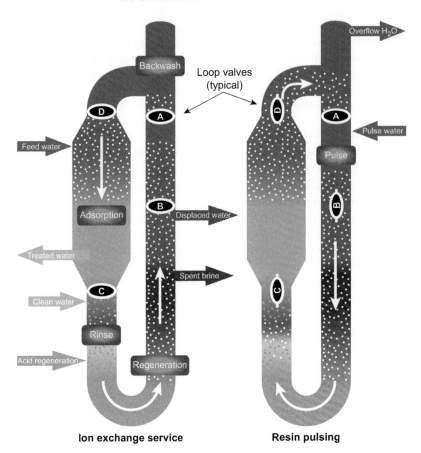

Ion exchange service   **Resin pulsing**

FIGURE 12.9 Schematic of EMIT "Higgins Loop" ion exchange treatment system. *Fischer (2009).*

When the resin in the adsorption zone becomes loaded with cations, the CBM flow to the Higgins Loop is momentarily interrupted to allow "pulsing" the resin bed through the loop in the opposite direction of the liquid flow. Liquid flows restart after the resin pulsing is complete.

Because 95% cation removal is not always needed to meet discharge limits, part of the CBM produced water flow bypasses the ion exchange treatment to be recombined with the treated flow downstream. This allows more water to meet discharge or beneficial use limits without having to increase the throughput of the treatment plant.

The Drake Process is the second ion exchange treatment system with a recirculating resin bed that has been pilot tested for the treatment of CBM water in the PRB (Drake Water Technologies, 2010). The patented Drake Process utilizes a loop consisting of three processing steps: (1) a fluidized bed contactor for loading cations from produced water onto the resin; (2) a separator to separate treated water from the loaded resin; and (3) a regenerating step where sulfuric acid is added to restore the resin for reuse. Pilot units capable of processing 361,000 gallons per day have been demonstrated using CBM produced water. Cation removal is reported to be 93% with a water recovery of 97%. A concentrated Na_2SO_4 brine or salt is also produced (Colorado School of Mines, 2009).

12.3.1.1.2 FIXED BED ION EXCHANGE

Information provided by the Wyoming Department of Environmental Quality (PCDL, 2009) indicated that three non-EMIT ion exchange treatment plants were treating CBM produced water in Wyoming in 2008. One of these plants uses the Eco-Tec's Recoflo compressed bed ion exchange system that is designed to treat 1.5 MGD (36,000 bpd) of CBM produced water (Colorado School of Mines, 2009). This system uses three separate compressed bed columns that contain a primary cation bed, an anion bed, and a polishing cation bed. H_2SO_4 and NaOH are required to regenerate the cation and anion ion exchange resins, respectively. The compressed bed technology employed by the Recoflo system features short bed height, small resin volume, low resin exchange loading, fine mesh resin, and short cycle times (PCDL, 2009). While conventional fixed bed ion exchange systems are typically onstream for 10–20 h and require several hours for regeneration, the Recoflo compressed bed technology is onstream for less than 30 min and regenerates in less than 7 min. Eco-Tec (Hausz) claims that operating in this manner utilizes only the most accessible exchange sites on the resin thereby maximizing exchange rates for both the ion exchange and regeneration processes. Recoflo compressed beds require less space and are more mobile than conventional ion exchange processes. Cation and anion removal is greater than 90%; water recovery has not been reported but is probably less than the 97% reported by cation-only ion exchange systems.

Rohm and Haas conducted a 6-month demonstration of an ion exchange treatment plant for CBM produced water in the PRB (PCDL, 2009). Although little information has been published about this demonstration for Anadarko Petroleum, the plant was thought to use Rohm and Haas Advanced Amberpack resins in a fixed bed configuration.

12.3.2.1 *Reverse Osmosis*

RO is a treatment process by which pressure is used to force liquid through a membrane with pore diameters less than 0.0001 μm. The membrane allows water molecules and other ions with small hydrated radii to pass through, but prevents or inhibits the passage of most CBM produced water constituents for which removal is desired. Because RO membranes are subject to scaling and fouling, RO is seldom a stand-alone treatment process. For CBM produced water treatment, RO is preceded by treatment processes that remove particulates, petroleum hydrocarbons, iron, and scale-forming cations. Chemicals are added to adjust pH, control scale formation, and suppress microbial growth.

Two RO plants for treating CBM produced water have been constructed in the PRB by Siemens for Petro-Canada. The Wild Turkey plant, which was built primarily to remove sodium, has a design capacity of 5 million gallons per day (119,000 bpd) and has been operating since May 2006. The Wild Turkey plant consists of: (1) influent ponds for aeration, solids settling, and surge capacity; (2) media filtration for removal of suspended solids; (3) primary RO treatment using six skids containing 6-m long cartridge filters; (4) secondary RO treatment (brine recovery) using six skids with 4-m long cartridge filters; and (5) a blending system that combines ratios of treated and raw water to obtain sodium concentrations that meet discharge criteria. Chemicals added include: (1) polymer injection for solids flocculation; (2) chlorine injection to control microbial growth; (3) bisulfate addition before RO to remove excess chlorine that can degrade RO membranes; (4) acid injection in RO stages to prevent $CaCO_4$ formation (scale); (5) antiscalant for control of silica scale; and (6) more biocides in the RO stages. Wastewaters generated during treatment are sent to evaporation ponds. The plant has been successfully meeting the permitted sodium discharge limit of 325 mg/L despite the occurrence of scaling problems when silica in the influent water exceeded design limit (Hook, 2007).

In 2008, Petro-Canada awarded Siemens the contract to treat CBM produced water at Mitchell Draw. The Mitchell Draw plant has a treatment capacity of 3.024 million gallons per day (72,000 bpd) and is similar to the Wild Turkey plant in design (Figure 12.10). However, the Mitchell Draw plant includes an ion exchange pretreatment to remove polyvalent cations (scale-forming ions) prior to RO treatment. The addition of ion exchange was expected to lessen the need for antiscalants and to eliminate acid feed. Eliminating acid injection enables the plant to operate at the naturally high pH of the influent water, which provides three potential treatment benefits: (1) residual organics are more soluble at high pH thereby reducing the potential for membrane fouling; (2) boron is present as borate (vs boric acid), a form

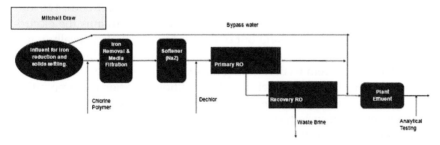

FIGURE 12.10 Schematic of the Siemens RO treatment system at Mitchell Draw. RO, reverse osmosis. *Hook (2007).*

better rejected by the RO membrane; and (3) silicate is more soluble at higher pH, decreasing the potential for silica scale formation on RO membranes (Welch, 2009). As of November 2009, no water had been treated at the Mitchell Draw plant so the benefits of replacing acid addition with ion exchange for scale control have not been substantiated by extended, full-scale operation.

12.3.2 Emerging Treatment Technologies: The Future of CBM Treatment?

New treatment technologies are being developed and demonstrated that appear to have application to the treatment of CBM produced water. Although no one technology can economically treat all types of CBM produced water, given the broad spectrum of water quality and beneficial uses, there are emerging technologies that will fill treatment niches not addressed by current technologies. For example, capacitive deionization may reverse the trend toward centralized treatment by offering portable, low cost treatment systems that can be operated at single well sites and then easily moved to meet treatment needs in other areas. Other technologies make use of low-grade waste heat to economically treat the saline/ hypersaline water produced from some coal basins. Previously, this water was thought to be untreatable and was deep injected at considerable expense. With new membrane distillation (MD) technologies, treatment costs may become competitive with the cost of deep injection, and the treat or inject decision may be swayed by the value of the treated water. Membrane use for treatment of CBM produced water is expected to increase in light of the torrid pace of advancements in membrane lifetime, flux, and selectivity. Described below are three types of treatment technologies that are expected to be evaluated at commercial scale for CBM produced water treatment.

12.3.2.1 Capacitive Deionization

In capacitive deionization treatment, CBM water flows between positively and negatively charged porous carbon electrodes. Charged ions in the treated water are attracted to the surfaces of oppositely charged electrodes. The ions remain fixed on the electrode surfaces while the water (minus the electrosorbed ions) flows through and out of the treatment cell. Periodically, when the electrode surfaces become loaded with ions, the water flow is stopped and polarization is removed, releasing the ions into a small volume of wastewater, which is then deep injected with or without additional concentration in evaporation pits. After cell regeneration, the polarity is reversed from the previous treatment cycle and the flow of treated water resumes.

In 2008, Aqua EWP demonstrated a 410 bpd capacitive deionization treatment system at a CBM site in Wyoming (Atlas and Wendell, 2008). The system (Figure 12.11), which consisted of two capacitive deionization

FIGURE 12.11   Picture of Aqua EWP capacitive deionization treatment system. *Welch (2009).*

cells connected in parallel, was operated for a period of 12 months, successfully removing 80% of the TDS with a water recovery of 83%. The authors mentioned that the CBM produced water from the gas field where the demonstration took place ranged from 1200 to 2500 ppm TDS. However, the authors did not specify the TDS or sodium content of the water tested. The only pretreatment needed was a 30-μm filter to remove particulates from the influent water. Posttreatment with calcium was required to the SAR requirement of 3.

Atlas and Wendell (Atlas and Wendell, 2008) speculate that higher TDS waters can be treated with capacitive deionization by adding addition cells in series (or by decreasing the flow rate through cells). The Aqua EWP Web site (AEWP) suggests that waters with TDS up to 35,000 ppm can be treated with capacitive deionization. However, Oren (Oren, 2008) disagrees, arguing that the decrease in differential charge efficiency at seawater salinities will make capacitive deionization uneconomical. Oren further points out that the cost of desalinating seawater with capacitive deionization would be greater than the cost of RO.

Capacitive deionization does have a niche in the treatment of CBM produced water where TDS values are often within the economic range of the treatment. Advantages of capacitive deionization over ion exchange and RO include: (1) low capital and O&M costs; (2) the system has a small footprint and is mobile, allowing it to be easily moved to meet changing treatment needs; (3) the system can be reconfigured to treat higher TDS water or to increase water recovery; (4) the system does not require chemicals except citric acid for semiannual cleaning of cells; and (5) the system needs minimal pretreatment and posttreatment.

12.3.2.2 Membrane Distillation

Currently, the treatment of produced water with salinities greater than that of seawater is limited to thermal distillation methods or RO with significant pretreatment and posttreatment. Current treatment for hypersaline water is expensive because of high capital costs and high O&M costs. MD may be the future treatment for hypersaline water because it operates at lower temperatures than conventional distillation and at lower pressures than conventional pressure-driven membrane separation processes (Oren, 2008). Further, the hydrophobic, microporous MD membranes offer near 100% rejection of ions, macromolecules, colloids, cells, and other nonvolatile compounds compared to the 90–95% rejection typical of RO membranes. Because the MD membrane is hydrophobic and not wetted by aqueous solutions, membrane fouling is less problematic. However, the supersaturation of some compounds can occur at the liquid/vapor boundary resulting in scaling.

MD can take place with a minimum temperature differential across the membrane of 20 °C (Oren, 2008). This allows inexpensive sources of low-grade heat (often waste heat from other processes), to be used, including solar energy. Because CBM production is often from areas where coal is mined and burned to generate electric power, thermoelectric power plants may be the source of low-grade heat needed for the MD of CBM produced water. Further, thermoelectric plants need water treatment for boiler and scrubber blowdown and for the supernatant from ash ponds. In the future, thermoelectric plants may need to treat condensate from the capture and compression of CO_2 as well. Therefore, synergies can be realized by colocating MD water treatment plants at the sites of thermoelectric plants.

12.3.2.3 Improvements to Pressure-driven Membrane Processes

RO treatment typically removes more than 99% of the contaminants in CBM produced water. Often this is more treatment than needed to meet discharge standards. In these cases, untreated water bypasses the treatment plant and is blended with treated water to achieve TDS or sodium values that meet mandated standards in the discharge water. This increases plant throughput. However, nanofiltration may offer an alternative. Nanofiltration membranes reject more than 99% of divalent cations but less than 90% of monovalent cations (Colorado School of Mines, 2009). The difference between the 99% ion rejection of RO and the 90% ion reduction by nanofiltration may be balanced by the lower energy costs and lower pressure requirements of nanofiltration. In the future, nanofiltration for CBM produced water treatment will probably grow especially if improved membranes continue to be developed.

References

Aqua EWP website. http://www.aquaewp.com/coalbedmethane.php.

ALL Consulting, 2003. Handbook on Coalbed Methane Produced Water. http://www.gwpc. org/elibrary/documents/general/Coalbed%20Methane%20Produced%20Water%20Ma nagement%20and%20Beneficial%20Use%20Alternatives.pdf.

Atlas, R., Wendell, J.H., April 2008. Low-power capacitive deionization method shows promise for the treatment of coalbed methane produced water. World Oil 229 (4).

Big Cat Energy website. www.bigcatenergy.com/ARID-Solution.aspx.

Beagle, D., 2009. CBM produced water management in the Powder River Basin of Wyoming and Montana, Montana. In: Presentation at the 16th Annual Petroleum and Biofuels Environmental Conference, Houston, TX, November 3–5, 2009. http://ipec.utulsa.edu/ Conf2009/Papers%20received/Beagle_CBM.pdf.

Cath, T.Y., 2009. Emerging technologies for the treatment of CBM produced water. In: Presentation at the 16th Annual Petroleum and Biofuels Environmental Conference, Houston, TX, November 3–5, 2009. http://ipec.utulsa.edu/Conf2009/Papers%20received/Cath_ Emerging.pdf.

Colorado School of Mines, 2009. An Integrated Framework for Treatment and Management of Produced Water, Technical Assessment of Produced Water Treatment Technologies, first ed. RPSEA Project 07122-12. http://www.aqwatec.com/research/projects/Tech_ Assessment_PW_Treatment_Tech.pdf.

Drake Water Technologies, 2010. Drake Process. http://www.drakewater.com/5401.html.

EMIT, 2010. Description of Higgins Loop Technology. http://www.emitwater.com/higgin s_loop.html.

Fischer, D., 2009. Summary of coalbed natural gas produced water treatment and management facilities, Powder River Basin, April 2009. In: Presentation before Powder River Basin Interagency Working Group, April, 2009. http://www.wy.blm.gov/prbgroup/.

Hausz, L. Evolution of counter-current ion exchange for industrial water treatment. http://www.eco-tec.com/pdf/Eco-Tec-%20Evolution%20of%20Counter%20Current%2 0Ion%20Exchange1.pdf.

Hook, T.A., 2007. Innovative technology for the treatment of coal-bed-methane produced water. In: Presentation at the 14th Annual Petroleum and Biofuels Environmental Conference, Houston, TX, November 6–9, 2007. http://ipec.utulsa.edu/Conf2007/Papers/ Hook_73.pdf.

IOGCC and ALL Consulting, 2006. A Guide to Practical Management of Produced Water from Onshore Oil and Gas Operations in the United States. http://iogcc.publishpath.com /Websites/iogcc/pdfs/2006-Produced-Water-Guidebook.pdf.

Lawson, K.W., Lloyd, D.R., 1997. Review: membrane distillation. Journal of Membrane Science 124, 1–25.

Oren, Y., 2008. Capacitive deionization (CDI) for desalination and water treatment – past, present and future (a review). Journal of Desalination 228, 10–29.

Personal communication with David Rectenwald of EPA UIC Program. November 2009.

Personal communication with Dennis Lamb, Wyoming Dept. of Environmental Quality, October 21, 2009.

Produced water management information website. http://www.netl.doe.gov/technologie s/pwmis/.

U.S. EPA. Frequently asked questions about UIC class V wells. http://www.in.gov/indot /div/pubs/waterway/Appendix_G-5_EPA_Frequently_Asked_Questions_About_UIC _Class_V_Wells.pdf.

U.S. EPA, 2009. Update on EPA's clean water act review of the coalbed methane (CBM) extraction sector. In: Presentation by C. Johnson at the 2009 International Petroleum and Biofuels Environmental Conference, Houston, TX, November 5, 2009. http://ipec.utulsa .edu/Conf2009/Papers%20received/Johnston_7.pdf.

Welch, J., 2009. Coalbed methane produced water – an evolution in treatment. In: Presentation at the 16th Annual Petroleum and Biofuels Environmental Conference, Houston, TX, November 3–9, 2009. http://ipec.utulsa.edu/Conf2009/Papers%20received/ Welch.pdf.

Wheaton, J., 2006. Evaluation of coalbed-methane infiltration ponds for produced water management. In: Biennial Report of Activities and Programs, July 1, 2004–June 30, 2006. Staff of the Montana Bureau of Mines and Geology. http://www.mbmg.mtech.edu/bien nial-2006/p73-74_br-2006.pdf.

13

Plugging In-Mine Boreholes and CBM Wells Drilled from Surface

Gary DuBois[1], Stephen Kravits[1], Joe Kirley[2], Doug Conklin[3], Joanne Reilly[3]

[1]Target Drilling Inc., Smithton, PA, USA, [2]Concrete Construction Materials, USA, [3]Coal Gas Recovery, LP, Affiliate of Alpha NR, Bristol, VA, USA

13.1 INTRODUCTION

The Mine Safety and Health Administration (MSHA) now requires a "Mine-Through Plan" as part of the Mine Operator's Ventilation Plan and mapping requirements that apply to methane gas wells or conventional gas wells (U.S. Department of Labor, 2005). A mine operator that utilizes either underground boreholes or surface CBM wells must describe in their Mine-Through Plan what material will be used to seal the boreholes and additional safety precautions to be taken prior to intersecting the borehole.

If a mine operator's ventilation plan does not permit water infusion of a horizontal degas borehole prior to mine through by their MSHA District Office then an alternate method must be used. Cement slurry has been predominantly used to plug underground horizontal degasification boreholes. However, pumping cement slurry in horizontal boreholes has proven ineffective filling the entire void of the borehole and the borehole sidetracks (Kravits et al., 2006). The ineffectiveness of cement slurry results from (1) dilution of the cement in the slurry caused by formation water in the borehole, (2) pressure buildup caused by heat and friction generated during the mixing and pumping of the cement slurry does not permit the entire borehole and sidetracks to be filled with the cement slurry, and (3) inherent shrinkage of the cement when it cures or sets (Aul and Cervik, 1979). Consequently, occasional face ignitions and production delays are possible with development and longwall mine through of boreholes plugged with cement slurry.

13.2 PLANNING THE PLUGGING OF BOREHOLES PRIOR TO MINE THROUGH

Inherent to any directionally drilled borehole or lateral drilled in coal or targeting the coal bed, the length and elevation changes of the borehole are the primary factors that must be considered to plug the borehole or lateral prior to mine through. Other factors are listed as follows in the form of questions that must be answered:

13.2.1 What is the Quantity, and Chemistry of the Water Produced from the Borehole or Coal Bed Lateral?

Water produced from the borehole or lateral should be measured and recorded during the gas production period and compared to water production just prior to plugging. This will determine if the coal bed was dewatered effectively. Very importantly, the water samples of the produced water prior to plugging must be collected and tested for water chemistry, no matter what plugging method or medium is used as it can adversely affect or prevent the plugging material to effectively plug the borehole or lateral.

13.2.2 Has the Borehole or Coal Bed Lateral Adequately Degassed the Coal Bed?

A coal bed contains methane at in situ reservoir pressure. As the coal bed is mined, or a borehole or coal bed lateral is drilled in the coal bed, a pressure sink is created and the methane gas is produced into the lower pressure borehole or mine entry. Consequently, the in situ gas pressure of the methane in the coal bed generally reduces within the effective drainage radius of the entry or borehole depending on the permeability of the coal bed and other reservoir parameters. It is important to measure the "shut-in" gas pressure of the borehole or coal bed lateral prior to plugging. Different types of plugging or sealing materials have different shear strengths. If the shut-in gas pressure of the borehole or coal bed lateral is still relatively close to the original shut-in gas pressure, the coal bed has not been effectively degassed. Consequently, the shear strength of the plugging material, when set or cured, must be able to withstand gas pressure in the coal bed after plugging or the material used to seal the borehole might be forced out of the borehole during mine through.

13.2.3 Were Drilled Intervals of Roof and Floor Eliminated?

Typically borehole stability is better in the coal bed compared to the surrounding rock strata. However, if the rock strata above or below the coal bed is weathered, weak sedimentary rock containing clay and shale, and it is intercepted by the borehole or coal bed lateral, the rock intervals could slough or collapse during and/or after drilling is completed. If rock intervals in the borehole collapse or slough, a plugging material with a lower viscosity is more likely to flow through the partially blocked or sloughed rock intervals.

13.2.4 If Sidetracks were Conducted to Eliminate Rock Intervals Drilled, Will the Sealing Material Fill the Abandoned Sidetracks?

The more extensively sidetracking was utilized to keep the horizontal borehole or coal bed lateral in the coal bed, the more difficult it becomes to adequately seal the borehole and/or coal bed producing laterals. A plugging fluid with a lower viscosity is more likely to flow into the roof and floor sidetracks than a sealing fluid with higher viscosity.

13.3 PRIMARY MEDIUMS USED TO PLUG DEGASIFICATION BOREHOLES

Water injection has been used to temporary plug degasification boreholes just prior to the mine through.

Advantages: Minimum cost; good availability; easy to pump and can be pumped from the well head eliminating the need of installing a pipe in the borehole.

Disadvantages: Inability to control the water's final destination as water seeks the path of least resistance including following the natural fracture system of the coal bed; quantity of water needed cannot be accurately determined because the path of the injected water is not controlled nor is the distance the water might travel; inability to maintain the water in the borehole during mine through, and after mining, thus allowing a conduit for methane flow. Depending on the elevation changes in the borehole and coal bed, water traps in the borehole can adversely affect mining conditions.

Cement slurry has been the primary sealing material used to plug degasification boreholes prior to mine through.

Advantages: When cement slurry cures it is a solid, with good compressive strength and high density.

Disadvantages: Cement slurry is difficult to pump long distances; may shrink up to 10% during curing; and might not cure or set depending on the volume of water and the chemistry of the water in the borehole.

Polymer gel has been the primary medium used by target drilling to plug degasification boreholes prior to mine through.

Advantages: The polymer gel can be pumped long distances through a relatively small diameter conduit from the vertical well to the in-mine horizontal borehole well head; cure time for the polymer gel can be adjusted, and the polymer gel can be pressurized and squeezed into cleats of the coal bed forcing gas and water back into the formation.

Disadvantages: The curing of polymer gel mixture can be adversely affected by the chemistry of the mix water in the borehole; the volume of water produced from the coal bed into the borehole just prior to plugging; water temperature causing the gel mixture to cure prematurely if the water temperature is too hot; mixing the gel chemicals is a tedious process requiring experience with the chemicals and the mixing and pumping equipment; and the cured polymer gel has minimal shear strength.

13.4 POLYMER GEL EVOLUTION

Target Drilling Inc. (TDI) has directionally drilled over 160 in-mine horizontal degas boreholes greater than 4000 ft and 20 CBM wells with horizontal laterals in the coal bed for Alpha NR's, Coal Gas Recovery, LLC (CGR) at Alpha's Cumberland Coal Resources, LP and Emerald Coal Resources, LP coal mines. As an alternative to water infusion or plugging with cement slurry prior to mine through, CGR and TDI solicited the technical support and experience of Concrete Construction Materials (CCM) to develop an alternative to water infusion or sealing these boreholes with cement slurry by using a polymer gel. Generally, the gel would be designed to have a fluid viscosity similar to water so that it could be pumped very long distances, and cure at a specifically designed time to a semisolid state, maintaining the semisolid state for more than 6 months. The gel was designed as a relatively thin liquid with viscosity of 1000 cp at shear so that it would flow like water through a 1.25 in plastic slick-line suspended from the surface with a steel cable down a vertical borehole and then horizontally to the underground horizontal borehole well heads. Pumping distances of the gel through the slick-line, from the surface to the underground horizontal well head could be up to 2 miles. As the borehole fills with gel and becomes pressurized, the gel would squeeze into and infiltrate the natural fracture system of the coal, forcing gas and water back into the natural fractures of the coal bed. Finally, as the water-soluble gel mixture cures into a semisolid, insoluble

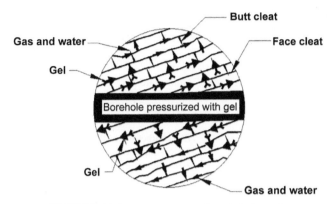

FIGURE 13.1   Cross-section of borehole in coal.

gel at the "designated cure time", a skin forms on the inside wall of the borehole reducing migration of gas and water back into the borehole (Figure 13.1).

13.5 CHEMICAL DESCRIPTION OF THE GEL

CCM developed a man made, proprietary inorganic, metal cross-linked polymer gel consisting of: (1) 97–99% water by weight; (2) 2–3% by weight liquid high molecular weight water-soluble, partially hydrated, poly acrylimide polymer product called VMA-007; (3) less than 0.5% by weight valiant chromium ion XLR-C used to "cross-link" or complex the poly acrylimide material (VMA-007) to change the polymer from water soluble to insoluble and; (4) less than 0.25% by weight liquid accelerator and conditioner to control reaction rate (Activator-M). In simple terms, the poly acrylimide, in excess of two million molecular lengths, is allowed to hydrate with water creating a mixture similar to a bowl of noodles. Then at a specific time based on the mix design, chromium III ions attach at bonding points in between the noodles creating extremely over cooked noodles. CCM hired an independent laboratory to develop specific concentrations of the individual components for the gel mix based on the desired times required for mixing, pumping, setting, and maintaining its semisolid state by conducting tests in their laboratory.

Table 13.1 is an information document for the metal cross-linked polymer gel mix structured as an material safety data sheet (MSDS), but it is *not* an actual MSDS sheet. Slight variations in mixing in the field make it difficult to develop an exact representative MSDS. Before the gel was used initially to seal underground degas boreholes, meetings were held

TABLE 13.1 Metal Cross-Linked Poly Acrylimide Gel Information Safety Data Sheet (Not an MSDS)

Section Number	Metal Cross-Linked Poly Acrylimide Gel
1. Product identification	Proprietary blend
2. Ingredients	Nonhazardous
3. Hazard identification	Light green to powder blue gel with no distinct odor
4. First aid measures	Eye contact-water flush; skin contact-soap and water; ingestion-physician; inhalation-fresh air; artificial respiration
5. National Fire Prevention Code	Health = 0; flammability = 0; reactivity = 0; special hazard = none
6. Accidental release measures	Wear personal protective equipment and remove with absorbent materials
7. Handling and storage	Keep away from heat and incompatible materials
8. Personal protection	Chemical resistant personal protective equipment
9. Physical and chemical properties	pH:6–9; SG:1.00–1.05 g/mL; insoluble to water; boiling point: >100 °C; freezing point: 0 °C; light gray to blue green gel with no odor
10. Stability and reactivity	Hazardous polymerization will not occur
11. Topological information	No information available
12. Ecological information	No information available
13. Disposal	Not an RCRA hazardous material. Discard according to regulatory agencies
14. Transportation	Primary hazard class/division: not restricted
15. Regulatory information	OSHA hazard communication status: nonhazardous
16. Other communication	HMIS ratings: health = 1; flammability = 0; reactivity = 0

RCRA, Resource Conservation and Recovery Act; OSHA, Occupational Safety and Health Administration; HMIS, Health Management Information System.

between representatives from the local work force's safety committee, mine management, MSHA District 2 representatives, an MSHA toxologist, CCM, and TDI. Those meeting concluded that the gel ingredients, mixed gel ingredients, and cured gel were non toxic or hazardous. Adequate training for personnel was conducted for safe handling, mixing, and pumping of the gel ingredients. The final gel mix either as a viscous liquid or a cured semisolid state would not be hazardous or jeopardize the safety and well-being of the local workforce if they came in contact. It is important to note, this product combination of gel mix was developed specifically for this application and to overcome the deficiencies of using a modified guar. The material is resistant to bacterial attack and stable to decay for long periods of time. Other natural gel systems (guar and low cost oil field viscosity agents) consist of food type materials and could present bacterial issues and long-term stability problems.

13.6 LABORATORY AND SURFACE TRIAL GEL TESTS

Laboratory tests and trial field tests were conducted prior to sealing the underground horizontal boreholes. The gel mix can be modified to fit different combinations of gel time, temperature, and required pumping and setting viscosity. The purposes of the laboratory tests were to simulate various formulas of the gel mix for sealing the horizontal underground boreholes. The primary criteria for successful laboratory tests, using water that was first tested for pH, were to develop: (1) a gel liquid viscosity of 500–1000 cp with high lubricate qualities for pumping thousands of feet; (2) a gel cure time of 5–8 h after mixing the ingredients and (3) a viscosity of about 20,000 cp for the cured gel. A Bradford programmable viscometer was used to plot the induction, build, gel and cure cycles for the gel tests.

Subsequently, the laboratory results were used to design a gel mix representative of gelling horizontal boreholes that was mixed and pumped into a 200′ by 3″ diameter high density polyethylene pipe (HDPE) laid on the ground, connected to 3″ wyes and shorter sections of 3″ HDPE pointing up to simulate abandoned sidetracks that intercepted the roof rock (Figure 13.2), capped off with closed 3″ valves. Sixteen-inch diameter holes were drilled periodically in the wall of the HDPE pipe to simulate the coal's face and butt cleat fracture system. Two gel mix designs were pumped into the HDPE pipe network on different days. The first gel mix cured in 7 h in the HDPE pipe with an ambient temperature of 60 °F, which was close to the

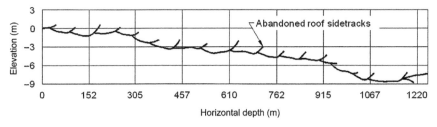

FIGURE 13.2   Vertical profile of coalbed methane borehole.

designed gel time of 5 h. The second gel mix pumped into the HDPE pipe was proportioned to gel or set in 1 h at 60 °F; however, the gel did not set for 12 h because the ambient air temperature was approximately 40 °F on a windy and rainy day. Consequently, it was decided to design the first gel mix with a set or cure time of 7–8 h to be used to seal horizontal boreholes.

13.7 GEL DESIGN FOR THE HORIZONTAL BOREHOLES

Table 13.2 provides the final dosage of gel ingredients designed for the horizontal boreholes based on the laboratory and HDPE trial gel tests with mixing gel batches with 1000 gallon of water. Due to potential underground problems independent of the gel mixing/pumping process, such as a mine fan outage, it was decided to mix 1000 gallon batches at a time because if pumping had to be terminated, up to 2000 gallons of gel mix would be lost, 1000 gallons being mixed while 1000 gallons were being pumped. If the entire borehole volume of gel was mixed, it would be possible to lose volumes exceeding 20,000 gallons. It was arbitrarily established to apply a Factor of Safety of 2 to the calculated volume of gel mixed to: (1) fill the boreholes, including sidetracks, (2) force gel into the fracture system of the coal displacing gas and water, and, (3) develop a maximum gel pressure of 100 psi at the horizontal well head to form a skin on the inner borehole wall reducing gas and water to flow back into the borehole. A very positive feature in the metal cross-link poly acrylimide gel mix was that the gel mix displays an affinity to attach or adhere to everything. The gel pressures in the boreholes were not allowed to exceed 100 psi because the collar pipes installed in the coalbed rib were water tested at 100 psi. *Trimming line was not installed in the horizontal boreholes* to transport the gel mix to the end of the horizontal boreholes. It was important that a sample of the water to be used to mix with the gel ingredients was tested for pH so that CCM could make final adjustments to the concentration of the gel ingredients as required.

TABLE 13.2   Gel Ingredients per Batch of Gel

Gel Ingredient	Volume by Weight		Volume by Liquid	
	Kilograms	Pounds	Liters	Gallons
Water	3846	8480	3785	1000
VMA-007 (tote)	78	173	76	20
XLR-C (blue pail)	155	41	15	4
Activator (round white pail)	9	21	7	2

13.8 GEL MIXING AND PUMPING PROCEDURES

After Target's staff reviewed the MSDS for the individual gel ingredients and the Information Safety Data Sheet for the gel, they were required to wear approved eye, hand, and foot personal protective equipment for the chemicals. Prior to mobilizing to each vertical vent borehole surface gel site, the 1.25″ HDPE slick-line was suspended down the vertical vent borehole with a telephone pager cable, both attached to steel cable. The telephone pager is a dedicated phone cable from the surface to pager phones at the bottom of the vertical vent borehole and at the horizontal well heads to maintain constant communication during the gelling process (Figure 13.3). The slick-line was then installed underground to the horizontal borehole well heads where it was hooked up with bypass wyes, two valves, and pressure gage. Water was then pumped from the surface through the slick-line to the horizontal well heads to verify that the slick-line did not leak. A sample of the water to be used to mix with the gel ingredients was tested to determine pH and if any other chemicals were present in the water that would require adjustments to the concentration of the gel ingredients.

The poly acrylimide polymer VMA-007 was supplied in totes; the chromium III metal cross-linker additive was supplied in 55 gallon drums and the Activator-M conditioner/accelerator is supplied partially hydrated in 3-gallon pails. As one batch of gel mix is being pumped, the next batch is being mixed. The preparation and mixing of a batch is timed to coincide with the batch being pumped so that a continuous pumping operation is achieved to keep the mixed liquid gel flowing into the borehole (Figure 13.4). Viscosity measurements were taken and recorded for samples of each batch of the mixed gel before pumping into the slick-line.

FIGURE 13.3   Slick-line suspended down vertical vent borehole with pager.

FIGURE 13.4   Simultaneously mixing a batch of mixed gel (left) while pumping gel mix (right).

13.8.1 Sealing Results of Underground Boreholes

Sixty-eight underground horizontal degasification boreholes drilled in four different longwall mines have been effectively sealed using metal cross-linked poly acrylimide gel. All of the boreholes were sealed by mixing and pumping the gel from the surface through plastic slick-line suspended down vertical vent holes and then traversed to the horizontal borehole well heads following the mixing and pumping procedures described earlier. The gel was pumped directly from the 1.25″ slick line into the 4″ diameter well head plumbing, then into the 3.75″ underground borehole.

The "second generation gel mixing machine" was used to mix and pump over 412,000 gallons of gel to seal the 68 underground horizontal boreholes. These 68 boreholes were drilled parallel to the gate road development section, so each borehole was intersected hundreds of times by the longwall before the longwall face advanced the entire length of the horizontal borehole (Figures 13.5 and 13.6). The gel proved to be very effective in sealing the borehole, thus almost eliminating (+99.0% of the time) production delays associated with methane liberation/accumulation from individual boreholes.

The first two boreholes sealed with the metal cross-linked poly acrylimide gel were B1-5 and B1-6 (Figure 13.7). The longwall face mined into the end of both boreholes to find the ends filled with the gel including the sidetracks resulting in zero methane caused longwall production delays. Cure time for the gel was designed for 7 h.

The actual volume of gel pumped into 68 underground boreholes was +412,000 gallons or 1.80 times the calculated volume of the 68 boreholes including all sidetracks. Pump pressure gradually increased as the boreholes filled with gel as expected. Pumping was stopped when the

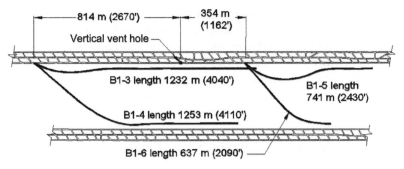

FIGURE 13.5   B1 longwall panel boreholes sealed with metal cross-linked poly acry-limide gel.

FIGURE 13.6   B1-5 mined into by longwall face without incident.

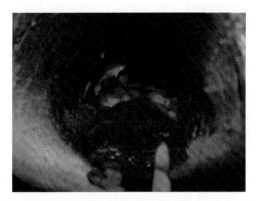

FIGURE 13.7   Cross-linked polymer gel forms skin on borehole wall.

horizontal well head pressure reached +50 psi and the volume was at least 1.50 times the calculated volume for each individual borehole. When the longwall intercepted the B1-5 borehole it was full of cured gel. Minimum sloughing of the cured gel in the borehole occurred after mine through.

Where the gel sloughed from the top of the inner borehole wall, the gel had formed a thin layer skin sticking to the inner wall. There was minimal gas and no water produced from the gelled B1-5 borehole when it was initially mined into by the longwall face. Sealing the B1-5 and B1-6 boreholes was considered a major success because the boreholes did not cause any longwall production delays. Consequently, it was decided to seal B1-3 and 4 boreholes with the gel several months later.

13.8.2 Results of Sealing the B2 Boreholes

The B2-5 and B2-6 longwall panel boreholes were sealed with gel pumping a total distance from the surface to the end of the boreholes over 3 km (10,000′) (Figure 13.8). The actual gel volume pumped for these two boreholes was 1.99 times the volume of the boreholes including sidetracks. Overlap of boreholes B2-3, B2-4, and B2-1 probably had an impact on the gel volumes actually pumped totaling 1.66 times the calculated borehole volume including sidetracks in addition to leakage through the coal ribs near the well heads at an average gel well head pressure of 53 psi. Likewise, only 1.14 times the calculated borehole volume with sidetracks for B2-1 of gel was pumped experiencing gel leakage at the rib near the well head with a well head gel pressure of 58 psi. The B2 longwall panel boreholes were either found to be full of gel or if not full, the gel had formed a skin on the inner wall of the borehole. Consequently, the B2 longwall panel was not interrupted by gas delays.

Modifications and improvements to the mixing and pumping procedures continued during the time frame to seal the balance of these 68 boreholes. If a borehole needed to be sealed during the hot summer months, they were usually completed at night to avoid the hot sun and heat of the summer daylight hours. The warmer summer temperatures reduced the time available to pump the gel before setting to a semisolid state. One improvement made during these 68 boreholes was to use potable water for mix water exclusively, rather than creek water that could contain high levels of bacteria. The results of B1 and B2 longwall panels continued for the balance of the 68 underground boreholes at both Alpha NR's Cumberland and Emerald Mines.

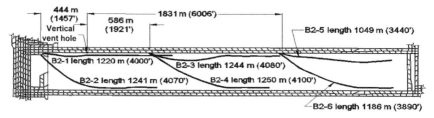

FIGURE 13.8   B2 longwall panel boreholes.

13.8.3 Sealing CBM Surface Wells, New Challenges

TDI was contracted by coal operator to seal three surface CBM horizontal wells utilizing the cross-linked polymer gel. These wells were both single wells and dual CBM wells and produced various amounts of gas and in situ water during their operating life but never completely degassed or dewatered the coal bed. TDI used the same gel formula and "second generation gel mixing machine" that had successfully sealed more than 73 miles of underground boreholes. These CBM wells were larger in diameter and therefore required more gel than a typical underground borehole. TDI assumed the gel mix and mixing pumping procedures that successfully sealed the underground boreholes would also work for surface CBM wells.

TDI took several samples for each batch of gel mixed indicating a majority of the samples cured or gelled. Unfortunately when the coal operator mined into these sealed CBM well laterals, they encountered ungelled slimy water that was pressurized by the in situ gas pressure and hydrostatic pressure of the ungelled water in the curve of the surface CBM well. The mine had a very difficult time dealing with the situation and experienced coal production delays before they were able to pack off the coal bed lateral, stop the flow of gas and resume coal production.

After numerous meetings and days of testing using the water produced from the coal bed laterals, it was determined the produced water from these CBM wells was very high in total dissolved solids (>20,000 mg/L). The high concentration of total dissolved solids (TDS) prevented the cross linking of the polymer gel chemicals resulting in the slimy water instead of cured gel. Although samples showed the majority of the batches gelled properly, these samples of cured gel indicated the "mix" water did not exhibit adverse water chemistry.

13.8.4 Reformulated Gel and "Third Generation Gel Machine"

After these shortcomings, TDI reevaluated the sealing process using the gel mixture and concluded that each individual underground borehole or surface CBM well needed to be evaluated as an individual entity, i.e., "one size does not fit all". Consequently TDI, CCM, and CGR established new protocol to seal underground boreholes or CBM well coal bed laterals. The protocol included: (1) measurement of the reservoir "shut-in pressure" for 48h prior to plugging, (2) measurement of the production rate of produced water at the well head, (3) water chemistry testing of produced water for total dissolved solids, pH and bacteria concentration and (4) additional testing to match the gel mixture curing characteristics to the borehole conditions.

From the results of the chemical analysis of the produced water, the water–chemical ratio will be adjusted, adding up to 50% more chemicals to the potable water used to mix with the gel chemicals. Extensive testing is completed in the laboratory to achieve the best results for each individual application, whether it is an underground borehole or a surface CBM well. Important factors that need to be considered when finalizing the chemical mix ratio are curing rate, mix and pumping viscosity, gel pumping conduit to the borehole or CBM well, volume of gel required to fill the borehole or coal bed laterals, estimated time to mix and pump the required gel, chemicals, water temperature, and ambient air temperature.

It was also apparent that a higher capacity "third generation gel mixing/pumping machine" (Figures 13.9 and 13.10) was needed to simultaneously mix and pump gel to seal future CBM wells that could require 5–10 times the quantity of gel than a single underground horizontal borehole. In addition, CBM wells required pumping through larger inner diameter drill string as compared to 1.25″ slick line and directly into the laterals of a CBM well. This allowed greater volumes of gel mix to be pumped in a given time. It was also apparent that the new "third generation gel mixing/pumping machine" needed to have the capability of finishing the sealing process in 8 h even if a mechanical failure occurred to any part of the gel mixing and pumping machine. The "third generation gel machine" was built with two completely independent gel mixing and pumping systems. These two systems can be used independently or used simultaneously with a pumping capacity of up to 180 gpm. The "third generation gel machine" is equipped with four 160-HP diesel engines, two to power two 6×5 centrifugal mixing pumps, and two engines to power two triplex pumps used to pump the mixed gel.

FIGURE 13.9   Second generation gel machine.

FIGURE 13.10   "Third generation gel machine" at DD-4.

13.8.5 Mine Techniques for Safe Mine through with Continuous Miner or Longwall

The mine through of both underground boreholes and surface CBM coal bed laterals has evolved in the last five years. The evolution has included water infusion, cement slurry and for the past five years, the use of the polymer gel to seal the degas boreholes and CBM wells. The mine operator is required to have an approved MSHA mine through plan, which are typically part of the mine's ventilation plan. MSHA may require a site-specific plan prior to mine through for any borehole or CBM well coal bed lateral.

Typically the mine operator will contract TDI to seal their boreholes or wells with polymer gel, 2–6 months prior to mine through. The best results for sealing the borehole or well with polymer gel occur when gas and water production rates have declined and the reservoir pressure has been reduced to <20 psi. However, sometimes this is not possible due to changes to the mining plan, delays accessing the surface property to drill, delays in obtaining permits, etc. Consequently, the underground boreholes or surface CBM wells may not have sufficient time to adequately degas and dewater the reservoir. Typically these underground boreholes and surface CBM wells need to be permanently isolated from the underground mine environment after they have been intersected with mining operations. TDI has used several different types of packers to make this isolation permanent.

- *Mechanical packers*: These packers use a rubber bladder sandwich between two pieces of 2″ steel casing. The casing and bladder are inserted into the borehole and mechanically squeezed to enlarge the bladder. They are also equipped with a flow-through tube to permit production of gas to continue after installation. These packers are relatively inexpensive and can be difficult and time consuming to install if there are substantial quantities of gas and water exiting the borehole. They are rated to withstand approximately 50 psi pressure depending on formation structure and length of the bladder. Importantly, they do not conform to the borehole inner diameter if it is not concentric which can cause the packer to fail when inflated.

- *Pneumatic packers*: These packers also use a rubber bladder element and can be purchased from numerous oil field or water well suppliers. The rubber bladders are inflated with compressed air or nitrogen on location. These packers are a little more complicated to install because a compressed air source is needed and must be plumbed to the pneumatic packer to inflate. The cost of these packers ranges from $1500 to $5000 depending on borehole diameter and pressure. These packers are safe if installed from a remote location but because they are inflated with 300–1000 psi, they can be dangerous if the bladder ruptures. The sudden release of the compressed air can be a violent event. These packers can also damage the coal structure if over inflated. This packer does not conform to the borehole inner diameter if it is not concentric or the diameter of the borehole exceeds the designed inflation diameter of the specific packer.
- *Poly Grout Bag Packer (Figures 13.11 and 13.12)*: TDI and CGR have developed a simple, cost-effective grout bag packer to pack off the intercepted boreholes or coal bed laterals permanently from the mine environment (Figure 13.11). They consist of a 1.0–2.0″ schedule 80 PVC pipe or casing that can be used as a flow-through tube for future water or gel injection. The PVC pipe is wrapped with small polyurethane bags of grout. The number of grout bags used is limited only by the length of PVC pipe utilized. The greater the packer length, the more contact area and the higher resistance the packer will have to the residual formation pressure acting on it. The bags and other components used are relatively inexpensive as a five-foot packer can be made for a few hundred dollars. The biggest advantage to this style of packer is, it will conform to an

FIGURE 13.11   Poly grout bag packer.

FIGURE 13.12   Activated grout bag packer.

eccentric or concentric borehole. These packers have been tested in the laboratory and underground in mined through boreholes. The poly grout bag packers have performed very well with formation pressures to 100 psi. Water actually ran out of the rib rather than forcing the poly grout bag packer out of the borehole or coal bed lateral. These packers are very easy to install and set up very quickly, stopping the flow of residual gas.

Other options available include attempting to intersect the borehole or coal bed lateral with an additional borehole drilled from underground. This can be a hit-or-miss proposition. If successful, it is possible to divert the flow of the gas away from the mined through borehole coal bed lateral. The second option is to attempt to seal the borehole or coal bed lateral through the new borehole that was drilled from underground. Water, cement, or polyurethane could be used to seal the borehole. However, the direction the grout flows cannot be always controlled due to the natural fracture system and other geologic and mining conditions.

13.8.6 Recent Results of Sealing Boreholes and CBM Wells with Reformulated Gel Mixed and Pumped with New Gel Machine

The reformulated gel and "third generation gel machine" has sealed three CBM wells and nine underground boreholes since January 2009. To seal these boreholes and CBM well coal bed laterals, 134,430 gallons of gel have been mixed and pumped. The nine underground boreholes have been successfully mined through without any safety incidents or coal production delays. Seven different coal bed laterals from three CBM wells have been successfully sealed with the reformulated gel and have been mined through with two continuous mining development sections.

13.8.7 DD-2 CBM Well (Figure 13.13)

The bottom 1000′of the right lateral was sealed using the reformulated gel by setting a packer at a measured depth of 4630′ in the right lateral. With the new gel machine, 1836 gallons of the reformulated gel was mixed and pumped through the drill string and through the packer into the final 1000′ of the borehole and pressurized to 400 psi. This lateral was successfully intercepted four times in a 3-entry development without a safety incident finding cured gel. DD-2 is still producing 18 months after the bottom 1000′ of the right lateral was sealed and has produced gas for nearly 48 months. The CBM well could produce for another 10 years.

13.8.8 DD-4 CBM Well

DD-4 was a three lateral CBM well and had produced methane gas for two years before it was sealed in July 2009 with polymer gel. The measured 48 h shut-in pressure leveled out at 23 psi, and produced less than 1 gpm water prior to sealing. Compressed air was pumped down DD-4's vertical production well while a short section of the 7″ casing was milled in the access well. A packer was set in both the vertical well and access well. Through the

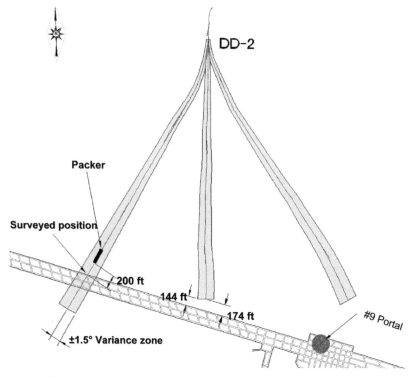

FIGURE 13.13   Reentered DD-2's right lateral and installed packer.

packer sealing all three coal bed laterals, 38,775 gallons of gel was pumped. The right lateral gained 63′ of elevation from beginning to end of the lateral. DD-4's all three coal bed laterals were mined through 19 different times by two different development sections. All but one mine through was successful. After each mine through was completed, pneumatic packers were installed in each lateral to isolate the lateral from the mine environment. When installing and inflating the pneumatic packer, the packer bladder ruptured, causing some coal debris to exit the lateral. This incident triggered the development of the poly grout bag packer. It had been discovered and confirmed underground that the compressed air injected in DD-4's lateral had been absorbed into the coal bed formation and artificially increased the formation shut-in pressure. This caused some difficulties in installing the packers when the mine through of DD-4 occurred.

13.8.9 DD-5 CBM Well

DD-5 was a four lateral CBM well and had produced methane gas for two years before it was sealed in December 2009. The 48 h shut-in pressure was 18 psi and had leveled out and was producing less than 1 gpm water. Compressed air was again pumped down the vertical well to help lift the cuttings from the milling of a short section of the 7″ casing in the access well that was cemented in the coal at the bottom of the curve. As an additional precaution, and before the packers were inflated, the vertical well was left open for three days to bleed the compressed air injected during the milling process, from the coal bed before sealing from the surface. The poly grout bag packer was used on all DD-5 mine intercepts. All four coal bed laterals were intercepted with the 3-entry development section. There have been 18 mine intercepts of four different DD-5 gas producing coal bed laterals. Into DD-5's gas producing coal bed laterals, 37,350 gallons of gel were mixed and pumped. These were all very successful, with minimum production loss and no safety incidents.

13.9 CONCLUSION

Numerous lessons have been learned over the nearly 10 years developing and using the polymer gel to seal in-mine boreholes and surface CBM well coal bed laterals. The most important lesson learned was that each borehole or CBM well is an individual entity and must be evaluated for its own unique parameters to effectively seal or plug with polymer gel. Following new protocol for sealing boreholes or coal bed laterals with polymer gel, these parameters must be evaluated and used to enhance the successful sealing using polymer gel resulting in the safe mine through. The development of the poly grout bag packer used to pack off the mined through in-mine borehole or coal bed lateral plugged with polymer gel has been proven very effective in replacing inflatable or mechanical packers. The poly bag grout packers have been easier and quicker to install, conform to the eccentric or concentric shape of the mined through borehole or coal bed lateral, have worked with residual gas pressures behind the packer of greater than 100 psi, and only cost a few hundred dollars. Perhaps, the polymer gel will not solve all potential technical problems associated with sealing degas boreholes or CBM wells coal bed laterals, however, the polymer gel has proven to be an excellent alternative as it is very effective to seal underground boreholes and CBM well coal bed laterals when the proper protocol is diligently followed. Lastly, the development of the polymer gel and the protocol developed will continue to be evaluated and analyzed to continually improve its performance. Our goals are to enhance the benefit of in-mine and surface CBM well degasification while facilitating the safe mine through for the safety and well-being of the mine workforce.

Acknowledgments

The authors wish to acknowledge the continued support and perseverance of Pennsylvania Services Corporation, Emerald Coal Resources, LP, and Cumberland Coal Resources, LP whose continued support is much appreciated. Special thanks is also given to Paul Henry, Coal Gas Recovery, LP's Alpha NR Degas Coordinator for his devotion to the development of the polymer gel and importantly the countless hours spent underground managing the mine through efforts of the gelled coal bed laterals and in-mine boreholes. Lastly, special thanks are given to Paul Henry and John Wood, TDI for their efforts in designing, testing, and managing the installation of the poly bag grout packers hands-on underground, proven to be more effective packing off the mined through coal bed laterals.

References

Aul, G., Cervik, J., 1979. Grouting Horizontal Drainage Holes in Coal Beds. U.S. Bureau of Mines RI 8843. 16 pp.
Kravits, S.J., Reilly, J., Kirley, J., DuBois, G., June 2006. Cross-linked polymer gel seals horizontal degas boreholes greater than 4000 feet long. In: 11th Annual U.S. Ventilation Symposium. Penn State University.
U.S. Department of Labor, Mine Safety and Health Administration, May 5, 2005. Program Information Bulletin No. P05–10. Ray McKinney, Administrator for Coal Mine Safety and Health.

Economic Analysis of Coalbed Methane Projects

Michael J. Miller[1], Danny A. Watson II[2]

[1]Cardno MM&A (Cardno Ltd.), Kingsport, Tennessee, USA, [2]Formerly
of Cardno MM&A (Cardno, Ltd.) Kingsport, Tennesee, USA, currently of
Gulfport Energy Corporation, Oklahoma City, Oklahoma, USA

14.1 INTRODUCTION

Economic analyses of coalbed methane (CBM) projects allow management teams to make informed decisions regarding financial investments, development, and operations of such projects. This can apply to either existing projects or to properties, which may have potential for future CBM development. Due to the dynamic nature of gas prices and the cyclical patterns of the energy industry, analyses are performed on a routine basis to update the financial outlook for specific projects. This paper provides the basic procedures, which are followed for most such analyses.

14.2 RESERVE CATEGORIES

When evaluating CBM projects, the estimated gas reserves are often defined as belonging to one or more reserve categories, developed by the Society of Petroleum Engineers (SPE), each of which represents a different likelihood of the reserves actually being produced. Capital costs are assigned to undeveloped reserve categories and varying degrees of risk are assigned to each category. The reserve classifications for CBM wells are the same as for other types of oil and natural gas wells.

Proved reserves are estimated quantities that geological and engineering data demonstrate with reasonable certainty to be recoverable in the future from known reservoirs under existing operating and economic conditions. Reserves categorized as *producing reserves* are expected to be recovered from completion intervals, which are open and producing at the time of the estimate. *Nonproducing reserves* include shut-in and behind-pipe reserves. *Shut-in reserves* are expected to be recovered from completion intervals open at the time of the estimate, but either had not started producing, were shut in for market conditions or pipeline connections, were not capable of production for mechanical reasons, or for which the timing when sales will commence is uncertain.

Undeveloped reserves are expected to be recovered from new wells on undrilled acreage or from deepening existing wells to a different reservoir. Undeveloped reserves may also be identified where a relatively large expenditure is required to either recomplete an existing well or install production facilities for primary or improved recovery projects. To be qualified as *proved undeveloped reserves*, reserves on undrilled acreage or improved recovery projects shall be limited to those drilling units offsetting productive units that are reasonably certain of production when drilled.

Probable reserves are less certain than proved reserves and can be estimated with a degree of certainty sufficient to indicate they are more likely to be recovered than not.

Possible reserves are less certain than probable reserves and can be estimated with a low degree of certainty, insufficient to indicate whether they are more likely to be recovered than not.

14.3 PROJECT AREA MAP

Development of one or more project area maps is a common and important first step in the economic analysis. Mineral ownership and lease positions are depicted so that the scope of potential future drilling on the property can be defined. The locations of core holes, from which geologic data have been obtained, and of existing CBM and other oil and gas wells are also indicated. Depiction of pipelines and related gas gathering facilities on or near the project area is also important to the analysis.

14.4 GEOLOGIC ASSESSMENT

Geologic assessment of CBM fields is accomplished by several means. Analyses of coal cores and geophysical logs obtained on or near the property can provide such information as coal thickness, coal density, and gas content. Interpretation is generally enhanced by mapping these data across the project area, as has been done for total coal thickness in Figure 14.1. Individual seam thickness maps and geologic cross sections to indicate the continuity of coal seams across the project area are also commonly prepared. The data are then used to conduct a volumetric analysis

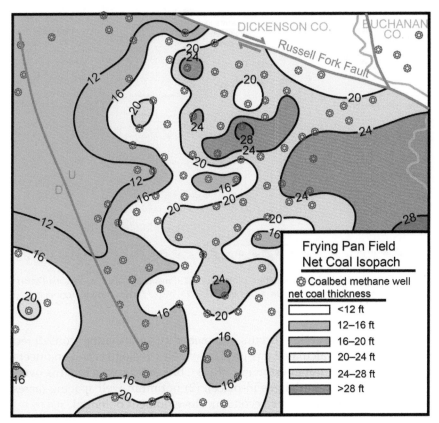

FIGURE 14.1   Total coal seam thickness map. *Cardno MM&A, with data provided by Virginia Department of Mines and Minerals, Division of Gas and Oil.*

to determine gas-in-place for the project area or individual well-spacing units according to the following formula.

$$GIP = Ah\rho G_c$$

where

GIP = gas-in-place, standard cubic feet
A = area, acres
h = coal seam thickness, ft
ρ = coal density, tons per acre-foot
G_c = coal seam gas content, standard cubic feet per ton

Permeability can be determined from analysis of core samples or, more reliably, from pressure transient analysis (e.g. injection-falloff testing) of coal seams. Geologic mapping of coal-seam structure based on coal core and geophysical log data often can provide an indication of where the

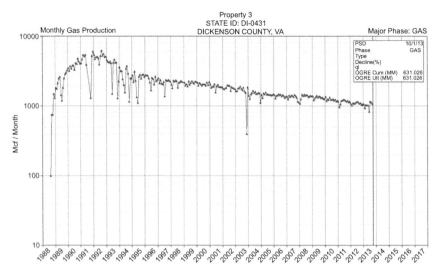

FIGURE 14.2   Historical production plot—DI-0431. *Production data from Virginia Department of Mines and Minerals, Division of Gas and Oil, 2013,* http://www.dmme.virginia.gov/d ivisiongasoil.shtml.

higher coal permeability occurs. Permeability information considered together with the calculated gas-in-place enables geologists and engineers to make informed initial estimates of (1) CBM recovery factors (recoverable fraction of the initial gas-in-place), (2) optimum well spacing (acres per well), and (3) project-area economic development potential.

14.5   FORECASTING FUTURE PRODUCTION

When available, historical production data are used to perform decline-curve analysis to estimate future production from CBM wells. The data are plotted on a semilogarithmic plot of production vs time. Figure 14.2 shows an example of this plot from well DI-0431 located in Dickenson County, Virginia.

The shape of the production profile for DI-0431 is typical of many CBM wells. Production increases throughout the dewatering phase, due to decreasing bottom-hole pressure and increasing relative permeability to gas, until reaching a peak production rate, followed by hyperbolic decline (represented by the curved section on the graph). Production eventually enters exponential decline (represented by the linear section on the graph) for the remainder of the life of the well. Future production is forecasted by fitting these historical production data with a trend line due to the relative predictability of natural gas production. Figure 14.3 shows DI-0431 with forecasted future production, which is indicated for a remaining well life of 40 years.

Type curves are developed from plotting historical production data from multiple wells to forecast future production for wells with little or no data,

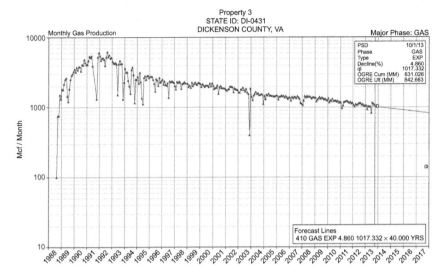

FIGURE 14.3   Decline-curve analysis—DI-0431. *Production data from Virginia Department of Mines and Minerals, Division of Gas and Oil, 2013,* http://www.dmme.virginia.gov/divisi ongasoil.shtml.

such as recently-drilled wells or undeveloped locations. A type curve is a representative production profile typical of wells producing in a given region. The profile is developed by gathering and plotting production data from a selection of analogous wells (on or near the project area) and then determining the average production volumes for the first month, second month, and so on. A trend line is fitted through the data and the resulting production profile is used to aide in forecasting production for wells with little or no historical production data. When forecasting individual wells, the type curves may be adjusted higher or lower, depending on the calculated gas-in-place from geologic assessment and calculated percentages of the gas-in-place, which is recoverable from the analogous wells. An example of a type curve developed from 398 CBM wells located in a portion of the Nora Field in Dickenson County, Virginia is shown below in Figure 14.4.

14.6 ECONOMIC EVALUATION MODEL

The economic model utilized to evaluate a CBM project incorporates the (1) gas-in-place derived from the geologic assessment, (2) forecasts of future gas production from existing and/or scheduled future wells, (3) schedule of future drilling and development, and (4) known or assumed economic parameters. Sophisticated reserve and economic evaluation software programs are often applied to efficiently perform the cash-flow assessment. These programs allow the forecasting of future production volumes (as discussed in previous sections) to be done by either manually fitting the established

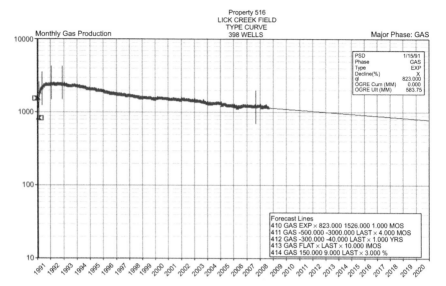

FIGURE 14.4 Production-type curve. *Cardno MM&A with production data from Virginia Department of Mines and Minerals, Division of Gas and Oil, 2013,* http://www.dmme.virginia.gov/divisiongasoil.shtml.

production trend for producing wells or allowing software to do so automatically. Proprietary OGRE® Systems software was utilized in preparing all graphical presentations and economic outputs included in this chapter.

The results of the decline-curve analysis are integrated into the economic model to generate cash flows on well-by-well, reserve category, and total project bases. Production from undeveloped well locations is forecasted based upon volumetric analysis of recoverable gas in the well units, the production-type-curve profile, and the scheduled turn-on-line dates of the undeveloped wells.

Economic parameters, which must be known or reasonably assumed in the economic model, are as follows:

Working interest: the percentage ownership in each well and, accordingly, the percentage of capital and operating costs which must be paid by an owner, which may also include the project operator.
Net revenue interest: the percentage of total gas sales revenues that will be received by the specific owner, a percentage that is typically reduced to less than the working-interest percentage by the royalty percentage paid to the mineral-estate owner.
Gas shrinkage: the expected percentage reduction of the produced gas volumes due to compressor and gas processing fuel, third-party pipeline retainage, and other losses.
Energy-content adjustment: the factor applied to convert from the produced volumetric gas volume, typically in thousands of cubic feet

(Mcf), to an energy-content amount, typically in dekatherms (dth), since most gas is sold on a dekatherm basis.

Gas sales price: the forecasted future gas prices.

Capital costs: the costs to drill, complete, and connect new wells to pipeline sales, and for installation of pipelines, gas compression, salt-water disposal, and other facilities costs.

Operating expenses: the per-well or per-volume-unit cost to operate CBM wells and other infrastructure, which includes employee labor, transportation, fuel, repair and maintenance, salt-water disposal, gas measurement, and other costs.

Production taxes: the severance and added-value taxes imposed by states, counties, and other jurisdictions, usually on a gas-volume basis.

Escalation rates: the assumed annual escalation percentages commonly applied to gas sales prices, capital costs, and operating costs.

Discount rate: the annual rate at which future net revenues are discounted to determine their present value.

Effective date: the date to which cash flows are discounted, usually near the date of the evaluation.

The assumption of future gas sales prices is a critical component in the economic evaluation of a property. When evaluating reserves under U.S. Securities and Exchange Commission guidelines, gas sales price is calculated as the average first-day-of-month price for the last 12 months prior to the end of the reporting period. This type of gas price scenario is referred to as "flat pricing" since the price is assumed to remain constant throughout the life of the project. A more common method, when values are determined for proposed property acquisitions and sales or for business planning purposes, is to apply the natural gas price futures published by the New York Mercantile Exchange (NYMEX). They are updated on a daily basis according to results of futures trading. The NYMEX prices are adjusted in the economic model by premiums and discounts as appropriate, depending on the proximity of the project to its gas market. When the gas from a CBM project is committed by contract to be sold at a specific price for a period of time, the contract price is of course applied in the economic analysis. An example of NYMEX-based gas price determination, including energy-content and market-basis adjustments is presented below in Table 14.1.

The schedule of future drilling and development is a particularly key component of the economic model for early-stage CBM projects. The most prospective well locations normally are assumed to be drilled first with Proved undeveloped locations followed by Probable locations and then Possible locations. Gas reserves normally are not assigned to undeveloped locations unless the expected rate-of-return (unrisked) exceeds a specified level, typically between 15% and 25%. The rate at which future drilling occurs depends on a number of factors, including the economic ranking of proposed drilling vs other capital opportunities, ability to obtain required

TABLE 14.1   NYMEX-Based Gas Price

Item	Units	Year 1	Year 2	Year 3	Year 4
NYMEX price	$/dth	$4.30	$4.18	$4.13	$4.15
Basis adjustment	$/dth	$0.06	$0.06	$0.06	$0.06
Adjusted price	$/dth	$4.36	$4.24	$4.19	$4.21
Btu adjustment	dth/Mcf	0.985	0.985	0.985	0.985
Net sales price	$/Mcf	$4.29	$4.18	$4.13	$4.15

NYMEX, New York Mercantile Exchange.

drilling and construction permits, availability of drilling rigs and other necessary services, available pipeline capacity, and confidence in the level of future gas prices. Calculated CBM project values normally are higher when an aggressive development schedule is assumed, which forecasts more gas sales in the early project years when the discounting of forecasted revenues is less.

14.7 ECONOMIC OUTPUT

The gross production of the wells is reduced by royalties and gas shrinkage to determine the net production, which is then multiplied by the gas sales price to determine net revenue. Net revenue is then reduced by production taxes and operating expenses to determine net operating income. Capital investments are deducted from net operating income to determine net before-income-tax (BTAX) cash flow and future cash flow is then discounted to estimate the present value. Projects may also be evaluated on an after-income-tax basis, which includes application of other factors, including determination of tangible vs intangible costs, depletion allowances, depreciation schedules, and the applicable federal and state income tax rates. An example BTAX economic output file for an undeveloped CBM well location is provided in Appendix A.

Economic factors that are determined by the model include the percentage rate-of-return on investment, the time to project payout (when cumulative cash flow is positive), the net income to investment ratio, and the total present value.

14.8 PROJECT RISK

A degree of risk is usually associated with both existing CBM wells and undeveloped well locations. A low amount of risk is associated

TABLE 14.2 Example Project Reserve Summary

Reserve Classification	Net Remaining Reserves (MMcf)	BTAX Cash Flow ($)	Unrisked BTAX Present Value ($)	Risk Adjustment (%)	Risked BTAX Present Value ($)
PDP	2400	5,190,000	1,850,000	95	1,757,500
PDNP	120	280,000	140,000	80	112,000
PUD	9600	14,500,000	4,280,000	60	2,568,000
PROB	19,200	31,500,000	7,560,000	35	2,646,000
POSS	24,000	42,400,000	8,480,000	10	848,000
Totals	55,320	93,870,000	22,310,000		7,931,500

PDP, proved producing reserves; PDNP, proved nonproducing reserves; PUD, proved undeveloped reserves; PROB, probable reserves; POSS, possible reserves, BTAX, before-income-tax.

with producing wells, as their gas production capabilities have normally been defined. However, higher degrees of risk are usually associated with nonproducing wells and undeveloped locations due to the absence of verifying production data, geologic uncertainty, and mechanical risk. As discussed previously, the estimated gas reserves of wells and prospective well locations are classified according to their inherent level of risk. An example project reserve summary is provided in Table 14.2, in which BTAX cash flow was discounted at 10% per year to obtain the unrisked present value, which was then risk adjusted by reserve category. The reserves can also be risked by applying various discount factors.

14.9 SUMMARY

Economic analyses of CBM projects, like analyses of other oil and gas production projects, can be very complex. There are many variables that affect the estimated value. For new project areas with little geologic data and few or no producing wells, the task can be very daunting and requires good professional judgment and reliance upon the best available analogies to other CBM projects. Even for projects where considerable historical development has already occurred and the geology and production characteristics are well defined, the uncertainties (market-based and political) surrounding future gas sales prices can lead to a significant range of possible future cash flows. For that reason, sensitivity analyses based on a range of gas-price forecasts are often conducted.

14.10 APPENDIX A: ECONOMIC OUTPUT FILE

RESERVES AND ECONOMICS

AS OF JANUARY 1, 2014

END-MO-YR	GROSS PRODUCTION OIL, MBBL	GAS, MMCF	NET PRODUCTION OIL, MBBL	GAS, MMCF	PRICES OIL $/B	GAS $/M	OPERATIONS, M$ NET OPER REVENUES	SEV+ADV TAXES	NET OPER EXPENSES	CAPITAL COSTS, M$	CASH FLOW BTAX, M$	10.00 PCT CUM. DISC BTAX, M$
12-14	.000	25.862	.000	21.271	.00	4.29	91.350	.000	27.931	325.000	-261.581	-264.532
12-15	.000	33.800	.000	27.801	.00	4.18	116.108	.000	29.478	.000	86.630	-189.442
12-16	.000	36.107	.000	29.698	.00	4.13	122.568	.000	28.771	.000	93.797	-115.531
12-17	.000	34.358	.000	28.259	.00	4.15	117.186	.000	28.418	.000	88.768	-51.942
12-18	.000	27.944	.000	22.984	.00	4.23	97.122	.000	25.515	.000	71.607	-5.310
12-19	.000	23.024	.000	18.937	.00	4.34	82.259	.000	23.309	.000	58.950	29.590
12-20	.000	19.537	.000	16.069	.00	4.50	72.334	.000	21.812	.000	50.522	56.781
12-21	.000	16.942	.000	13.935	.00	4.64	64.649	.000	20.758	.000	43.891	78.256
12-22	.000	14.937	.000	12.286	.00	4.76	58.451	.000	19.999	.000	38.452	95.359
12-23	.000	13.343	.000	10.975	.00	4.89	53.619	.000	19.446	.000	34.173	109.177
12-24	.000	12.046	.000	9.908	.00	5.03	49.870	.000	19.044	.000	30.826	120.509
12-25	.000	10.972	.000	9.024	.00	5.21	47.021	.000	18.757	.000	28.264	129.954
12-26	.000	10.068	.000	8.281	.00	5.43	44.944	.000	18.195	.000	26.749	138.080
12-27	.000	9.297	.000	7.647	.00	5.43	41.503	.000	17.716	.000	23.787	144.650
12-28	.000	8.631	.000	7.099	.00	5.43	38.529	.000	17.302	.000	21.227	149.980
12-29	.000	8.053	.000	6.624	.00	5.43	35.951	.000	16.942	.000	19.009	154.319
12-30	.000	7.543	.000	6.204	.00	5.43	33.671	.000	16.625	.000	17.046	157.856
12-31	.000	7.092	.000	5.833	.00	5.43	31.658	.000	16.345	.000	15.313	160.745
12-32	.000	6.691	.000	5.503	.00	5.43	29.867	.000	16.096	.000	13.771	163.107
12-33	.000	6.330	.000	5.206	.00	5.43	28.255	.000	15.871	.000	12.384	165.038
12-34	.000	6.006	.000	4.940	.00	5.43	26.811	.000	15.670	.000	11.141	166.617
12-35	.000	5.705	.000	4.692	.00	5.43	25.465	.000	15.483	.000	9.982	167.903
12-36	.000	5.420	.000	4.458	.00	5.43	24.195	.000	15.306	.000	8.889	168.944
12-37	.000	5.149	.000	4.235	.00	5.43	22.985	.000	15.137	.000	7.848	169.780
12-38	.000	4.891	.000	4.023	.00	5.43	21.834	.000	14.977	.000	6.857	170.444
12-39	.000	4.647	.000	3.822	.00	5.43	20.743	.000	14.825	.000	5.918	170.965
12-40	.000	4.414	.000	3.631	.00	5.43	19.707	.000	14.680	.000	5.027	171.367
12-41	.000	4.194	.000	3.450	.00	5.43	18.724	.000	14.543	.000	4.181	171.671
12-42	.000	3.984	.000	3.277	.00	5.43	17.785	.000	14.413	.000	3.372	171.894
12-43	.000	3.785	.000	3.113	.00	5.43	16.895	.000	14.289	.000	2.606	172.051

12-44	.000	3.596	.000	2.958	.00	5.43	16.054	.000	.000	14.172	1.882	172.154
12-45	.000	3.416	.000	2.810	.00	5.43	15.251	.000	.000	14.060	1.191	172.213
12-46	.000	3.245	.000	2.669	.00	5.43	14.486	.000	.000	13.953	.533	172.237
12-47	.000											
12-48	.000											
S TOT	.000	391.029	.000	321.622	.00	4.72	1517.850	.000	.000	609.838	325.000	172.237
REM.	.000	.000	.000	.000	.00	.00	.000	.000	.000	.000	.000	172.237
TOTAL	.000	391.029	.000	321.622	.00	4.72	1517.850	.000	.000	609.838	325.000	172.237
CUM.	.000	.000										
ULT.	.000	391.029										

	CUM.	ULT.
NET OIL REVENUES (M$)	.000	.000
NET GAS REVENUES (M$)	1517.850	1517.850
TOTAL REVENUES (M$)	1517.850	1517.850

PROJECT LIFE (YEARS)		33.000
DISCOUNT RATE (PCT)		10.000
GROSS OIL WELLS		.000
GROSS GAS WELLS		1.000
GROSS WELLS		1.000

--------PRESENT WORTH PROFILE--------

DISC RATE	PW OF NET BTAX, M$	DISC RATE	PW OF NET BTAX, M$
.0	583.012	30.0	-62.072
2.0	456.125	35.0	-89.121
5.0	318.852	40.0	-110.920
8.0	222.047	45.0	-128.860
10.0	172.237	50.0	-143.876
12.0	130.835	60.0	-167.585
15.0	80.444	70.0	-185.444
18.0	40.316	80.0	-199.366
20.0	17.825	90.0	-210.515
25.0	-27.603	100.0	-219.641

BTAX RATE OF RETURN (PCT)	21.96
BTAX PAYOUT	11/30/2017
BTAX PAYOUT (DISC)	02/26/2019
BTAX NET INCOME/INVEST	2.79
BTAX NET INCOME/INVEST (DISC)	1.53

References

OGRE® Systems, Oil and gas reserves evaluation software.

Society of Petroleum Engineers (SPE).

Virginia Department of Mines and Minerals, Division of Gas and Oil, 2013, http://www.dmme.virginia.gov/divisiongasoil.shtml.

Legal Issues Associated with Coalbed Methane Development

Sharon O. Flanery[1], Ryan J. Morgan

Steptoe & Johnson PLLC, Charleston WV, USA

15.1 INTRODUCTION

Historically, coalbed methane was considered to be a danger to coal miners and a hindrance to coal mining in general. It was not until the latter part of the last century with emerging technology that coalbed methane began to be viewed as a valuable resource that could be developed independently. This revelation, and the resulting commercial production of the mineral, has raised several legal issues not present with the normal development of minerals such as coal or natural gas. These issues have required statutory guidance, judicial interpretation, and greater attention to written agreements among the parties seeking to develop the methane.

Foremost among these legal issues is the uncertain ownership, and resulting right to develop, the methane gas located within the coal seam. This answer to this question has been answered differently in various coal-producing jurisdictions throughout the United States and remains an unresolved issue in many jurisdictions. Some states have developed statutory guidelines to follow in developing coalbed methane. These statutes govern permitting requirements, as well as the pooling process, which allows for the development of the resource in the absence of control of 100% of the minerals in a given area. In other states the courts have determined who owns coalbed methane, and consequently who has the right to develop it.

---

[1] The author would like to thank Erin F. Anderson, Jonathan R. Ellis, Andrew Graham, Daniel B. Kostrub, Timothy M. McKeen, and Kathy Milenkovski for their contributions to this chapter.

Other legal issues raised by the development of coalbed methane include the ability to mine-through coalbed methane wells, as well as the determination of which party is responsible and liable for the cost of plugging the well and the loss of any coal reserves from the drilling or stimulation of such well or the plugging of such well. In addition, there is a potential loss of coalbed methane that may be caused by the mining activities. Because many coalbed methane wells are drilled using relatively new horizontal techniques, there are still uncertainties as to the viability of safely plugging those wells for subsequent mine-through. Finally, on the environmental front, new national and international regulatory schemes are being proposed that would limit greenhouse gas emissions in an effort to reduce global warming.

15.2 THEORIES OF OWNERSHIP

In the area of property law, there is always the temptation to extract a bright-line rule from a case that hinged on a particular factual scenario, a case that might have yielded a different result with a slightly different set of facts. Courts have avoided making blanket declarations that coalbed methane is a part of the coal estate, a part of the gas estate, or a separate mineral estate altogether. Instead, the cases that have presented the issue of ownership have been decided on a case-by-case basis by interpreting specific language of specific legal documents, typically a deed or a lease, in an attempt to ascertain whether the parties to such documents intended to grant or reserve the right to produce coalbed methane. Courts have also considered other factors, including dictionary definitions, safe mining operations, the economic value of coalbed methane, the chemical makeup of coalbed methane, the physical relationship between coal and coalbed methane, and the legal theories relating to the ownership of oil and gas, in reaching their decisions. As a result, the ownership cases cannot be easily divided into categories, although some discussions of these cases refer to a dichotomy between coal owners and gas owners or Eastern states and Western states. Despite this legal murkiness, a few general principles have emerged from the cases.

Some courts have held that the coalbed methane is part of the coal estate so long as the coalbed methane is adsorbed or present in the coal strata[2]; however, these same courts have also held that, once the coalbed methane escapes the coal strata, it may be produced by the owner of the gas estate. Thus, under this holding, gas produced from gob wells, which are wells

---

[2] *U.S. Steel Corp. v. Hoge*, 468 A.2d 1380 (Pa. 1983); *NCNB Texas Nat'l Bank v. West*, 631 So.2d 212 (Ala. 1993); *Continental Resources of Ill., Inc. v. Illinois Methane, LLC*, 847 N.E.2d 897 (Ill. App. Ct. 2006).

drilled into the gob area and not the coal seam, would belong to the gas owner even though it is produced as a result of longwall mining operations. Other courts have held that, as a general proposition, the owner of the gas estate has the exclusive right to drill for and produce coalbed methane, even if the coalbed methane remains adsorbed to the coal in the coal seam.[3] These courts have also held that the owner of the coal estate has a coequal right to vent the coalbed methane away from the coal seam in preparation for mining. Finally, some states have examined the issue but refused to answer the ultimate question of ownership—instead, the courts have adopted a process of elimination ruling in a particular case that a given party does not have the right to develop the coalbed methane without addressing who does have that right.

15.2.1 Coalbed Methane is Part of the Coal Estate

The earliest of the coalbed methane ownership cases is *U.S. Steel Corp. v. Hoge*, decided by the Supreme Court of Pennsylvania in 1983. In *Hoge*, the court held that a deed, executed in 1920, granting all of the coal of a certain stratum together with certain mining rights, but reserving to the grantor the right to drill and operate through the coal for oil and gas, did not reserve the coalbed methane as a part of the gas estate. Rather, the coalbed methane belongs to the owner of the strata in which it may be found. As a result, coalbed methane that is still contained within the coal seam belongs to the coal owner, whereas coalbed methane that has escaped from the coal seam (e.g., gob gas) becomes part of the gas estate. While the *Hoge* court gave consideration to the physical characteristics of coalbed methane and the history of its development as a mineral, the court focused on the dangers presented to the coal mining industry by the presence of coalbed methane in coal seams. The *Hoge* court went so far as to state in its holding that it "strain[ed] credulity to think that the grantor intended to reserve the right to extract a valueless waste product with the attendant potential responsibility for damages resulting from its dangerous nature".[4] This

---

[3] *Carbon County v. Union Reserve Coal Co.*, 898 P.2d 680 (Mont. 1995); *Harrison–Wyatt, LLC v. Ratliff*, 593 S.E.2d 234 (Va. 2004); *Newman v. RAG Wyoming Land Co.*, 53 P.3d 540 (Wyo. 2002).

[4] The *Hoge* decision greatly influenced the Supreme Court of Alabama in deciding another early ownership case, *NCNB Texas Nat'l Bank v. West*, 631 So.2d 212 (Ala. 1993). In a more recent decision, the Illinois Court of Appeals held, in *Continental Resources of Ill., Inc. v. Illinois Methane, LLC*, 847 N.E.2d 897 (Ill. App. Ct. 2006), that coalbed methane gas found in coal seams or in mine voids is owned by the coal estate owner. In its decision, the court, much like the Supreme Court of Pennsylvania in *Hoge*, gave special attention to the fact that "[h]istorically, coalbed methane was considered a 'dangerous waste product of coal mining.'"

logic was subsequently adopted by the highest court in Alabama, an intermediate appeals court in Illinois, as well as a trial court in Kentucky.[5]

In 1993, the Supreme Court of Alabama twice considered whether certain conveyances included the right to produce coalbed methane. In the first case, the court held that coalbed methane was included in a grant of "all the coal and other minerals", where such language appeared in mineral leases executed in 1898 and 1902, respectively.[6] The court specifically held that "an express grant of 'all coal' necessarily implies the grant of coalbed methane gas, unless the language of the grant itself prevents this construction."[7] In this first decision, however, the court did not address the question of whether coal and coalbed methane could be treated as separate mineral estates under Alabama law. Nevertheless, the court addressed that very same issue later in 1993, when it held that a reservation of "all gas" includes the right to coalbed methane that migrates out of the coal seam, but not coalbed methane within the coal seam. Further, the court held that, absent the clear intent of the parties, a reservation of "coalbed methane gas" does not include coalbed methane contained within the coal seam, and that, consequently, the right to produce such coalbed methane remains with the coal owner.[8]

More recently, lower state courts have reached the same conclusion, although such lower court decisions do not generally have the precedential value as a state's highest court. In 2006, the Appellate Court of Illinois held that oil and gas leases executed in the 1980s, which granted "the right to produce from those lands oil, all gases, liquid hydrocarbons, and their constituent products", did not include the right to produce coalbed methane found in coal seams or mine voids.[9] Instead, coalbed methane in coal seams or mine voids is controlled by the coal owner. Likewise, in 2009, a Kentucky state trial court concluded that a deed executed in 1924 that conveyed "[a]ll the veins and beds of coal in and underlying [the subject property]" also conveyed the right to produce coalbed methane from within those coal seams, although the court recognized that, under Kentucky's rule of capture, the gas owner has the right to produce any coalbed methane that migrates out of the coal seam.[10] As discussed in

---

[5] In general, the opinions of the highest court in each state have precedential value and are binding on the other courts within the state while, in general, the opinions of a lower court do not have precedential value and are not binding on other courts.

[6] *Vines v. McKenzie Methane Corp.*, 619 So.2d 1305 (Ala. 1993).

[7] *Id.*

[8] *NCNB Texas Nat'l Bank*, 631 So.2d at 212.

[9] *Cont'l Resources of Ill.*, 847 N.E.2d at 897.

[10] *Bowles v. Hopkins County Coal, LLC*, Civil Action No. 06-CI-395 (Hopkins Cir Ct. 4th Judicial Cir.).

Section 15.2.2 below, however, a Kentucky federal district court in an earlier decision reached a different conclusion as to which party controls the right to produce coalbed methane from within the coal seam, making the issue of coalbed methane ownership in Kentucky even more clouded.

15.2.2 Coalbed Methane is Part of the Gas Estate

Not long after the Alabama decisions, a split of authority emerged when, in 1995, the Supreme Court of Montana held, in interpreting a coal severance deed executed in 1974, that coalbed methane is not a constituent part of the coal estate and further held that the owner of the gas estate, through its lessee, has the right to drill for and produce coalbed methane, even though the owner of the coal estate has a "mutual, simultaneous right to extract and to capture such gas for safety purposes, incident to its actual coal mining operations".[11] The court did not decide how, or even if, the gas owner would be compensated by the coal owner for the gas captured for mine safety purposes. Unlike the Supreme Court of Pennsylvania, the Supreme Court of Montana relied heavily on definitions of the terms "coal", "mineable coal", and "gas" from sources such as the American Heritage Dictionary and the Montana Code. The Supreme Court of Montana also gave special attention to the fact that the coal severance deed provided for a per ton royalty for the mined coal, but made no provision for a royalty on the coalbed methane. Just as the *Hoge* court found it unbelievable that anyone would reserve the coalbed methane in a coal severance deed, the Supreme Court of Montana found it unbelievable, or at least an "unreasonable assumption", that anyone would convey away the coalbed methane without receiving some kind of compensation.

At least one federal district court has held that the gas owner owns the coalbed methane where a deed contained the following language: "[granting] all the oil, gas, and all other minerals and mineral rights of every kind and character except the coal and coal rights". The United States District Court for the Western District of Kentucky held that the grant included the coalbed methane because, under Kentucky law, the use of "minerals and mineral rights of every kind and character" means that the grantor intended to grant "all organic and inorganic substances that can be taken from the Earth".[12] The court further relied on the only United States Supreme Court decision to address coalbed methane ownership, which occurred in a 1999 opinion involving coalbed methane

---

[11] *Carbon County*, 898 P.2d at 680.

[12] *Michael F. Geiger, LLC v. U.S.*, 456 F. Supp. 2d 885 (W.D. Ky. 2006). The court's conclusions regarding Kentucky law are at odds with the conclusions of the Kentucky trial court in *Bowles*, discussed hereinabove.

under federal land patents in Utah, *Amoco Prod. Co. v. So. Ute Tribe*.[13] As discussed in more detail in Section 15.2.3, the *Amoco* court held that at the time of the land grant, "coal" meant solid coal. Likewise the Western District of Kentucky court held "coal" meant "solid fuel resources", and could not be interpreted to include the coalbed methane.[14] As a result, the court held that the gas owner controls the production of coalbed methane, even if it remains within the coal seam. The court noted that the coal owner retains the right to capture and vent coalbed methane as part of its mining operations, although it can only vent the coalbed methane and it is prohibited from marketing the captured coalbed methane.

15.2.3 Negative Rulings—Rulings on Who Does Not Own Coalbed Methane

The courts in some states have begun a trend of what can be called "negative rulings"—a ruling that examines the relevant documents and finds that a particular party is not the owner of coalbed methane without making a clear decision among the remaining possible owners as to who in fact owns the coalbed methane. In these states, while the ownership issue has been pursued in the judicial system, the courts have resisted issuing opinions that would clearly resolve the ultimate question of who owns coalbed methane.

In a 2004 decision, the Supreme Court of Virginia held certain deeds executed by the fee owner (i.e., the owner of the surface and all the minerals) in 1887 that conveyed "all the coal", did not convey the coalbed methane.[15] Much like the Supreme Court of Montana, the Supreme Court of Virginia relied upon definitions in its analysis; however, instead of using modern-day definitions as found in dictionaries or statutes, the Supreme Court of Virginia considered the definition of "coal" as it appeared in *The American Cyclopedia: A Popular Dictionary of General Knowledge*, published in 1873, and the Ninth Edition of *Encyclopedia Britannica*, published in 1877.[16] The court did not believe that the term "coal", as used in the latter part of the nineteenth century, could mean anything other than "a solid rock substance used as fuel", and that, as a result, coalbed methane could not be included in a grant of coal in that time period.

---

[13] *Amoco Prod. Co. v. So. Ute Tribe*, 526 U.S. 865, 119 S.Ct. 1719, 144 L.Ed.2d 22 (1999).

[14] *Michael F. Geiger, LLC*, 456 F. Supp. 2d at 888.

[15] *Harrison–Wyatt*, 593 S.E.2d at 234.

[16] Note that the Supreme Court of Virginia considered these sources because they were cited in the coal owner's brief.

The analysis used by the Supreme Court of Virginia was heavily influenced by the 1999 United States Supreme Court opinion in *Amoco*, where the court interpreted certain land patents from which the United States had reserved "all coal".[17] Much like the Supreme Court of Virginia, the United States Supreme Court turned to dictionaries contemporary to the patents and determined that the term "coal" as used in land patents granted in the early twentieth century referred only to the solid mineral, and did not include coalbed methane.

Like the Supreme Courts of Montana and Virginia and the United States Supreme Court, the Supreme Court of Kansas, in a more recent case, held that certain coal severance deeds, executed between 1924 and 1926, that conveyed "all coal without reference to quality or quantity, ... together with the right to mine and remove the same", did not convey the right to produce coalbed methane, but instead conveyed only "the solid mineral coal".[18] Similar to other courts that have considered the issue, the Supreme Court of Kansas highlighted the dangerous nature of coalbed methane, as well as the commercial viability, and relative economic value, of coalbed methane production at the time of the coal severance deeds. But, unlike the Supreme Court of Pennsylvania, the Supreme Court of Kansas found that all of these factors weighed against the coal owner's right to produce coalbed methane, even though the court recognized that the coal owner would have the right to ventilate coalbed methane from a coal seam as part of its mining operations. This right, however, is not the same as ownership, and the Supreme Court of Kansas specifically noted that it could not discover in the language of the coal severance deeds any intention on the part of the grantors to permit the coal owner to produce coalbed methane without "exercising the right to mine and remove the coal".

The West Virginia Supreme Court of Appeals avoided the final ownership issue in a 2003 case by making the determination that the parties to an oil and gas lease executed in 1986, before commercially viable operations for producing coalbed methane were known in the leasing area, did not intend to grant the right to produce coalbed methane.[19] The Court of Appeals of Indiana took a similar position in a more recent case, when it held that the parties to a 1976 oil and gas lease would not have contemplated that such a lease included the right to produce coalbed methane.[20]

---

[17] *Amoco Prod. Co.*, 526 U.S. at 865.

[18] *Central Natural Resources, Inc. v. Davis Operating Co.*, 201 P.3d 680 (Kan. 2009).

[19] *Energy Dev. Corp. v. Moss*, 591 S.E.2d 135 (W. Va. 2003).

[20] *Cimarron Oil Corp. v. Howard Energy Corp.*, 909 N.E.2d 1115 (Ind. Ct. App. 2009).

15.2.4 Ownership is Case and Fact Specific—the Wyoming Example

It is not uncommon for courts to hold that it is necessary to review the particular facts in reaching a decision, and under this analysis no clear rule can be provided. Wyoming is perhaps the best example of this fact-specific, case-by-case nature of court decisions relating to the ownership of coalbed methane. Since 2002, the Supreme Court of Wyoming has considered the ownership issue four times, with varying results in three of the four cases.[21] In its earliest decision, in 2002, the court held that a 1974 grant of the surface and "coal and minerals commingled with [the] coal", that also reserved the "oil, gas, and other minerals" did not convey coalbed methane as a part of the coal, for reasons similar to the reasons given by the Supreme Court of Montana as set forth above. In its second case, in 2003, which presented a similar grant from roughly the same period in time, the Supreme Court of Wyoming reached the same conclusion. But, in its third decision, in 2004, the Supreme Court of Wyoming was presented with a different set of facts, namely that in a deed executed in 1975, the grantor conveyed "all of grantor's undivided interest in and to the coal…together with all of grantor's undivided interest in and to all other minerals, metallic or nonmetallic, contained in or associated with the deposits of coal…, subject to the reserved royalty hereinafter provided". This language was sufficiently different from the grants presented in the first two cases for the Supreme Court of Wyoming to reach a different conclusion from its first two ownership cases. The court gave special emphasis to the phrase "all other minerals, metallic, or nonmetallic" and found it to be critical in its conclusion that the grantors had intended to reserve no mineral interest of any kind, thus conveying the coalbed methane with the coal. The fourth case, in 2005, presented a twist on the now familiar fact pattern. Here, the Supreme Court of Wyoming affirmed a trial court ruling that the term "oil rights" as used in deeds executed in the 1940s did not encompass the right to produce gas, including coalbed methane.

Even though the various courts that have addressed the issue of coalbed methane ownership have analyzed any number of factors in reaching their respective decisions (e.g., the chemical characteristics of coalbed methane, the adsorption of coalbed methane into coal seams, the dangers coalbed methane pose to coal mining operations), the single most important factors appear to be the time period of the grant or reservation of the coal or the gas and the intention of the parties. When addressing this issue, courts have relied upon traditional approaches to deed or lease

---

[21] *Mullinix LLC v. HKB Royalty Trust*, 126 P.3d 909 (Wyo. 2006); *Caballo Coal Co. v. Fidelity Exploration & Prod. Co.*, 84 P.3d 311 (Wyo. 2004); *McGee v. Caballo Coal Co.*, 69 P.3d 908 (Wyo. 2003); *Newman*, 53 P.3d at 540.

interpretation to reach their conclusions and have resisted the temptation, as one West Virginia case puts it, "to wave a wand and declare coalbed methane to be either 'coal' or 'gas'".[22]

15.3 THE RESPONSE TO UNCERTAINTY—ISSUES AND DEVELOPMENT IN VARIOUS STATES

15.3.1 Background of Permitting and Pooling

Historically, coalbed methane was considered a nuisance and a hindrance to coal mining. For many years, and in some places to this day, coalbed methane is vented from the coal seams ahead of mining operations. In West Virginia, a large coal-producing state, it is public policy that venting coalbed methane is an approved means of making the mining of coal safer.[23] Ventilation of coalbed methane from the coal mines "was accomplished by the use of large fans and wells to discharge the coalbed methane (CBM) into the atmosphere".[24] During the latter part of the last century, it became evident that coalbed methane could be a valuable resource. Since that realization, some companies have begun to capture and sell the coalbed methane instead of venting it into the atmosphere. Due to advancements in drilling techniques over the past several years, companies are now drilling for, and economically producing, vast quantities of this once valueless commodity.

Due to the increased interest in producing and selling coalbed methane, a few states have developed specific statutory schemes for companies to follow,[25] or as in Virginia, which does not have a specific coalbed methane statute, the regulation can be found in the Virginia Gas and Oil Act.[26] These acts and statutes set forth detailed, yet somewhat convoluted, procedures that outline the necessary requirements to obtain a permit to drill coalbed methane wells. Some statutes and acts also set forth standards to follow if coalbed methane wells are drilled and require pooling.

15.3.2 The Permitting Process

The process to permit coalbed methane wells is an arduous task. Initially, a company will need to have available coal reserves in which to drill the wells. Geologists will review coal formations and mine plans to place

---

[22] *Moss*, 591 S.E.2d at 143.

[23] W. VA. CODE § 22-21-1.

[24] *Harrison–Wyatt, LLC*, 593 S.E.2d at 234.

[25] KY. REV. STAT. ANN. § 349.00, *et seq.*; W. VA. CODE § 22-21-1, *et seq.*

[26] VA. CODE § 45.1-361.1, *et seq.*

potential coalbed methane wells in the most advantageous arrangement. A large amount of the efforts are in the planning and laying out of well units. Once a company has an idea of where the wells will be located in relation to the coal reserves, the next consideration is the topography of the land. The importance of topography is whether there is sufficient area to drill a horizontal well or if the only option is a vertical well. Once these preliminary steps are complete, it is necessary to obtain title examinations, which will provide information regarding the owners of the various mineral estates. This is necessary due to the notice requirements in the statutes or acts. Generally, those who are provided notice of the coalbed methane well will be the surface owner at the well drill site, the oil and gas owners, and the coal owners and operators. When title is complete, and the well layouts are set, there should be sufficient information to complete the permit application.

The requirements for permitting coalbed methane wells will vary state by state. Most states, however, have some requirements in common. The following are some common requirements necessary to include in a permit application: notice along with information relating to the recipients' ability to object to the permit and where to send the objection, consent of coal owners or operators whose coal will be penetrated, whether there is a plan to stimulate the well and if so, consent of the coal owners to such stimulation technique, and a description of the well. An important point to remember is anyone who is supplied notice of the potential coalbed methane well may have the ability to object to the well based upon the controlling statutes or regulations. If someone raises an objection, generally a hearing will be scheduled and there will be a review of the permit application and the objection.

Stimulation or fracturing of a coal seam generally requires consent of the coal owner or operator because of the potential damage to the coal seam and its roof, and in turn, damage to the coal reserves and safety issues. While in recent years certain fracturing techniques have become technically viable and are used regularly in vertical wells, the statutory schemes vary with respect to addressing a coal owner's objection or lack of consent to such stimulation. For example, in Virginia, a coalbed methane well operator must obtain a signed consent from the coal operator of each coal seam, which is located within 750 horizontal feet of the proposed well location (1) which the applicant proposes to stimulate or (2) which is within 100 vertical feet above or below a coal-bearing stratum that the applicant proposes to stimulate. However, in West Virginia, a process is set out to address the well operator's failure to obtain the coal owner's or operator's consent in wells which will be stimulated, which process includes a hearing and the establishment of an additional bond by the well operator.[27]

---

[27] *Id.* § 45.1-361.29; W. VA. CODE § 22-21-13.

There is at least one other common statutory theme that is unique with respect to the interplay between coal operations and coalbed methane operations, and that is the requirement that the coalbed operator plug a coalbed methane well to allow for mine-through upon notice by the coal operator of mining operations. In Virginia and West Virginia, the coalbed methane operator is not compensated for loss of reserves or plugging, while Kentucky's statute sets forth specific provisions in certain instances that allow for the plugging and mining through of a well at the well operator's expense with no liability to the well operator by the coal operator.[28] The determination of whether the coalbed methane operator is compensated under the Kentucky statute depends on if the well was located within the boundaries of any area for which a mine permit has been issued or for which a mine permit or a mine permit modification application has been filed. In these instances no compensation is due the coalbed methane operator.[29]

15.3.3 The Pooling and Unitizing Process

Pooling is a process entirely separate from permitting, although the processes can be completed simultaneously. According to Blacks Law Dictionary, pooling is "the bringing together of small tracts of land or fractional mineral interests over a producing reservoir for the purpose of drilling an oil or gas well". Pooling and unitizing (i.e., the bringing together of small tracts of land or fractional mineral interests over a part of a reservoir—as opposed to the entire reservoir—for the purpose of drilling an oil and gas well) are terms that are used interchangeably. An operator utilizes pooling and unitization when the operator does not control all the interests in all the tracts encompassed in an area from which a well will drain. For example, consider a well that will drain 100 acres and there are four separate 25 acre property tracts that make up this 100 acres. Further consider that the operator controls through leases or ownership three of the four 25 acre tracts, but the operator is unable to lease or purchase the oil and gas interests on the remaining 25 acre tract. The operator can utilize pooling and unitization to bring this additional 25 acre tract into the 100 acre unit if the operator shows that such pooling is necessary to prevent waste, avoid excessive drilling, and protect correlative rights. The pooling and unitization process generally provides to the owner or lessee of the outstanding 25 acre tract the right to elect to participate in the unit as an investor, to participate after the well operator recovers 2–300% of its

---

[28] Ky. Rev. Stat. Ann. § 349.030; W. Va. Code § 22-21-22. Virginia's statute is silent on the issue of mine-through.

[29] Ky. Rev. Stat. Ann. § 349.030.

costs, or to not participate and lease its interests in exchange for a designated royalty, usually one-eighth of the production from the well attributable to its proportionate share of the unit (i.e., 25/100 in this case). If the owner or lessee of the remaining 25 acre tract fails to make an election, the pooling and unitization process provides that the interest is deemed to be leased to the well operator in exchange for a royalty that is the same as the voluntary lease election option. Note that pooling and unitization are generally recognized to be a constitutional exercise of the state's police power. The goals associated with this process are to prevent waste of this natural resource, maximize efficient recovery, avoid unnecessary and excessive drilling, and protect correlative rights.[30]

State statutes or rules will dictate specifically what must be included in a pooling application. Most pooling applications include a detailed description of the unit. A map showing the size and location of the unit generally accompanies the application packet. The application also includes a listing of everyone who is a potential claimant in the well. This includes owners and lessees of fractional interests of the oil and gas as well as the coal. Similar to the permitting process, the pooling process has a notice requirement. Anyone with a potential claim to coalbed methane is required to receive notice of the potential pooling. The notice should include a copy of the pooling application along with information regarding the time and place for the pooling hearing. Once a pooling application is complete and submitted, there is generally a hearing before an administrative or regulatory board. The duties of the board are to approve the size and shape of the unit, assure the proper parties are noticed, designate the operator of the well, and approve the pooling application. Once the pooling application is approved, individuals who have not leased their claimed coalbed methane interest to the well operator have a specified period of time to elect to either become a participating or nonparticipating owner in the well, and in lieu of making an election, those individuals will be deemed to be leased, as discussed above.

Because of the ownership uncertainty of coalbed methane, the coalbed methane pooling and unitization statutory process may differ from that of conventional oil and gas wells in three primary ways: (1) identification of coal owners in addition to oil and gas owners for each tract within a unit; (2) escrowing of royalties if the coal ownership differs from that of the oil and gas owner for a tract; and (3) a provision that provides protection from claims of trespass for coalbed methane development.

As to identification of owners, in conventional wells, a well operator pools or unitizes the fractional interests belonging to owners of the

---

[30] Jeff L. Lewin, "Coalbed Methane: Recent Court Decisions Leave Ownership 'Up in the Air,' But New Federal and State Legislation should Facilitate Production," W. Va. L. Rev. 631 (1994).

oil and gas within each property tract in an area that will be pooled or unitized. In coalbed methane wells, however, a well operator pools or unitizes the fractional oil and gas interests, as well as the fractional coal interests belonging to those mineral interests' owners within each property tract in an area that will be pooled or unitized. The purpose of pooling all the potential ownership interests in a coalbed methane well is to ensure that all potential claimants to the ownership of coalbed methane have the opportunity to be heard in the process and ultimately have the ability to participate in the unit or lease their interests, whether voluntarily or involuntarily, to the well operator. Thus, when a well operator receives a pooling order from the jurisdictional board at the end of the pooling process, the order extends to the interests of all potential claimants to ownership of the coalbed methane. This limits the well operator's exposure from claims from either the oil and gas owner or the coal owner that its interests are not subject to the pooling order.

The next significant difference of the coalbed methane pooling process driven by ownership uncertainty is the escrowing of royalties that are associated with parties that through the pooling process either voluntarily lease, or are deemed to have leased, their mineral interests. In conventional oil and gas wells, if a party with a fractional interest in a property tract within a unit elects to lease, or is deemed to have leased, its oil and gas interests to the well operator during the pooling process, the fractional oil and gas interests owner is paid a royalty, typically one-eighth, by the well operator on the fractional oil and gas interest owner's proportionate share of production from the total unit. Because the coal owner or the gas owner both may have claims to the coalbed methane, royalties associated with leased interests, or interests that are deemed to be leased, are escrowed if there are different owners to the coal estate and oil and gas estate for the property tract at issue. The escrowed amounts are not disbursed until: (1) a judicial determination of ownership of the coalbed methane, or (2) the coal owner and the gas owner reach a voluntary agreement with respect to the escrowed royalties.

The final major difference of the coalbed methane pooling process driven by ownership uncertainty is that the statutory scheme generally provides an explicit provision that development of coalbed methane shall not be subject to trespass claims where the well operator was acting under color of title from either the gas owner or the coal owner. Because trespass claims potentially expose the operator to punitive as well as compensatory damages, this explicit provision provides additional protection to the coalbed methane operator from various claimants to ownership of the coalbed methane alleging trespass by the well operator who may not own or control both the coal and the oil and gas estate.

15.4 COAL MINE METHANE AS A COMMODITY— ALLOWANCES AND OFFSETS

Emissions of greenhouse gases,[31] including methane associated with coal mining activities, are being blamed by many for climate change. While some continue to question the science underlying these claims, various regional, national, and international regulatory schemes designed to control greenhouse gas emissions are being developed. The most popular approach to regulating greenhouse gases is referred to as "cap and trade". Under a cap and trade program, a "cap" or over-all limit on the number of tons of greenhouse gas emissions allowed per year is established, and "allowances" to emit greenhouse gases are then allocated by various measures (e.g., auctioned, given away) to those industries that emit them. The "cap" gets tighter as time goes on, meaning the pool of allowances shrinks, so that companies must find ways to reduce their emissions or acquire additional allowances from other companies who have more than they need. The idea is that the marketplace will drive emission reductions; those who can reduce emissions most efficiently will do so and can then sell their excess allowances to those who cannot reduce emissions in a cost-effective manner.

In the United States, no federal greenhouse gas legislation exists as of the publication date of this book, although various cap and trade propos-als have been introduced in Congress, with one passing the House of Rep-resentatives for the first time in 2009.[32] Some state and regional programs exist, and while one—the Regional Greenhouse Gas Initiative (RGGI)[33]— is up and running, most are still in different stages of planning and devel-opment. These include the Western Climate Initiative (WCI),[34] California's

---

[31] The primary greenhouse gases are carbon dioxide (CO_2), methane, nitrous oxide, and fluorinated gases such as hydrofluorocarbons, perfluorocarbons, and sulfur hexafluoride.

[32] H.R. 2454, *American Clean Energy and Security Act* (Waxman-Markey), passed the House of Representatives in June 2009.

[33] The states of Connecticut, Delaware, Maine, Maryland, Massachusetts, New Hamp-shire, New Jersey, New York, Rhode Island, and Vermont are signatory states to the RGGI agreement. These 10 states have capped CO_2 emissions from the power sector, and will require a 10% reduction in these emissions by 2018.

[34] The WCI membership is comprised of Arizona, California, Montana, New Mexico, Oregon, Utah, and Washington, as well as the Canadian provinces of British Colum-bia, Manitoba, Ontario, and Quebec. WCI's goals are to reduce greenhouse gas emis-sions by 15% below 2005 levels by 2020. Phase I of the WCI program is planned for January 1, 2012. Phase II, which is much broader, is planned to launch January 1, 2015.

Global Warming Solutions Act,[35] and the Midwestern Greenhouse Gas Reduction Accord.[36]

In addition to any mandatory emission reduction programs that exist or may be enacted, a number of voluntary emission reduction programs exist for those who want to proactively reduce their carbon footprint. The Chicago Climate Exchange (CCX), established in 2003, was the first voluntary greenhouse gas trading program to be established in the United States. Members of the CCX make legally binding commitments to reduce their greenhouse gas emissions or to purchase carbon offsets from third parties.

A carbon offset is a financial instrument that can be bought or sold in the marketplace. One carbon offset represents the reduction of one metric ton of carbon dioxide or its equivalent (CO_2e) in other greenhouse gases.[37] Common projects used to create offsets include the use of renewable energy as an alternative to fossil fuels, energy efficiency projects, the destruction of industrial pollutants, agricultural by-products, or landfill methane, and forestry projects, and depending on the trading program, projects that collect and/or destroy coal mine methane, which would have vented to atmosphere.

There are two markets for carbon offsets—the compliance market and the voluntary market. The compliance market, which is by far the larger of the two, includes those entities subject to mandatory caps on carbon emissions through some regulatory program. In addition to reducing their emissions, or in lieu of doing so, entities can buy offsets to comply with their cap requirements. Companies, governments, or even individuals not subject to a regulatory cap on their greenhouse gas emissions can voluntarily purchase offsets for moral or business reasons—such as to counter the emissions they generate by driving their personal vehicles.

Each program has its own protocols governing what types of projects can create offsets, as well as how those offsets are tracked and traded. There are, however, elements common to all. All the programs require,

---

[35] The Global Warming Solutions Act, AB 32, was signed into law in 2006. The Act caps California's GHG emissions at 1990 levels by 2020, with the reduction to be accomplished through an enforceable statewide program that will be phased in starting in 2012. The Act directs the California Air Resources Board (CARB) to develop appropriate regulations, adopt market-based compliance mechanisms including cap and trade and establish a mandatory reporting system to track and monitor global warming emissions levels.

[36] The participating states and provinces are Iowa, Illinois, Kansas, Manitoba, Michigan, Minnesota, and Wisconsin.

[37] For example, one ton of methane is equal to 21 CO_2 equivalents, which reflects the fact that methane has 21 times the global warming potential of CO_2.

in some form, the submission of an application to qualify a proposed offset project as eligible under the applicable regulations, the notion of "additionality", and the verification of emission reductions. Additionality addresses whether the greenhouse gas reductions would have occurred in the absence of the offset program. Activities that are required by law, that represent common practice in an industry, or that are so financially desirable that they would have occurred even without the revenue generated by the creation of offsets are usually not considered additional and thus cannot be used to create offsets. Verification is the process of ensuring the emission reductions that have actually occurred. Verification commonly requires the involvement of a third party who has been approved to serve as an independent verifier.

Currently, there are a few exchanges operating in the United States where carbon offsets can be bought or sold, with more in the planning stages. As noted above, the first was the CCX. In addition, in 2008, the New York Mercantile Exchange (NYMEX), began listing a suite of environmental futures and options contracts, based on commonly traded instruments in the European and U.S. Markets: European Union Carbon Allowances, U.N.-certified carbon offsets, RGGI carbon allowances and other noncarbon emission allowances. Thirteen major financial institutions, calling themselves the Green Exchange venture, have lodged an application with the U.S. Commodity Futures Trading Commission to operate as a stand-alone exchange and to list the environmental contracts that are currently traded on NYMEX. Until such approval is granted, NYMEX will continue to trade these instruments.

Outside the United States, the market for carbon offsets is more fully developed. The Kyoto Protocol, a 1997 treaty, binds most of the world's developed nations to a cap and trade system for greenhouse gases.[38] The European Union Emission Trading Scheme is the largest multinational, greenhouse gas emissions trading scheme in the world and was created in conjunction with the Kyoto Protocol.[39]

The eligibility of greenhouse gas emission reduction projects at coal mines to generate offsets depends on the program. Under RGGI, only emission reductions from five defined project categories can create offsets: landfill methane capture and destruction; sulfur hexafluoride reductions from the electric transmission and distribution sector; carbon sequestration attributable to afforestation; reduction or avoidance of CO_2 emissions for natural gas, oil, or propane end-use combustion due to energy efficiency improvements in the building sector; and avoided methane

---

[38] The United States did not ratify the Kyoto Protocol and thus is not bound by its emission reduction requirements.

[39] *See* http://ec.europa.eu/environment/climat/emission/index_en.htm.

emissions from agricultural manure management operations.[40] Thus, projects that reduce coal mine methane emissions are not be eligible for CO_2 offsets allowances under RGGI at this time.

The CCX allows projects that collect and/or destroy coal mine methane, which would have vented to atmosphere to generate offsets.[41] Under the CCX protocol, offsets are issued at a rate of 21 metric tons of carbon dioxide per ton of methane collected and destroyed.[42] Project owners must demonstrate clear ownership of the greenhouse gas mitigation rights.[43] Methane produced before and after mine-through from predrainage wells will only be eligible after the well is mined through. All methane produced before and after mine-through from predrainage wells from within a −50 m to +150 m vertical range of the mined coal seam will become eligible when the well is mined through. Methane produced outside the vertical limit can become eligible if the candidate project demonstrates sufficient analytical evidence that connects methane generated outside the established vertical range to the mined seam in question. All methane produced from qualifying wells at abandoned coal mines will be eligible. Coalbed methane produced from coal seams unrelated to mining activity is ineligible.[44] Under the CCX protocol, "coalbed methane" is defined as "methane that resides within or is produced from coal seams. For the purposes of this protocol, coalbed methane refers to methane produced from coal seams unrelated to mining activities". Conversely, "coal mine methane" is defined as follows: "As part of the mining process, methane contained in the coal and surrounding strata may be released. This methane is referred to as coal mine methane since its liberation is a result of mining activity. In some instances, methane that continues to be released from the coal-bearing strata once a mine is closed and sealed may also be referred to as coal mine methane (or abandoned mine methane) because the liberated methane is associated with past coal mining activity."[45] Thus, under the CCX, some coal mine emission reduction projects are eligible to generate offsets.

[40] *See* http://www.rggi.org/offsets/categories.

[41] *See* http://www.chicagoclimatex.com/content.jsf?id=1021.

[42] *Id.*

[43] See Chicago Climate Exchange General Offset Program Provisions (Issued 8/20/2009), available at http://www.chicagoclimatex.com/docs/offsets/CCX_Gener al_Offset_Program_Provisions_Final.pdf. For more detailed information, see Chicago Climate Exchange Offset Project Protocol: Coal Mine Methane Collection and Combustion, available at http://www.chicagoclimatex.com/docs/offsets/CCX_Coal_Min e_Methane_Collection_and_Combustion_Final.pdf.

[44] *Id.*

[45] *Id.*

The market for CO_2e offsets is still in its infancy in the United States and could change radically if a federal cap and trade program is enacted.

15.5 ISSUES TO CONSIDER WITH REGARD TO COALBED METHANE WELL PLUGGING AND SUBSEQUENT MINE-THROUGH

15.5.1 Plugging Coalbed Methane Wells in General

Well plugging is an important issue that requires careful coordination and negotiation where the developer of the coalbed methane well is someone other than the owner of the affected coal seams. Specific state jurisdictions may have different oil and gas, as well as mining, requirements, and procedures regarding plugging. In addition, there are federal requirements and procedures promulgated by the Mine Safety and Health Administration (MSHA) that may also apply. It is important to understand the regulations and laws that will control the plugging process.

One issue to consider before beginning any coalbed methane drilling project is to determine whether the coalbed methane well operator will be required to plug the wells on demand or whether the well operator has any recourse when a mine operator wants to mine-through an existing coalbed methane well. As noted above, certain state coalbed methane statutes generally require that a well operator plug coalbed methane wells for mining operations; however, if there is not a controlling statute to address this issue, the parties must negotiate with respect to superior property or contractual rights to develop the coal seam(s) over the coalbed methane gas, the notice that will be given, and the method and time to complete the plugging process.

Regardless of whether the coalbed methane well will be plugged on demand by the coal operator, the well operator and coal owner or operator should discuss and, if possible, reach an agreement in advance on certain issues relating to plugging for safe mine-through of the affected coal seams. A plan should be developed setting forth the specific technique to be utilized for the plugging or abandonment for mine-through. Both the well operator and coal operator will want to know that the agreed-upon plan will be deemed sufficient to meet all applicable state and federal statutory requirements.

Issues that may arise, if not addressed statutorily, between the coalbed methane well operator and the coal owner or operator include: (1) the allocation of financial responsibility in the event that a well operator plugs a well for mine-through and subsequent state or federal regulatory changes require a different technique be used for such plugging; (2) the party that is liable for any failure to adhere to the plan and any resulting damage that occurs—either because the plan was not followed or the plan did not work as anticipated; (3) compensation for the well operator who may lose income from future production. As set forth in Section 15.3.2, the West Virginia statute that requires a coalbed methane operator to plug its coalbed methane wells does not have this requirement for financial compensation.[46] If not addressed statutorily, the coalbed methane well operator will likely want compensation for any recoverable reserves that will be lost due to the mining activity while the mining company may seek to avoid that cost based upon its long-standing right to vent methane in conjunction with its mining activity. The ability to negotiate payment for this lost income will relate directly to the issue of whether the mine operator has a right to demand the well be abandoned or if the mine operator has to obtain agreement and consent from the well operator prior to plugging the coalbed methane well; (4) a coal owner's or mine operator's right to receive compensation if any of the coal is sterilized—i.e., cannot be mined due to plugging or safety issues; (5) liability for issues that develop in the future during mining that may be due to coalbed methane operations such as fracturing or improper plugging; and (6) reclamation costs.

Most state and federal regulatory agencies have not promulgated any specific technical regulations relating to coalbed well plugging techniques for mine-through, thus underscoring the need for private agreements to regulate these aspects of the interplay between the coalbed methane developer and coal owners and mine operators until the concerns arising from regulatory uncertainty are resolved.

15.5.2 Technical MSHA Regulations for Plugging or Abandoning for Mine-Through

The major concern with regard to plugging or abandoning the coalbed methane well located in a mineable coal seam is assuring subsequent mine-through so that coal reserves can be recovered. There is currently uncertainty in that regard, particularly with respect to horizontally drilled wells. Note that in addition to permanent plugging, other nonpermanent plugging alternatives for horizontal wells such as "water infusion" and "bentonite slurry infused at hydrostatic or higher pressures, have been noted as potential abandonment technologies to allow for mine-through" and, while these alternative are currently being evaluated, they have been used in wells.[47]

Applicable mining regulations promulgated by the MSHA expressly prohibit mining within 150 ft of any well; however, that standard may be

---

[46] W. VA. CODE § 22-21-22.

[47] MSHA Program Information Bulletin P08-20 (2008).

modified to allow an operator to mine-through gas wells where the operator can show that its proposed method provides at least the same level of protection to miners.[48] Section 101(c) of the Mine Act allows coal operators to file petitions for modification of a mandatory standard, including the 150-foot prohibition just cited, where "an alternative method of achieving the result of such a standard exists, which will at all times guarantee not less than the same measure of protection afforded the miners of such mine by such standard".[49]

In the context of vertically drilled coalbed methane wells, this petition for modification process has greater certainty and reliability as the MSHA routinely approves petitions to modify the 150-foot barrier requirement for vertical coalbed methane wells where the well is plugged with expanding cement from the bottom of the hole (or at least 50 ft below the lowest mineable seam) to the surface (or at least 50 ft above the highest mineable seam).

The concern, however, arises where coalbed methane wells are drilled horizontally. Though the technology leads to far greater coalbed methane recovery, the MSHA stated in 2008 that "[d]irectionally drilling coalbed methane wells…is a relatively new technology and plugging or other methods that will protect coal miners have not been established".[50] In that 2008 program information bulletin (2008 MSHA PIB), the MSHA expressed concerns that plugging is more difficult with horizontal wells due to "undulati[ons]…, water cutting, gas cutting, numerous branches, length of the holes, and…potential [hole] caving".[51] According to the MSHA, the resulting voids could become "pressurized methane reservoirs".[52] Unfortunately, the need to obtain approval for subsequent mine-through is even more important in the context of horizontal coalbed methane wells with multilateral branches as the 150-foot barrier around those branches would be extensive, and significant amounts of coal reserves could be sterilized if the multilateral branches could not be subsequently mined-through.

This uncertainty raises issues as to the viability of plans to subsequently mine-through horizontal coalbed methane wells and is something that the coalbed methane well operator should be aware of as it will be of concern not only to the well operator but also the coal owner or operator who will mine-through the well. At this time, it is not clear what the criteria for approval of such petitions for modifications under

[48] 30 C.F.R. § 75.1700.

[49] 30 U.S.C. § 811.

[50] MSHA Program Information Bulletin P08-20 (2008).

[51] Id.

[52] Id.

Section 101(c) of the Mine Act will be or whether the MSHA will even grant petitions for certain mine operations prior to a more definitive determination on the effectiveness of plugging horizontal wells. Although the MSHA has not categorically approved plugging or other postproduction alternatives for horizontal wells, it has, as noted above, approved certain potential nonpermanent abandonment technologies such as "water infusion" and "bentonite slurry infused at hydrostatic or higher pressures" in certain mining operations to allow for subsequent mine-through.[53]

The MSHA itself has changed positions on the issue of plugging horizontal wells. In 2005, the MSHA issued a PIB (2005 MSHA PIB) exempting horizontally drilled wells from the 150-foot barrier requirement because it found that there "are no functional differences between degasification holes and [coalbed methane] holes or methane drainage systems and [coalbed methane] production systems".[54] Accordingly, after the 2005 MSHA PIB, the MSHA required only that the horizontal wells be included and addressed in the mine's ventilation plan: "[Coalbed methane] wells are subject to the ventilation plan...requirements that apply to methane degas holes."[55] Until the 2008 MSHA PIB, the MSHA district offices routinely approved plans to mine-through horizontal wells as part of the mines' ventilation plans; however, issues with mining through of some horizontal wells prompted the MSHA to look more closely, and to regulate more firmly, in this area. The 2008 MSHA PIB recognized a change in enforcement policy from the 2005 MSHA PIB: "MSHA has determined that the barrier specified in 30 C.F.R. §75.1700, i.e., 300 feet in diameter, must be maintained...around all vertical and horizontal legs and branches of the gas well that are within the coal seam...[and that a] petition for modification will be required to mine-through."[56] Thus, at the time of this publication, horizontal coalbed methane wells are subject to the provisions of 30 Code of Federal Regulations Section 75.1700 (i.e., the barrier specified of 300 ft in diameter, must be maintained around all vertical and horizontal legs and branches of the well) and may be mined through only if the MSHA grants a petition of modification of that standard. While petitions for modification may be granted, the petition process is significantly longer than the mine ventilation plan approval process and is somewhat less certain.

---

[53] *Id.*

[54] MSHA Program Information Bulletin P05-10 (2005).

[55] *Id.*

[56] MSHA Program Information Bulletin P08-20 (2008).

15.6 IMPORTANT ISSUES TO ADDRESS IN LEGAL AGREEMENTS

There are many factors to consider when parties such as a well operator and a coal owner or operator are entering into a coalbed methane agreement for the development of coalbed methane from the coal seams. It is extremely important that the parties' entire agreement be reduced to writing. Most written agreements are drafted at the beginning of the process when the parties enjoy a cordial relationship with each other. As the transaction carries out its course, unexpected circumstances can arise. Because the parties may not always agree how to handle such unforeseen situations when they arise, there is a great benefit to considering and addressing these possible circumstances in a written agreement at the beginning of the development process. Enforcing verbal agreements will be difficult at best in the court system and it is unlikely that both sides will have the same understanding of the details of any oral agreement. Therefore, all parties' rights and responsibilities should be considered and included in the final written document.

As set forth above, the ownership of coalbed methane gas is uncertain in many jurisdictions. Although coalbed methane is found in the coal seams, it is in fact a gas, and different jurisdictions have reached different conclusions regarding the legal ownership of coalbed methane gas. Unless development will occur in a jurisdiction that has established a bright-line rule regarding ownership, you must consider the claims of all possible owners and when practical should negotiate an agreement to address the ownership uncertainty and the division of revenues and royalties. Therefore, in addition to the normal agreements between mineral owners and mineral developers, you must consider the possibility of drafting agreements between parties with conflicting claims of ownership as well. In some jurisdictions, statutes and regulations address a number of issues that would normally be addressed in an agreement. Among the most common provisions of such a statute are the ability to develop coalbed methane without the consent of all possible owners and the allocation or escrow of revenues when ownership is uncertain.

Most written agreements typically address the issue of how revenues and profits will be shared among the parties, but, in addition, development costs and liabilities also need to be allocated. There are other rights that should be considered in the allocation process. For example, coalbed methane production could possibly generate carbon credits, other tax credits, or government incentives that could be allocated in a written agreement. Coalbed methane development also involves significant water usage and disposal. The parties should consider who should bear the responsibilities for the water issues relating to development and disposal.

After coal mining operations extract all of the coal in a given area of a seam, coalbed methane gas may continue to exist and be subject to capture. A good agreement should anticipate this possibility and determine who will be entitled to control and capture the coalbed methane gas at that point in time. When a coal seam is mined out, the remaining area is often referred to as "container space". Legal issues can arise when a coal seam owner or operator extracts all of the coal as to whether the coal owner has the right to continue to use that container space to transport coal or other minerals, whether the coal owner or operator may inject substances into such space, and whether the coal owner or operator continues to have the right to control the coalbed methane, which may be found in that space. Even those courts that have held that coalbed methane is part of the coal owner's estate have conditioned their ruling upon the coalbed methane being adsorbed to, or part of, the coal seam. Once that gas has been released and migrates into other strata or a gob zone, the ownership becomes even more difficult to resolve. Therefore careful negotiating and agreement in advance is advised to resolve potential conflicting claims.

As the previous issues make clear, one of the primary reasons for conflicts and the resulting need for written agreements comes from the fact that often the ownership of the coal, gas, surface, and other estates is not vested in one party. As a result it is important to understand whose rights are superior and whose rights are subservient based upon the jurisdiction and any existing documents affecting this issue—e.g., severance documents in the chain of title, leases, development agreements, and the like. Regardless of which party is in a superior bargaining position, it is important for all parties to enter into a written agreement that considers these issues before development begins.

Litigation can arise in any number of the previously described situations. From disputes with surface owners regarding drilling to disagreements with other mineral owners, coalbed methane development can create the potential for litigation. Written agreements should allocate liabilities for litigation costs and potential money damages. Indemnification for various damages should also be considered.

In conclusion, there are numerous issues to consider when drafting written agreements relating to coalbed methane development. It is important that these issues be considered and the entire agreement reduced to writing. A comprehensive written agreement will serve all parties well throughout the life of the transaction.

15.7 CONCLUSION

Recent developments in technology have allowed for the development of coalbed methane, and this new era of commercial production has required adjustments and adaptations in the legal world to account for various issues not present with the conventional development of other minerals, such as oil and gas. Issues as simple as ownership of coalbed methane are still in flux in certain states; however, a few states have enacted statutes that allow for production of the mineral even in the face of this uncertainty. A coalbed methane well operator needs to be aware of and operate under the provisions of these statutes. Additionally, these statutes may provide protection from adverse claims through use of the pooling process and through any trespass exclusion provisions. The well operator also must be aware of any well plugging requirements with respect to its coalbed methane operations when mining operations will occur. These plugging issues must be taken into account when reviewing and assessing the coalbed methane operation as a whole. Like ownership, issues with well plugging requirements and greenhouse gas emissions remain unsettled and subject to change. In light of this uncertainty, the best defense for parties wishing to develop the resource is a written agreement addressing all contingencies.

CHAPTER
16

Permitting Coalbed Methane Wells

Joanne Reilly

Coal Gas Recovery, LP, Affiliate of Alpha NR, Bristol, VA, USA

16.1 INTRODUCTION

Prior to drilling any coalbed or coal mine methane (CBM) wells in the Northern Appalachian Basin, a series of permits must first be obtained from the regulatory agency for the state in which the well will be drilled. Additional permits or plans may be required by various Federal agencies, the Army Corp of Engineers, local counties, or townships. Permits can take months to obtain and cover not only the drilling of the well but water withdrawals from streams, pipeline installations, building installations, and zoning and planning regulations.

Permitting wells in the Northern Appalachian Basins has become a rigorous exercise, due in no small part to the development of the Marcellus Shale. States including Pennsylvania and New York have placed restriction on the use of water from streams and the disposal of drilling and production fluids (Pennsylvania and West Virginia). While this chapter does not claim to be a complete guide to coalbed methane (CBM) permitting in all states, this chapter will attempt to outline the major permits required for the installation of a CBM well.

16.2 ZONING

Zoning restrictions are generally local, and require the proposed use of a property to fit into a restricted use category such as R1 (residential, single family), residential multifamily, commercial and extractive industry. Zoning regulations can be obtained from the local township or county. If the proposed well location does not fit into the particular category, then an

Coal Bed Methane
http://dx.doi.org/10.1016/B978-0-12-800880-5.00016-4

exception must be granted by the zoning authority. Most requests for exceptions involve the attendance at several meetings of the zoning board and perhaps the township or county supervisors meetings. Zoning boards may require extensive mapping, notification to surrounding land owners within a certain distance from the property boundaries, and a public meeting where past practices and future effects can be discussed by the general public. Often special conditions may be placed on the zoning exception such as fencing, noise, or light restrictions. Additionally, zoning boards may require that the project plans be reviewed by their consulting engineer at a cost not controlled by the well developer. These additional plan reviews may be required even if the well developer's plans are sealed by a professional engineer. It should be kept in mind that the zoning regulations may apply to any exploration to be completed in the political subdivision as well as the actual installation of the well, the compressor station and the pipeline.

It should be noted that in Pennsylvania, the Pennsylvania Supreme Court has ruled in favor of townships limiting the location of gas wells through zoning. The individual townships cannot require those wishing to drill wells to abide by restrictions that only the state may impose the regulation which states:[1] "…all local ordinances and enactments purporting to regulate oil and gas well operations regulated by this act are hereby superseded…The Commonwealth, by this enactment, hereby preempts and supersedes the regulation of oil and gas wells as herein defined." However the townships may limit the location of wells to certain zoned areas. As of this writing, the issue of zoning in Pennsylvania and its townships is still unsettled with various drilling companies in litigation with townships over zoning ordinances.

16.3 PLANNING COMMISSIONS

The International Building Code has been adopted by many states (the code may be obtained through any bookstore) and is used in conjunction with comprehensive development plans for the political entity. Often local entities, such as county planning commissions are tasked with enforcing the Code and the comprehensive plans. Any proposed building plans (such as treatment plants or compressor housing) must be reviewed and approved by the enforcing agency. Building plans sealed by a professional architect or engineer are required to be submitted and the enforcing agency may require the plans to be reviewed and approved by their consulting engineer at a cost to the well developer. Often the enforcing agency will piggyback with other local agencies such as conservation districts or taxing agencies. These agencies may also require storm water drainage plans, parking areas, access road paving, and landscaping, depending

---

[1] Act 223, Oil and Gas Act, Sec. 601.602

on the comprehensive plans developed for the political entity. One item worth noting is that even though the plans presented to the political entity are prepared and sealed by a registered professional engineer, the political entity may require the plans to be reviewed by the political entity's own consulting engineer, at an additional cost to the well developer.

16.4 ROADS: LOCAL, STATE, HIGHWAY

Roads in the Appalachian Basin can be divided into limited access (interstate) highways, state maintained roads, and local county, township or city maintained roads. Access to the roads for specific projects including driveway access, heavy hauling, and pipeline installation require separate permits from different agencies or governing bodies.

Local roads may be posted for relatively low weight limits. Hauling permits accompanied by bonds guaranteeing the restoration of the road surface may be required. It is advised that the producer documents the condition of the road before hauling large loads as disputes regarding responsibility may arise.

State maintained roads may also be posted with low weight limits. Permits will be required. Enforcement by police can be intimidating with fines in some cases approaching $15,000 for hauling on state roads without a permit to exceed the weight restrictions.

16.5 WATER WITHDRAWAL

If a CBM project will be drilled using quantities of surface-obtained water, the well operator should be aware of the water registration requirements for the state in which the well is drilled. Many states including Pennsylvania, Ohio, West Virginia, Virginia, and Tennessee require surface water withdrawal registration when a specific gallons per day threshold is reached. In Pennsylvania,[2] Water Resources Planning became effective November 15, 2008. The purpose of this chapter is to establish "the registration, monitoring, record keeping, and reporting" in order for Pennsylvania to plan the use of its water resources. The Pennsylvania water resources are broken into several river basins (for example the Susquehanna River Basin and the Delaware River Basin), each with its own Commission that has jurisdiction over the planning, development, and regulation of the water resources within the basin. While at the time of this writing there was not one for the Monongahela River Basin in southwestern Pennsylvania, there was some discussion that one may be formed. There is, however, an Ohio River Basin Commission.

[2] 25 PA Code Chapter 110

In Pennsylvania, each entity that withdraws in excess of an average rate of 10,000 gallons per day in any 30-day period from a watershed is required to register with the Department of Environmental Protection (DEP). Tennessee has a similar threshold of 10,000 gallons per day. Registration forms are available through the various state Web sites and may include a registration fee. West Virginia, according to the Department of Natural Resources (DNR) Web site,[3] is trying to determine the status of the law in regard to water withdrawals.

16.6 WELL DRILLING PERMITS

Well drilling permits vary in complexity by state. Those states in which the Marcellus shale is being developed have opted to increase the complexity of the permitting procedure.

West Virginia requires operators to obtain a drilling permit through the DEP, Office of Oil and Gas. Located on the Web site is a permit application package specific to CBM wells. The application requires the applicant to notify in writing coal owners, lessees, and operators, natural gas owners, lessees and operators, surface property owners, and the Oil and Gas Inspector. Also required are a public notice, a location plat, an erosion and sedimentation control plan, a construction and reclamation plan and water well testing of wells within 1000 ft of the CBM well. West Virginia requires an annual production report.

Pennsylvania's drilling permit applications have been changing rapidly due to the increased activity in the Marcellus shale. The Pennsylvania DEP's oil and gas regulations will continue to change to meet the increased interest in drilling Pennsylvania gas as new proposed regulations affecting all aspects of drilling and plugging wells were introduced in January 2010. Currently, standard drilling permit application requirements include a written notice to the surface owner, coal owner, operator and lessee, coal mine operator within 1000 ft of the well, gas storage operator, and surface owner with water or water purveyor within 1000 ft of the well. A map showing the coal lease area is also required. Details of the drilling and casing plans must be submitted with the application including a map illustrating any laterals to be drilled, their direction and length (alternate casing plans may be submitted). If the disturbed area of the project (including access road and pipeline) is greater than five acres, an Earth disturbance permit application must be submitted. A Pennsylvania Natural Diversity Index (PNDI) search with negative results must be included. Stamped and sealed plats of the well location are required. New to the plat requirement is that as of 2010, the plats must be submitted in NAD 83. Pennsylvania also has a new permit fee structure. Permit

[3] http://www.dcnr.state.pa.us/

application that formerly cost $350 to submit can now total thousands of dollars depending on the depth and total lateral footage. Pennsylvania DEP has a permit fee calculator on their Web site. Pennsylvania requires that the wells be bonded. Pennsylvania also requires an annual report of the water and gas production for each well.

Virginia has developed their own regulations for CBM well permitting.[4] CBM wells are under the jurisdiction of the Department of Mines, Minerals and Energy (DMME). Virginia's regulation underwent a review and revision in the fourth quarter of 2009 to bring the regulations up to the standards currently being used by the industry. Permit applications must include proof of notice to all affected parties, a plat developed using the Virginal Coordinate System of 1983, an operations plan (including groundwater characterization and a list of proposed drilling fluids), an Spill Prevention Control and Countermeasures (SPCC) plan, an erosion and sedimentation control plan, a casing plan, proof that the well conforms with local mining operation plans, safety plans for wells located within 200 ft of a mine (sealed or active) and a geologic column identifying coal, oil, gas, and groundwater strata. It should be noted that no oil-based drilling fluid can be used without the permission of the DMME Director. Virginia requires a well completion report. CBM well operators are also required to submit monthly reports of the amount of produced water for the month and cumulatively for each well.

Ohio oil and gas regulations are established under two separate state codes.[5,6] Ohio requires submittal of a permit application that includes a signed and sealed plat (specifically by an Ohio registered surveyor), a well bond, proof of insurance, a brine, and other residual waste storage and disposal plan, a sworn statement that all the requirements of political subdivisions have been met (including zoning), a road hauling plan for heavy equipment, a reclamation plan, a permit application fee ($250), and names and addresses of all royalty holders. Ohio requires an oil, gas, and brine production annual report. A completion report for each well is required.

16.7 PENNSYLVANIA NATURAL DIVERSITY INDEX

PNDI refers specifically to the Pennsylvania Natural Diversity Inventory, a web-based interactive interface that allows the well developer to "one stop shop" for potential impacts from the development of the proposed site. The site was designed for well developer access.[7] Through this site the well

---

[4] Virginia Administrative Code 4VAC25-150

[5] Ohio Administrative Code 1501:9

[6] Ohio Revised Code Chapter 1509

[7] http://www.gis.dcnr.state.pa.us/hgis-er/login.aspx

developer can enter well plans and locations. Four agencies will report back regarding any natural heritage impacts at the proposed site: PA Department of Natural Resources, PA Game Commission, PA Fish and Boat Commission, and the US Fish and Wildlife Services. The report that is produced must be submitted with numerous permit applications in Pennsylvania including the Well Drilling Permit Application and the Erosion and Sedimentation Control General Permit-1 (ESCGP-1) Permit Application. This search can oftentimes produce some disturbing results including plant or animal species that can only be identified during one or two months each year. If a potential impact is reported, the Department of Conservation and Natural Resources (DCNR) requires, for example in the case of plant species, a certified botanist to evaluate the proposed location prior to site construction. The botanist must either confirm or deny in a written report that the identified species exists or not in that particular location. In 2009, proposed site locations in Southwestern Pennsylvania were affected by the location of an Indiana Bat hybernarium. Developers were not allowed to cut rough-barked trees within the time period that the bats were raising young in the trees.

16.8 COALBED METHANE REVIEW BOARD

In February 2010, legislation was passed in Pennsylvania the intent of which was to provide an alternative to litigation regarding the location of CBM wells and access roads. The legislation establishes a Coalbed Methane Review Board to which surface property owners may appeal if the surface property owner objects to the location of a CBM well or access road. Written notification must be provided to the surface property owner prior to submitting the drilling permit application as specified in the legislation. A drilling permit will not be issued by the DEP unless the notification requirements are satisfied. If an objection to the well location is filed, a conference between the parties and the Board must be held within 10 days of the filing. Any agreement reached at the conference must be filed with the DEP in writing within 10 days of the completion of the conference. If the parties fail to agree on a location, the Board will determine a location for the well, which will be delivered in writing to the surface owners.

16.9 STREAM CROSSINGS, WATER OBSTRUCTIONS, AND ENCROACHMENTS

Some states (including West Virginia and Pennsylvania) consider all waters to be under title to the state. As such, stream-crossing permits are required for all streams. Stream crossings are required for access road crossings, temporary or permanent, and utility line crossings. Projects that involve water encroachment into navigable waters or lakes and reservoirs may require joint permitting with both the state and the Army Corp of Engineers.

Pennsylvania's permit application for water obstruction and encroachment is filed with the Pennsylvania DEP, but is jointly reviewed and approved by the US Army Corp of Engineers and the Pennsylvania Fish and Boat Commission. This application is used when any "activity changes, expands or diminishes the course current or cross section of a watercourse, floodway or body of water" (Commonwealth of Pennsylvania Joint Permit Application instructions are available on the Pennsylvania Web site).

However, Pennsylvania DEP has a series of general permits, including general permits for utility line stream crossings, minor, and temporary road crossings, which may ease the permitting burden. Utility line stream crossings are addressed under one set of statutes.[8] A general permit, GP-5 is required for utility lines, including gathering pipelines. Also included in the general permit category for Pennsylvania is GP-7 for minor road crossings (constructed across a wetland, less than 100 ft long and less than 0.1 acre in area) and GP-8 for temporary road crossings (in place for less than one year). The general permits may be used if the applicant agrees to the general permit conditions. All general permit application forms and requirements are available on the Pennsylvania DEP Web site.[9]

Virginia requires stream-crossing permits for projects encroaching on waters situated in watershed having an area of 3000 or more acres or in wetland areas. This permit involves not only the Virginia Department of Environmental Quality, but also the US Army Corp of Engineers and the Virginia Marine Resource Commission.

West Virginia requires a Stream Activity application filed through the West Virginia Division of Natural Resources, Office of Land and Streams. The Office of Land and Streams can provide permits for road and pipeline crossings. Section 10 (navigable waters) may require joint permitting between the Office of Land and Steams and the US Army Corp of Engineers. Applications for Stream Activity permits are available.[10]

16.10 EROSION AND SEDIMENT CONTROL PLANS AND PERMITS

The Environmental Protection Agency (EPA) may not require an National Pollution Discharge Elimination System (NPDES) permit for

---

[8] Pennsylvania's DEP Watershed Management PA Code, Title 25, Chapter 105

[9] www.dep web.state.pa.us

[10] www.wvdnr.gov

"discharges of storm water runoff" from oil and gas operations.[11] The EPA or the primacy state does however require the development of storm water erosion and sedimentation control plans for well sites encompassing 5 acres or more. The 5 acres includes not only the drilling site, but any access roads and pipelines to the site.

In Pennsylvania the permit required is known as the ESCGP-1. Requirements for the permit include site maps, maps illustrating earth disturbances, runoff calculations, best management practices, and a $500 permit fee to the Clean Water Fund. Each drawing (for an expedited permit) must be signed and sealed by a registered professional engineer, geologist, or surveyor. Pennsylvania is also requiring those professionals preparing the permit applications to have attended a recent DEP regulatory training class. The plans must be submitted to the district Oil and Gas Management office for approval. The only exception to this is for transmission lines. These may be submitted to the County Conservation Office for review.

For West Virginia, oil and gas activities are specifically exempt from the West Virginia general NPDES permit for construction activities.[12]

The Virginia Erosion and Sediment Control Law also specifically exempts oil and gas activities.[13] It is worth noting, however, that some Virginia counties require erosion and sedimentation control plans.

16.11 CALL BEFORE DIGGING

Alternately known as "Call Before You Dig" or "One Call", states have put in place nonprofit organizations that field calls by well developers who want to locate utility lines before breaking ground. The One Call organization will notify the various utilities in the specified area. The utilities will then mark the buried lines. Some One Call organizations require membership and charge a small fee for each call. Note that if the one call is not placed and a line is ruptured, fines to the contractor breaking the line may total several thousands of dollars. It is suggested that the prudent well operator with gathering or transmission lines should join the One Call system in their respective states.

---

[11] Section 402(I)(2) of the Clean Water Act

[12] Section 10 of General Permit WV0115924

[13] Title 10.1, Chapter 5, Article 4 of the Code of Virginia as amended

16.12 SPCC PLANS

An SPCC Plan is required under Federal regulations.[14] Managed by the EPA, SPCC plans are required for all onshore facilities storing 1320 aggregate gallons or more of oil in above ground facilities (the amount of oil stored is only subject to that contained in 55 gallon containers). The purpose of the plan is to minimize the occurrence of any oil discharge into the navigable waters of the United States. The plan must contain the following:

- description of the facility and its layout;
- type and capacity of oil contained on site;
- discharge prevention measures;
- discharge or drainage controls;
- countermeasures for discharge, discovery, response, and cleanup;
- disposal of recovered materials;
- contact list and telephone numbers for the facility response coordinators;
- information and response for reporting a discharge;
- prediction of the direction, rate of flow, and total quantity of oils that could be discharged;
- a description of the containment and diversionary structures or equipment to prevent a discharge;
- plans for inspections and tests and retainment of records;
- personnel training in discharge prevention procedures;
- emergency response procedures;
- site security including valves and lighting;
- loading and unloading procedures;
- facility drainage;
- bulk storage containers; and
- transfer operations.

Drawings of the site and the various parts of the facility must be included in the plan.

The plan must be reviewed and certified by a professional engineer as to its adequacy and applicability to the facility. The plan must be available on site and available for any inspection. It must be reviewed and updated at least every 5 years or more frequently.

[14] 40 CFR 112

16.13 PPC PLANS

Preparedness, Prevention, and Contingency Plans are prepared in response to the PA DEP requirements in order to prevent environmental accidents and spills. Pennsylvania has opted for a format that combines the requirements of both the State and Federal emergency response programs into one. All commercial and industrial operations (including oil and gas well drilling) that have the potential for causing accidental pollution of the water, air, or land are required to develop and implement a PPC plan. The PPC plan must include the following parts:

- A description of the operation and its location. In rural areas it is helpful to contact the county emergency management office in order to have an address assigned to the well site. In the event of an accident it will be easier to find the site if it has an address, rather than describing the area as so many miles past the intersection of road A and B.
- A list of the chemicals used on site and the wastes generated. Material Safety Data Sheet information should be included in this section.
- A narrative regarding the possibility of an accidental discharge and how the discharge would be contained.
- Site security and protection from factors such as severe weather and vandalism.
- Preventative maintenance and inspection and monitoring of possible failure points.
- Training programs.
- Waste disposal methods including hauler information.
- The accidental discharge response measures to be taken by the site operator or any contractors, communications and alarm systems, evacuation plans, and any necessary emergency equipment.
- Accidental discharge reporting procedures including notifications to local emergency agencies, the DEP regional office, the EPA National Response Center, Fish and Boat Commission, and the downstream water users. The reporting procedures should include a list of contacts and telephone numbers.

A copy of the plan must be on site and available for inspection. A copy of the plan should be submitted to the DEP with any other permit applications and plans, including the drilling application. Additionally the plan should be submitted to county and local emergency management agencies for those sites containing storage tanks with greater than 21,000 gallons (500 bbl) above ground storage. If there are any changes to the plan, it should be revised.

16.14 ENVIRONMENTAL IMPACT STATEMENTS

As of this writing (end of 2009), environmental impact statements are not required for wells drilled in Pennsylvania.

16.15 BRINE WATER DISPOSAL

Most states require the reporting of brine production from oil and gas wells (CBM wells fall into this category). While most reporting includes the oil and gas production as well as the brine production, some states such as Pennsylvania also require a separate report of the brine as a residual waste. The brine must be tested for its constituents and the laboratory analysis submitted with the production.

The eventual disposal of the brine is dependent upon individual state regulations. Some states prohibit disposal in surface waters but allow spreading the brine on fields and roads as long as a permit has been obtained (Ohio, West Virginia, Pennsylvania). West Virginia issued a revised general permit application effective January 1, 2010 (GP-WV-1-07). This permit provides for the land application of produced water from CBM wells. Other states allow brine disposal in publicly or privately owned treatment facilities (Pennsylvania, West Virginia) while others only allow disposal in disposal wells (Ohio). Well developers should consult the disposal regulations for the state in which the well is located.

Well developers should anticipate produced water disposal regulation to be tightened following the 2008 and 2009 total dissolved solids and golden algae problems encountered in the Monongahela River and Dunkard Creek. Pennsylvania has proposed new source limits for Total Dissolved Solids (TDS) of 500 mg/L as a monthly average.

16.16 MINE SAFETY AND HEALTH ADMINISTRATION

As an outside contractor (not affiliated with the mine) working in a mine, the contractor is required to obtain an Independent Contractor Identification Number. The number is obtained from the Mine Safety and Health Administration (MSHA) Web site[15]; the site administrator assigns the number by e-mail. The contractor is required to register this number with any mine operators the contractor is servicing.

Once the contractor ID Number has been obtained, the contractor is required to submit to MSHA the quarterly 7000-2 report, detailing the

---

[15] www.msha.gov

amount of hours worked by the contractor in the mines. Also required is the quarterly report 7000-1, Mine Accident, Injury and Illness Report. Both can be submitted on line.

For the well developer, if the project is associated with an active mine, the developer should contact the MSHA District Office before initiating the project. The development plans should be presented and discussed. MSHA will require a ventilation plan submittal for any underground project that impacts the mine.

16.17 OCCUPATIONAL SAFETY AND HEALTH ADMINISTRATION

Surface operations such as drilling, pipeline installation, and gas treatment facilities are under the jurisdiction of the Occupational Safety and Health Administration (OSHA). OSHA conducts inspections of workplaces in an effort to eliminate workplace injuries. According to the OSHA Web site, "Top priority are reports of imminent dangers-accidents about to happen; second are fatalities or accidents serious enough to send three or more employees to the hospital. Third are employee complaints. Referrals from other government agencies are fourth. Fifth are targeted inspections-such as the Site Specific Targeting Program, which focuses on employers that report high injury and illness rates, and special emphasis programs that zero in on hazardous work such as trenching or equipment such as mechanical power presses. Follow-up inspections are the final priority."

Employers are required to track and post for inspection work-related accidents resulting in injuries and any work-related illnesses. Any accident resulting in death or injury to three or more people requires notification to OSHA.

16.18 CONCLUSION

Permitting in today's regulatory environment can be a daunting task. In addition to the above listed permitting and plan requirements, and depending on the project complexity, other agencies that may require permits or plans or reporting are Air Quality (for internal combustion engines including compressors), Department of Homeland Security,[16] EPA's Risk Management Plan[17] statutory program designed to prevent chemical

---

[16] CFR Part 27, "Chemical Facility Anti-Terrorism Standards" for large quantities of specific chemicals

[17] CAA 112(r)

accidents and releases, EPA's Toxic Substances Control Act,[18] EPA's SARA (Superfund Amendments and Reauthorization Act),[19] Tier II reporting of hazardous chemical inventories, and Department of Transportation[20] pipeline safety. Due to the proliferation of shale gas harvesting, Federal and state regulations regarding oil and gas have changed and continue to change since the writing of this chapter. The well operator should examine all the Federal and State programs to determine their applicability to their project.

[18] USC Title 15 Chapter 53 Subchapter 1 Section 2601

[19] SARA Title III Section 312 Tier II

[20] DOT Title 49

17

United States Lower 48 Coalbed Methane—Benchmark (2010)

Stephen W. Lambert

Schlumberger Data and Consulting Services, USA, Retired

17.1 INTRODUCTION

Beginning in the mid-1970s, research spear-headed by the U.S. Bureau of Mines began exploring the possibility of applying oil field technologies to remove gas from coal seams. These approaches involved the drilling and completion of vertical wells from the surface into coal seams designed to remove methane ahead of mining operations. Initial attempts using these techniques were applied to operating coal mines in the Black Warrior and Appalachian basins, and to unmined areas within the San Juan Basin, USA. The subsequent success of these efforts showed not only that gas could be readily drained from coal seams in advance of mining, but also demonstrated that recovery rates were high enough to be potentially commercial. Commercial sale of methane followed in 1977 from the San Juan Basin, New Mexico, then from the Black Warrior Basin, Alabama at the end of 1980. To date, there have been some 30 so-called CBM "play areas" across the U.S. Lower 48 (Figure 17.1). In this chapter, a play area can refer geographically to a coal basin, or a specific coal-bearing region within a coal basin.

CBM development and production increased from fewer than 100 wells in the early 1980s followed by rapid expansion during the late 1980s through 2006, as shown on Figure 17.2. At the start of 2007, new well production start-ups began to slow somewhat in almost all basins, primarily due to lower product pricing and reduced average per well production rate. The decrease in overall rates over the last several years is principally due to the slowdown of new well development in the Powder River (Figure 17.3), and also to fewer start-ups of exceptionally high-rate wells within the San Juan (Figure 17.4). Nevertheless, at the beginning of 2007, the International Exploration and Production database reported a total of 36,000 producing wells selling approximately 4.9 US billion cubic feet of CBM daily.

311

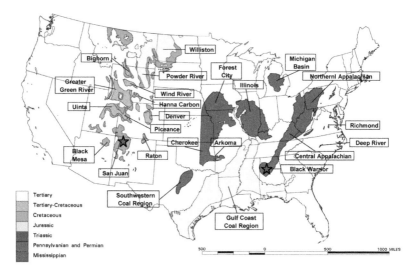

FIGURE 17.1   Geological age of coalbed methane gas play areas within the U.S. Lower 48 States.

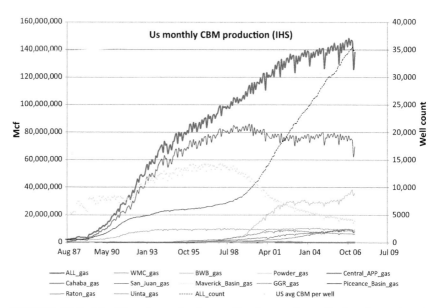

FIGURE 17.2   Monthly sales production profiles from major coalbed methane plays in the U.S. Lower 48 States, as reported by the IHS International Exploration and Production Database.

At last count, IHS identifies13 so-called "major" CBM producing play areas reporting data. Although estimations have varied over recent years as more information becomes available, a current compilation of reports estimating the amounts of in-place gas indicates 483 trillion cubic feet

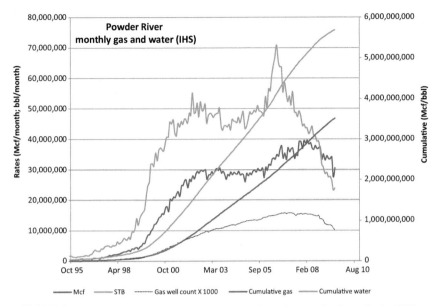

FIGURE 17.3 Monthly sales production profile from the Powder River Basin (IHS).

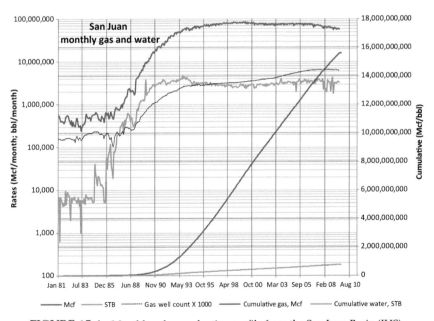

FIGURE 17.4 Monthly sales production profile from the San Juan Basin (IHS).

TABLE 17.1   Summary of Estimated U.S. Lower 48 Coalbed Methane Gas-in-place; Total Coalbed Methane Considered Recoverable and Produced; and the Estimated Total Acreage Proven by Drilling

Number of major coalbed methane play areas	13
Reported estimated gas-in-place, trillion cubic feet	483
Gas-in-place ultimately recoverable, trillion cubic feet	88
Total coalbed methane produced, trillion cubic feet	24
Remaining gas-in-place recoverable, trillion cubic feet	64
Average gas-in-place per well unit, thousand cubic feet	1,070,175
Total gas-in-place area proven by drilling and production, acres	3,418,160
Average coal thickness within gas-in-place area, ft	31
Average midpoint depth of coal targets, ft	2847
Average well spacing used in this analysis, acres	87
Average number of coal seams	17

(trillion cubic feet) contained within these 13 CBM play areas, of which 88 trillion cubic feet are considered recoverable. Of this amount, 24 trillion cubic feet of methane gas have been reported produced, leaving an estimated 64 trillion cubic feet of coal gas yet to be developed within these 13 major play areas. Appendix A includes all 30 U.S. Lower 48 CBM play areas' gas-in-place estimations and shows the calculated total acreage required to account for this gas-in-place. Much of this information is summarized on Table 17.1 specifically for the major play areas. Comparing the sum total CBM produced with the total volume of gas-in-place believed recoverable, the remaining potential of CBM exploitation is huge.

17.2 PRODUCTION

17.2.1 Rate Profile

Most CBM production profiles (for gas and water) are characterized by an initial steep decline of water rates while gas rates increase. Gas rates peak at some point, then exhibit a very slow decline over the longer term. On a regional basis, this more or less typical profile can be best observed by normalizing the production data so that all wells within a specified play area all "start-up" at hypothetically the same time. Such zero-time analyses are typically used to formulate area type-curves, which can then be used to forecast future CBM production reserves. An example of such a type-curve analysis is shown on Figure 17.5, using all data from the Black Warrior Basin. From

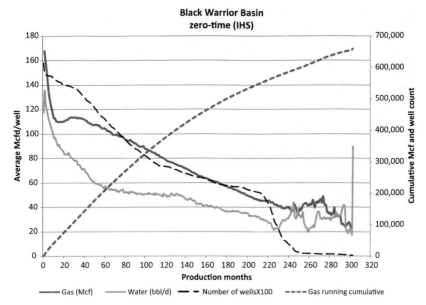

FIGURE 17.5   Zero-time type-curve analyses of all Black Warrior coalbed methane wells.

these results, the average well rate profile may be anticipated and also the longer-term cumulative volume of gas to be produced can be projected. The number of wells reporting during any given month is a critical factor in these analyses because as the number of wells decrease within the data population, the authenticity of the type-curve deteriorates.

For reference purposes, similar zero-time analyses for each basin/play area reporting data from the IHS are provided in Appendix B at the end of this report.

17.2.2 Regional Variations

The principle drivers of CBM well production variability are both regional specific (basin, play area, fairway, etc.) and location specific. Well performance is affected over relatively broad regional areas by gas-in-place concentrations and by existing in situ stress and hydraulic pressure gradients. Such regional controls on CBM production can be easily inferred and appreciated in map view as presented on Figure 17.6. It is evident how regional production differs greatly between the western Mid-Continent region which contains thin coals, lower gas-in-place, and low reservoir pressure gradients in contrast to the much thicker coals, higher gas-in-place values and pressures gradients contained within the San Juan Basin region.

The San Juan Basin itself is also a good example of where distinctly different production drivers exist, as indicated on Figure 17.7. In this case,

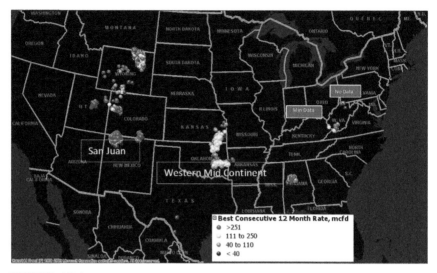

FIGURE 17.6   U.S. Lower 48 coalbed methane production distribution by basin/play area.

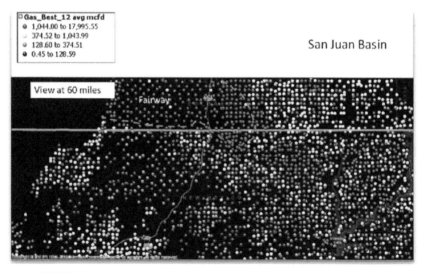

FIGURE 17.7   Map view of San Juan Basin production rate distribution.

the basin is generally divided into two distinct CBM gas production areas, i.e., the "fairway" and "nonfairway" productive areas. The fairway represents about 15% of the total productive area, yet this area accounts for over 75% of the total coal gas produced from this basin. Coal reservoirs are thickest in the fairway, locally exceeding 90 ft in cumulative thickness. This area is also set apart by highly (over)pressured reservoirs, higher

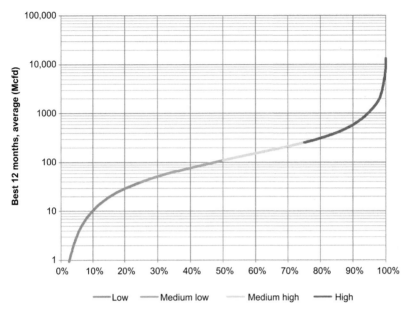

FIGURE 17.8 Distribution of all US coalbed methane well production based on the best consecutive 12-month daily average.

permeability (20–100 mD), and greater coal seam gas concentrations. Outside of the fairway region, coals are generally thinner (20–40 ft) having generally lower permeability (1–30 mD) and exist under lower (normal hydrostatic or less) reservoir pressure settings.

The distribution of all U.S. CBM well production shown graphically on Figure 17.8 is based on the best consecutive 12-month daily gas rate average. As indicated, the historical *median* well production rate of all U.S. CBM wells reporting data has been approximately 110 thousand cubic feet daily during its best 12 months of performance; while the mathematical *average* rate of these 36,000 wells (not shown on Figure 17.7) is considerably higher, at approximately 250 thousand cubic feet daily, mainly because of the very high individual well rates within the San Juan fairway region skew these data significantly to the high end.

Although production rates vary, all small or large CBM plays show very similar distributions. A good example of a large mature play area production distribution is that of the Black Warrior Basin, shown on Figure 17.9.

An example of a much less developed play is shown on Figure 17.10 for the Piceance Basin, where the typical CBM production rate distribution is just beginning to take shape.

The production distributions for all basin/play areas reporting IHS data are included in Appendix C of this report.

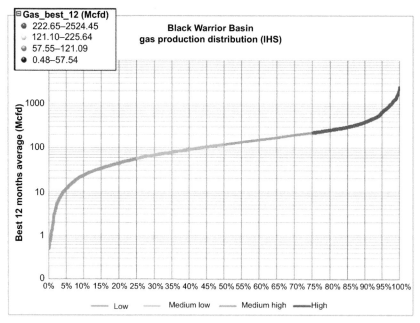

FIGURE 17.9   Production distribution of the Black Warrior Basin.

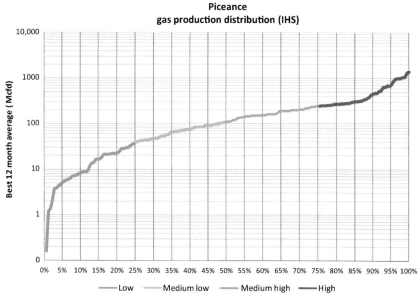

FIGURE 17.10   Production distribution of the Piceance Basin.

17.2.3 Field-Specific Variations

Because of the complex reservoir mechanisms that control gas and water flow in coals, production from CBM wells tends to have rather complex characteristics. Initial production from most coal reservoirs is dominated by water with small amounts of gas. As the water moves out of the natural fracture system, hydrostatic pressure is reduced, gas desorbs from the internal surface of the coal. As gas saturation (within the cleat and fracture system) increases, relative permeability to gas increases while relative water permeability to water decreases. With stabilization of the gas and water relative permeability, peak gas production occurs. From this point, gas and water production slowly decline, controlled not only by the key reservoir parameters (especially permeability) but also by the synergetic affects created by the pumping of adjacent wells. Thus, one of the most basic approaches applied historically to the large-scale production of CBM is to establish patterns of wells that work to "shield" incoming water recharge, and/or "interfere" by increasing well pressure drawdown efficiencies.

One striking example of the "shielding" effect that CBM wells can have on intercepting water inflow before reaching interior wells can be observed in the Gurnee Field, Cahaba Basin, Alabama. Figure 17.11 indicates

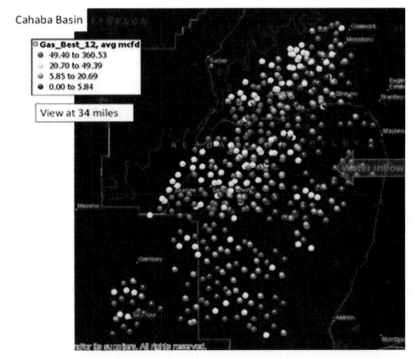

FIGURE 17.11 "Shielding" effect that coalbed methane wells have on intercepting water inflow before reaching interior wells is that observed in the Gurnee Field, Cahaba Basin, Alabama.

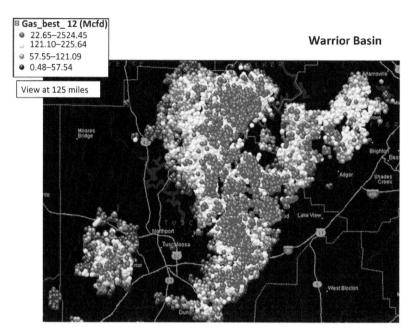

FIGURE 17.12   Black Warrior Basin showing synergetic effects of coalbed methane well pattern "interference".

significant recharge originating from the eastern edge of the basin, and that a number of low gas-producing wells are probably a result from a lack of ability to draw water pressure down over a large enough area.

Although many factors influence the productivity of CBM wells, including well construction, well spacing, coal thickness, gas content, and stress conditions and the abundance and open of natural fractures, virtually all field areas within maturely developed show the synergetic effects of CBM well pattern "interference", as shown on Figure 17.12, a production rate map from the Black Warrior Basin.

17.2.4 Location-Specific Variations

Changes in reservoir permeability appear to be very well location specific due primarily to heterogeneities in the natural fracture system within the reservoir (relative amounts of open cleat and natural fractures and their spacing and aperture width). Permeability in coals have also been shown to be highly in-situ stress sensitive, leading to an order of magnitude changes in permeability from location to location within producing fields. Examination of numerous producing wells across extensively developed CBM plays indicate very large variations in well performance. As an example, Figure 17.13 shows the best

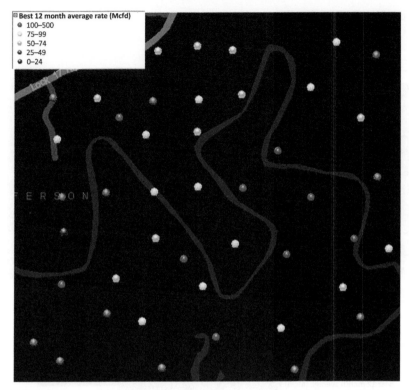

FIGURE 17.13 Best 12-month gas production rates from a tightly spaced (1000 ft spacing) pattern of producing coalbed methane wells in the Oak Grove Field, Black Warrior Basin, Alabama.

12-month gas production from a tightly spaced pattern of producing CBM wells in the Oak Grove Field, Black Warrior Basin, Alabama. All of the wells were drilled and completed nearly identically, with only slight well-to-well variations in coal thickness, gas content, and other reservoir parameters. All these wells were on the same small spacing—approximately 1000 ft apart—on a square grid. Hence, only variation in reservoir permeability can explain the large variation in well production across this field.

Extensive studies of CBM production data from highly developed CBM plays have also indicated that a high degree of variability exists across producing plays, and for smaller areas (down to field scale) within these producing plays, as demonstrated by a portion of the Cherokee Basin CBM area on Figure 17.14; and within the much less developed Greater Green River region, Figure 17.15. This high degree of variability within the coal reservoirs has significant implications for evaluating prospective CBM areas.

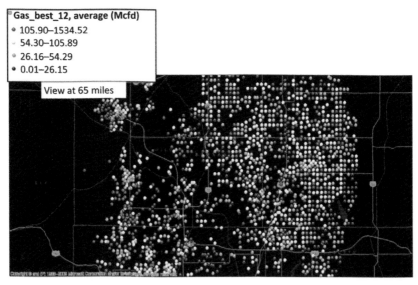

FIGURE 17.14   Variability across a portion of the Cherokee Basin coalbed methane area.

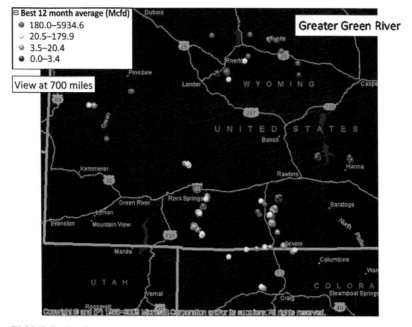

FIGURE 17.15   Variability across the greater Green River coalbed methane region.

17.2.5 Completion Variations

With recent improvement in downhole technology and the associated reduction in costs, horizontal drilling is becoming an attractive alternative in certain reservoir settings to vertical CBM wells. The first large-scale application of horizontal wells in coal occurred in the mid-1990 in the Hartshorne coal of the Arkoma Basin in Oklahoma, USA. As of August 2009, of the 1919 CBM wells reported producing by IHS, approximately 342 of these were horizontal completions, almost exclusively in the Hartshorne coal. An analysis of the production rate distributions, comparing the performance with Arkoma vertical wells is shown on Figure 17.16. Drawing from this analysis, it can be seen that horizontal wells in the Arkoma have outperformed vertical wells by roughly 100%; and that the probability of obtaining a higher production rate is significantly improved compared to vertical well production.

There is no doubt that location-specific completions have significant impact on CBM productivity. In fact, finding and developing area-specific completions have played the most vital role within the industry's past 35-year development history, starting from simple single-zone openhole and cased hole multiple zone completions. The rapid development of the Powder River Basin, for example, was based on a nontraditional hydraulic stimulation operation involving relatively small volume high-rate injections of nondamaging (water) via openhole. During the mid-1980s, it was

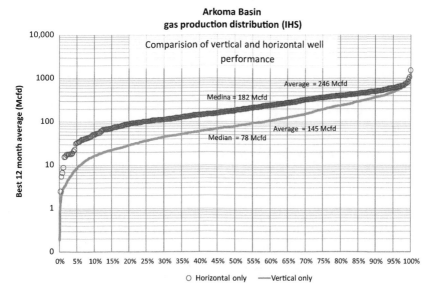

FIGURE 17.16 Production rate distribution of vertical and horizontal wells within the Arkoma Basin.

a "cavitations" completion method that ultimately proved most effective in the "fairway" region of the San Juan Basin, and so on.

17.3 WHAT THE DATA SUGGEST

The naturally occurring complexities of the coal gas reservoir have never been hostage to even the best engineered completion design. As earlier mentioned, these natural complexities far outweigh the effect of any type and recipe used to complete CBM wells, as the data continue to reflect highly variable production results in every coal basin/play area worldwide. For example, the mapped distribution of horizontal well production in the Arkoma Basin, shown in map view on Figure 17.17, presents as much *variability* if one were to compare these data with the same data distribution from vertical well production only, Figure 17.18.

The CBM data are consistent with an overall reservoir system where production is driven principally by preexisting natural fractures that can contain varying amounts of free and adsorbed gas. The source rock for this gas is believed to be numerous coal and other carbonaceous layers present within this system, and not necessarily confined only to the coal targeted by completion. In some locations, one or more of these source

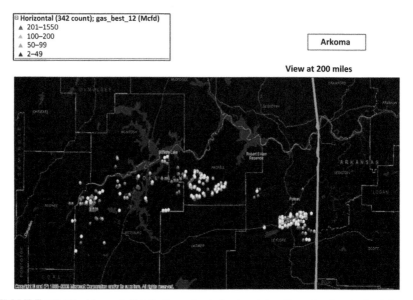

FIGURE 17.17 Mapped distribution of only horizontal well production in the Arkoma Basin.

rock layers are likely in direct contact with natural fractures; and at other locations, these source rock layers may not contain any significant natural fractures, as portrayed in Figure 17.19.

In vertical well completions, hydraulic stimulation treatments are likely most effective when induced fractures from the wellbore result in the intersection of these fractures. Therefore, in vertical CBM well scenarios,

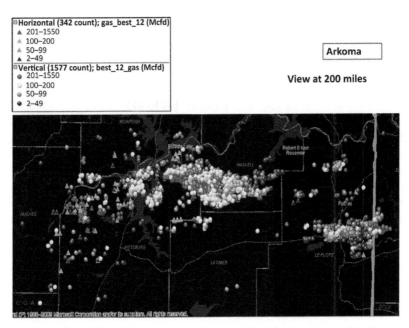

FIGURE 17.18 Mapped combined distribution of horizontal and vertical well production in the Arkoma Basin.

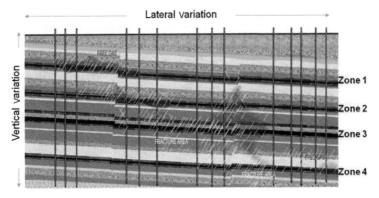

FIGURE 17.19 Conceptual reservoir system consistent with coalbed methane production data characteristics.

the probability of higher rate production is more or less coincidental with where both source rock and naturally occurring fractures coexist. These types of connections produce large variations of amounts and rates of fluid (water and/or gas) production, and many times display difficult to explain "disconnects" between anticipated amounts of produced water and gas rates, and are contrary to modeled responses to pressure drawdown. In wells where most or all selected target completions are not coincidental with natural fractures, production results are believed to present more characteristic "normal" low permeability coals, displaying typically lower gas and water production rates that are more easily explained by reservoir modeling.

17.4 FUTURE TREND

Starting from a little known research idea, the CBM industry has expanded greatly over the last 40 years. However, the complexities of describing site, field, or even regional CBM reservoir characteristics required to design the most effective drilling, completion, and production operations remain elusive in many areas. The relatively low rate (but very long term) productivity of coal formations continue to dictate the need for new and cost-effective technologies in order for the CBM industry to reach full potential in the USA. Improved borehole imaging and geochemical logging methods, for example, are now being developed to aid in the identification and high grading of coal reservoirs. Applications of horizontal well technology appear to be much more consistent with the complex nature of the coal reservoir, greatly increasing the probability of intersecting an increased number of naturally occurring fractures within the system (Figure 17.19). Also, driven by improved steering capability, reduction in formation damage, and by an overall reduction in drilling cost, horizontal well technologies may well "unlock" gas-bearing coals that have yet to demonstrate consistent attractive results. For example, the application of horizontal well technologies appear to be much more consistent for producing shallow (<1000 ft) less gassy but permeable coal seams within heretofore difficult (to complete) large potential play areas, such as in the Illinois and Northern Appalachian basins.

Major Basin/Play Areas

Basin Play/Area	Cahaba	Black Warrior Basin	Central Appalachian	Greater Green River (Less Sand Wash Basin)
State(s)	AL	AL	WV, VA, KY, TN	WY
Primary coal formation(s)	Pottsville Formation	Pottsville Formation	Norton Formation, New River Formation, Lee Formation, Pocahontas Formation	Fort Union, Lance, Mesa Verde Group, Frontier Formation
Topographical setting	Hilly, moderate to dense vegetation	Moderate—hilly, heavy vegetation	Moderate—hilly, heavy vegetation	Fair to rugged, extremely large area/some very remote areas, semidesert to forested
Structural style	Foreland thrust; gentle to moderate folding and faulting	Triangular-shaped, deepens to S–SW, moderate faulting	Foreland thrust, elongate NE–SW, numerous anticlines and synclines	Laramide, intermontane, numerous subbasins, simple to complex structure, faulting and folding prevalent
Estimated methane original gas-in-place Dry Ash-Free (DAF), Tcf	1.75	19.00	10.00	117.50
Estimated methane original gas-in-place ultimately recoverable (DAF), Tcf	0.35	8.00	3.14	2.50
Basin play area, acres	224,000	22,400,000	14,624,000	9,472,000
Total basin play acreage required to account for original gas-in-place estimation (i.e., "Original Gas In-Place (OGIP) area")	159,040	1,904,000	1,213,792	6,440,960

(Continued)

— cont'd

Basin Play/Area	Cahaba	Black Warrior Basin	Central Appalachian	Greater Green River (Less Sand Wash Basin)
Percentage of total OGIP acreage required to account for original gas-in-place estimation	71.00%	8.50%	8.30%	68.00%
Total CBM produced (recorded or estimated), Tcf	0.02	2.15	0.03	0.06
Remaining estimated recoverable, Tcf	0.33	5.85	3.05	2.43
Average original gas-in-place per well unit, mcf	879,722	799,159	646,052	2,908,693
Total OGIP area proven accessible by drilling and production activity, acres	22,800	456,000	167,760	35,520
Total OGIP area not proven by drilling and production activity, acres	136,240	1,448,000	1,046,032	6,405,440
Average gross coal footage within OGIP area	22.5	21.0	11.3	56.3
Average midpoint depth of coal targets, representative of OGIP area	3500	3000	2500	4000
Unit size well spacing applied to this analysis, acres	80	80	80	160

Basin Play/Area	Illinois (IL)	Maverick (TX)	Northern Appalachian	Piceance	Powder River
State(s)	IL	TX	PA, WV, OH, KY, MD	CO	WY, MT
Primary coal formation(s)	Carbondale Group	Olmos Formation	Allegheny Group	Williams Fork Formation, Iles Formation	Wasatch Formation, Fort Union Formation
Topographical setting	Flat, open, farmland	Flat, arid, fee surface	Moderate to rugged, dissected terrain, heavy vegetation, populated	Moderate rugged, semiarid, variable vegetation	Flat, rolling hills, arid, remote
Structural style	Interior Cratonic Basin, numerous anticlines, monoclines, minor to severe structure and faulting (especially in SE)	Foreland basin, NE–SW depositional strike, anticline structures	Foreland thrust; elongated NE–SW syncline, minor structures, moderate faulting and folding	Laramide, intermontane basins; elongated NE–SE, asymmetrical syncline; minor faulting and folding, very deep at axis	Laramide, foreland basin; elongate N–S, asymmetrical syncline (dips westward), some faulting
Reported estimated methane original gas-in-place (DAF), Tcf	13.0	1.40	61.00	81.00	62.00
Estimated methane original gas-in-place ultimately recoverable (DAF), Tcf	2.00	0.14	10.57	8.39	15.00
Basin play area, acres	33,920,000	536,960	19,392,000	4,288,000	16,512,000
Total basin play acreage required to account for original gas-in-place estimation (i.e., "OGIP area")	Not reported	354,394	13,225,344	3,087,360	8,866,944

(Continued)

— cont'd

Basin Play/Area	Illinois (IL)	Maverick (TX)	Northern Appalachian	Piceance	Powder River
Percentage of total OGIP acreage required to account for original gas-in-place estimation	Not reported	66.00%	68.20%	72.00%	53.70%
Total CBM produced (recorded or estimated), Tcf	0.004	0.001	0.001	0.12	3.52
Remaining estimated recoverable, Tcf	Not reported	0.14	10.40	8.29	11.47
Average original gas-in-place per well unit, mcf	Not reported	314,667	363,214	2,103,722	559,174
Total OGIP area proven accessible by drilling and production activity, acres	Not reported	1200	19,600	23,520	1,429,040
Total OGIP area not proven by drilling and production activity, acres	Not reported	353,194	13,205,744	3,063,840	7,437,904
Average gross coal footage within OGIP area	Not reported	22.5	15.0	37.5	56.3
Average midpoint depth of coal targets, representative of OGIP area	Not reported	1900	1750	4000	1500
Unit size well spacing applied to this analysis, acres	Not reported	80	80	80	80

Basin Play/Area	Raton (CO)	San Juan	Uinta	Western Mid-Continent
State(s)	CO, NM	CO, NM	UT	OK, KS, AR, MO, IA, NE
Primary coal formation(s)	Raton Formation, Vermejo Formation	Fruitland Formation, Menefee Formation	Blackhawk Formation, Ferron Sandstone (SS)	McAlester Formation, Hartshorne Formation, Cabaniss Formation, Krebs Formation, Marmaton Group, Cherokee Group
Topographical setting	Moderate—rugged, semiarid to forested	Moderate—rugged, semiarid, forested	Gently rolling—rugged, semiarid, remote areas	Flat—moderately hilly, farmland
Structural style	Laramide, asymmetrical, numerous igneous sills and dikes, minor structures and faulting	Laramide, asymmetric, monocline; minor structures and local faulting	Laramide, intermontane; asymmetrical syncline; structurally complex	Elongated trough E-W, very complex in areas, faulting and folding in eastern portions of basin Interior Cratonic Basin; dips westward, localized folding, limited faulting Interior Cratonic Basin; oval, elongated NE-S, possible faulting, little subsurface detail
Estimated methane original gas-in-place (DAF), Tcf	10.00	84.00	10.00	12.40
Reported estimated methane original gas-in-place ultimately recoverable (DAF), Tcf	6.04	18.50	8.88	4.70
Basin play area, acres	1,408,000	4,800,000	9,248,000	54,328,320
Total basin play acreage required to account for original gas-in-place estimation (i.e., "OGIP area")	830,720	2,904,000	1,895,840	6,452,091

(Continued)

— cont'd

Basin Play/Area	Raton (CO)	San Juan	Uinta	Western Mid-Continent
Percentage of total OGIP acreage required to account for original gas-in-place estimation	59.00%	60.50%	20.50%	11.88%
Total CBM produced (recorded or estimated), Tcf	0.91	15.62	0.95	0.67
Remaining estimated recoverable, Tcf	5.13	2.88	7.90	4.02
Average original gas-in-place per well unit, mcf	963,455	2,313,427	420,534	570,276
Total OGIP area proven accessible by drilling and production activity, acres	226,720	523,120	67,200	445,680
Total OGIP area not proven by drilling and production activity, acres	604,000	2,380,880	1,828,640	6,006,411
Average gross coal footage within OGIP area	24.0	67.5	15.0	21.8

	Bighorn	Black Mesa		Deep River/Richmond	Denver	Eastern Anthracite Fields
Average midpoint depth of coal targets, representative of OGIP area	2750	3500		3900	1867	
Unit size well spacing applied to this analysis, acres	80	80		80	80	

Minor Basin/Play Areas

Basin Play/Area	Bighorn	Black Mesa	Coosa	Deep River/Richmond	Denver	Eastern Anthracite Fields
State(s)	WY, MT	AZ	AL	NC	CO	PA
Primary coal formation(s)	Fort Union, Mesa Verde Group	Mesa Verde Group		Cummock Formation, Tuckahoe Formation	Cretaceous Laramie Formation	Peach Mountain, Mammoth, Priimrose, Buck Mountain
Topographical setting	Flat, arid, very remote			Flat to rolling hills, farmland, some forested areas	Flat, few rolling hills, semiarid, grass and range land	Moderate to rugged, dissected terrain, heavy vegetation, populated
Structural style	Laramide, intermontane, elongate N–S, some faulting			Triassic-age, isolated coalfields, half-graben, structurally complex in areas	Laramide, foreland basin, asymmetric syncline, dips westward, minor structures and faulting	Steeply dipping and discontinuous, folding and faulting frequently distorts and displaces the coal beds; complex stratigraphy

(Continued)

— cont'd

Basin Play/Area	Bighorn	Black Mesa	Coosa	Deep River/Richmond	Denver	Eastern Anthracite Fields
Estimated methane original gas-in-place (DAF), Tcf	3.00	2.00	0.50	0.50	2.00	4.00
Estimated methane original gas-in-place ultimately recoverable (DAF), Tcf	0.83	0.18	0.05	0.05	0.30	0.05
Basin play area, acres	1,920,000	2,048,000	64,000	160,000	4,800,000	640,000
Total CBM produced (recorded or estimated), mcf	16,000	1000	150,000	4000	5000	6000

Basin Play/Area	Williston Basin	Hanna-Carbon	Henry Mountains	Kaiparowits	Michigan Basin	North Central Coal Region
State(s)	ND, SD, MT	WY	UT	UT	MI	MT
Primary coal formation(s)	Sentinel Butte Formation, Bullion Creek Formation, Slope Formation, Cannonball Formation, Ludlow Formation	Hanna Formation, Ferris Formation, Medicine Bow Formation, Almond Formation	Masuk Formation, Ferron SS	Straight Cliffs Formation	Saginaw Formation	Fort Union, Mesa Verde

Topographical setting	Flat, grass and range land	Flat to rolling hills, grassland		Dissected mesa, very diverse		Flat to moderate terrain, very remote areas, variable vegetation cover
Structural style	Flat, gently dipping, shallow	Laramide, intermontane basins, synclinal axis, numerous faults and some folding		Colorado Plateau, dissected mesa, northerly trending folds, anticlines, and monoclines, Peripheral faulting	Small, circular basin, shallow and gently dipping toward center, little structural activity	Laramide, intermontane basins, and coalfields, simple to complex structure, thick glacial cover in areas
Estimated methane original gas-in-place (DAF), Tcf	10.00	15.00	1.10	10.00	1.00	3.70
Estimated methane original gas-in-place ultimately recoverable (DAF), Tcf	0.50	4.37	0.11	0.20	0.10	1.20
Basin play area, acres	20,480,000	768,000	288,000	1,056,000	3,200,000	12,800,000
Total CBM produced (recorded or estimated), mcf	200,000	8000	4000	1000	1000	40,000

Basin Play/Area	Over Thrust Belt	Sand Wash	Southwest Coal Region	Southwest Washington (Chehalis, Centralia)	US Gulf Coast—Central Louisiana Coalbed Methane (Central Louisiana)
State(s)	NM, WY	CO	TX, OK	WA	LA
Primary coal formation(s)	Evanston/Adaville	Mesa Verde Group, Fort Union Formation	Harpersville Formation, Mingus Formation	Skookumchuk Formation	Wilcox equivalent
Topographical setting		Flat to moderate hills, semiarid	Flat to moderately hilly, variable vegetation		
Structural style	Very complex, series of over thrusts many faults, erosion unconformities	Laramide, intermontane basins	Eastern shelf of midland basin, Forth Worth Basin, fairly simple to moderate structurally	Unknown	Flat, shelf area, dips to south
Estimated methane original gas-in-place (DAF), Tcf	5.00	39.50	29.00	0.50	2.00
Estimated methane original gas-in-place ultimately recoverable (DAF), Tcf	0.50	3.95	5.80	0.05	0.20
Basin play area, acres	1,920,000	3,200,000	19,200,000	640,000	3,200,000
Total CBM produced (recorded or estimated), mcf	310,000	80,000	40,000	1,000	360,000

APPENDIX B: BASIN/PLAY AREA ZERO-TIME ANALYSES

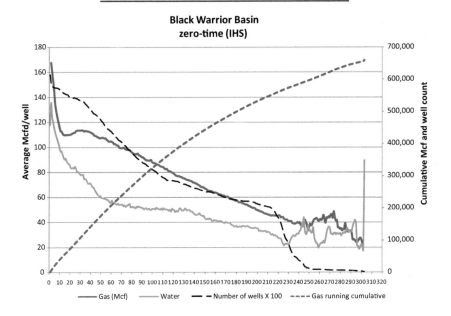

Black Warrior Basin zero-time (IHS)

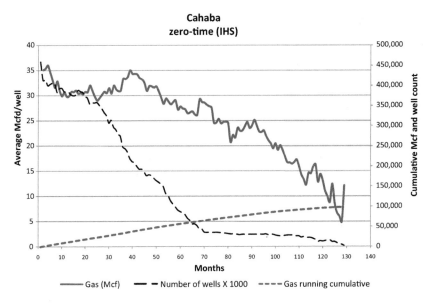

Cahaba zero-time (IHS)

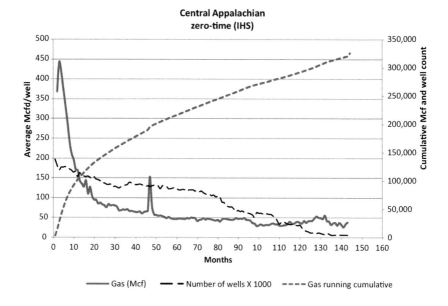

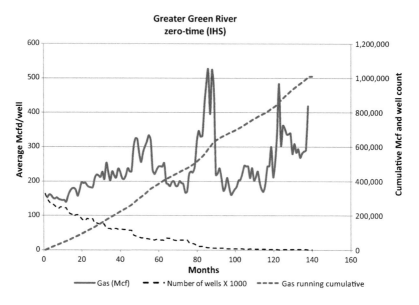

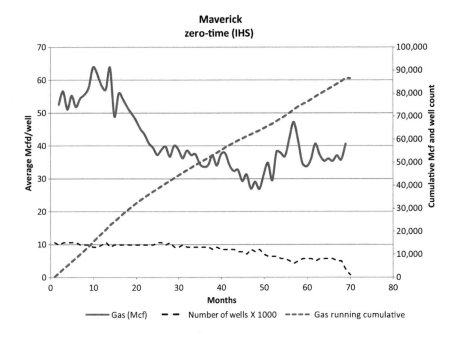

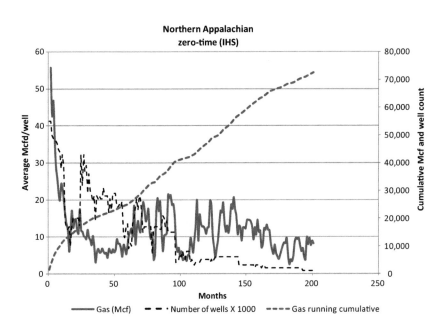

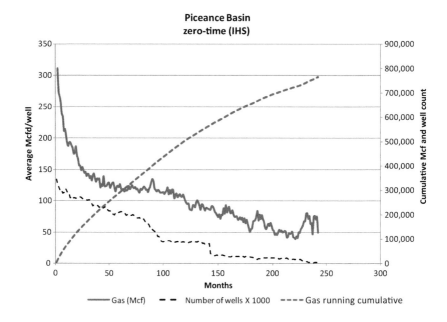

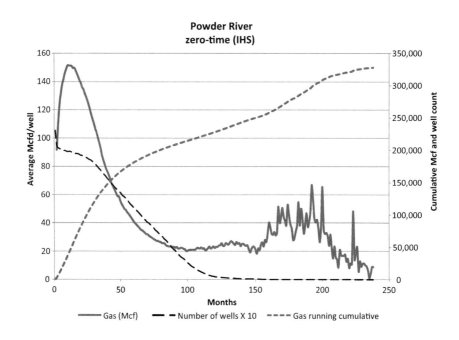

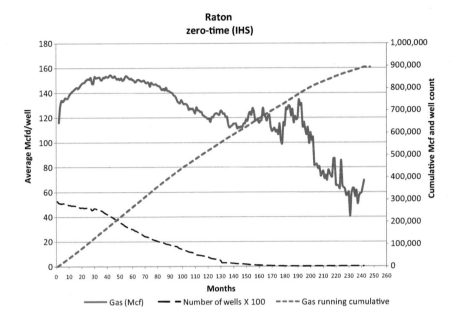

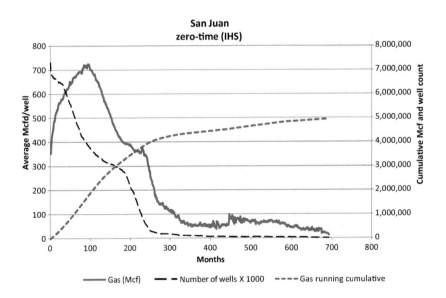

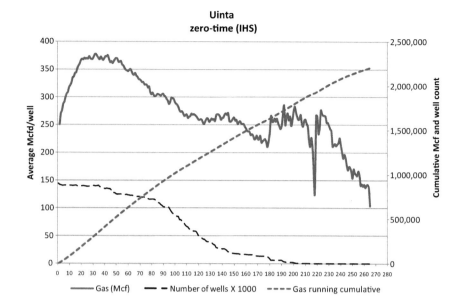

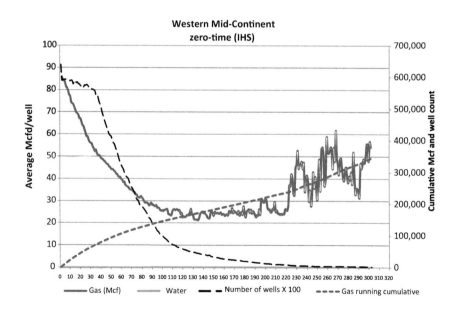

APPENDIX C: BASIN/PLAY AREA PRODUCTION DISTRIBUTIONS

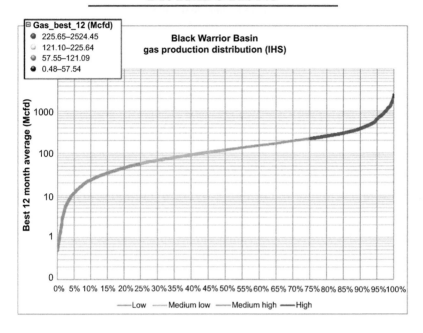

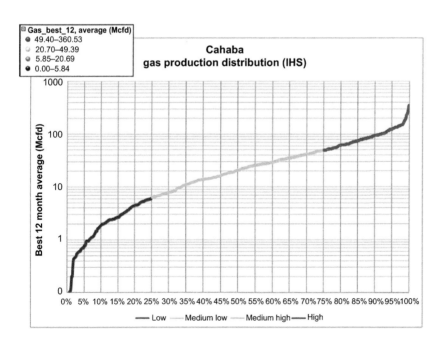

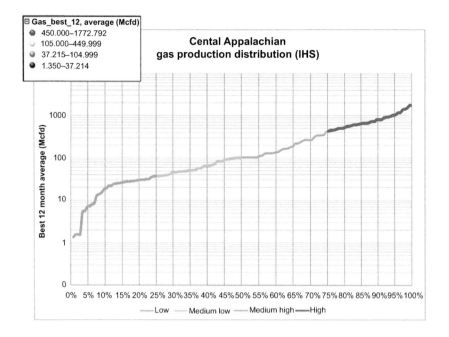

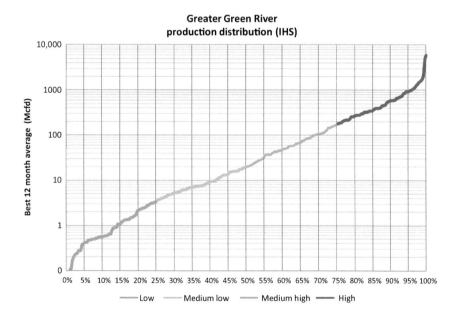

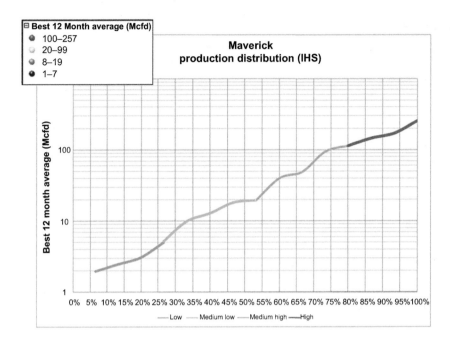

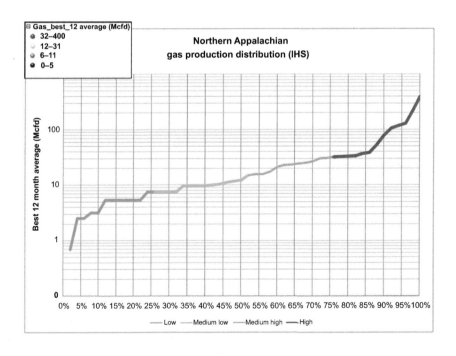

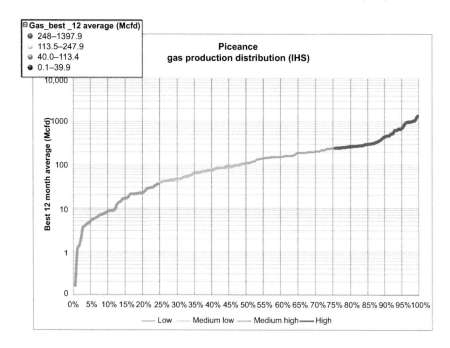

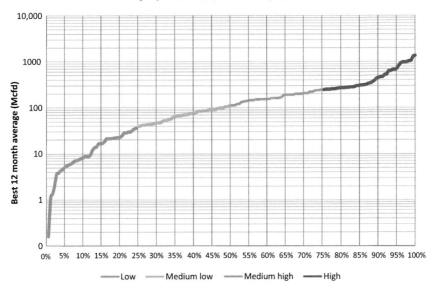

Piceance
gas production distribution (IHS)

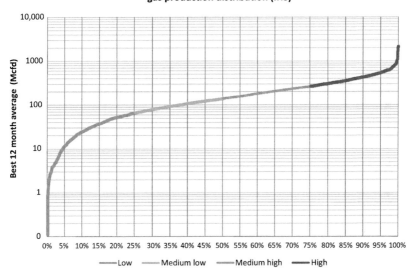

Raton Basin
gas production distribution (IHS)

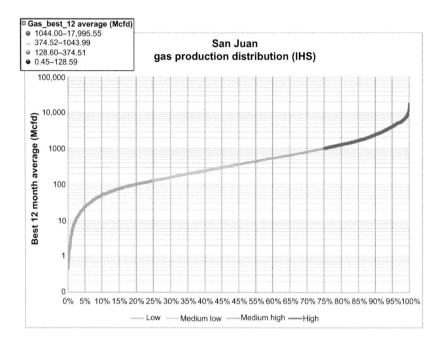

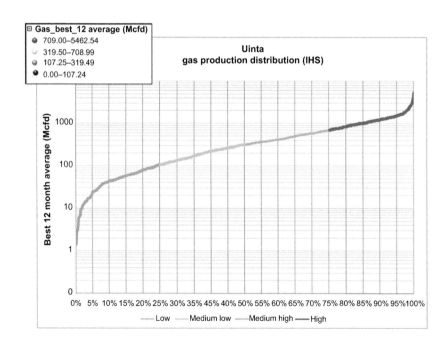

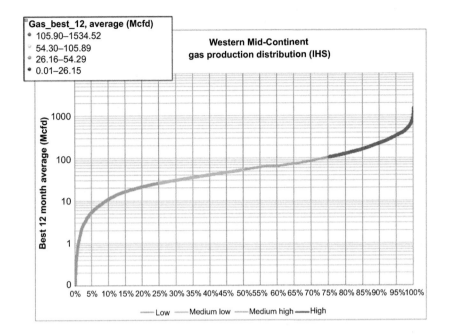

Worldwide Coal Mine Methane and Coalbed Methane Activities

Charlee Boger, James S. Marshall, Raymond C. Pilcher

Raven Ridge Resources Inc., Grand Junction, Colorado, USA

18.1 INTRODUCTION

Since the 1980s, coalbed methane (CBM) has gained attention as a saleable natural gas resource. CBM can be extracted from coal seams, which will never undergo mining (CBM), or it can be produced as a part of the coal mining process as coal mine methane (CMM). The largest CBM reserves lie in the former Soviet Union, Canada, China, Australia, and the United States as shown in Figure 18.1 below, but much of the world's CBM remains untapped. In 2006 it was estimated that of global resources totaling 143 trillion cubic meters, only 1 trillion cubic meters has been recovered (WCI, n.d.). In the United States, CBM supplies about 10% of the market for natural gas, a position largely made possible by the nonconventional fuels federal tax incentives of 1980–2002. The United States has to date produced about 600 US billion cubic meters of CBM (BP, 2007). Worldwide, CBM exploration and production is underway.

Currently, over 200 CMM projects developed in 14 countries are capturing 3 billion cubic meters per year (OECD/IEA, 2008). The remaining methane is vented to the atmosphere, representing the loss of a valuable energy resource.

There are three primary reasons for recovering CMM. The first is increased mine safety. Worldwide, there have been thousands of recorded fatalities from underground mine explosions in which methane was a contributing factor. Using methane drainage systems, mines can reduce the methane that must be removed by their ventilation air, ultimately reducing mine explosions.

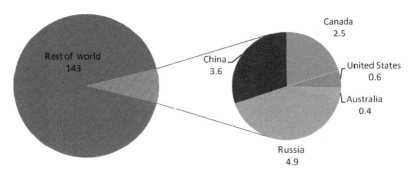

FIGURE 18.1   Global coalbed methane reserves (trillion cubic meters). *WCI (n.d.), AAGA (2009), BCMEM (2008), NSDOE (2009), GOA (2010), EIA (2009b), Gazprom (2008).*

The second reason is to improve mine economics. By reducing emissions and preventing explosions and outbursts, methane drainage systems can cost effectively reduce the amount of time that the coal mine must curtail production. Moreover, recovered methane can be used either as fuel at the mine site or sold to other users to generate revenue for the mine or a developer.

The third reason for CMM recovery and use is that it benefits the global and local environment. Methane is a major greenhouse gas and is second in global impact only to carbon dioxide; methane thus is detrimental to the environment if vented to the atmosphere. Each additional ton of methane released to the atmosphere is as much as 21 times more effective in potentially warming the Earth's surface over a 100-year period than each additional ton of carbon dioxide.

18.2 STATUS OF CMM AND CBM IN SELECTED COUNTRIES

There are many opportunities for CBM production as well as decreasing CMM emissions by increasing recovery of this abundant fuel. The following section examines the status of CBM and CMM recovery and use in key countries worldwide. Additional countries not covered in this report that show potential for CMM and CBM projects include Argentina, the Republic of Korea, New Zealand, and Nigeria.

18.2.1 Australia

Australia produces coal primarily from the eastern portion of the continent in Queensland, New South Wales (NSW), and Victoria. Total coal reserves are estimated at nearly 76.2 billion metric tons (EIA, 2009a). About 396 million metric tons of coal were produced from 118 privately

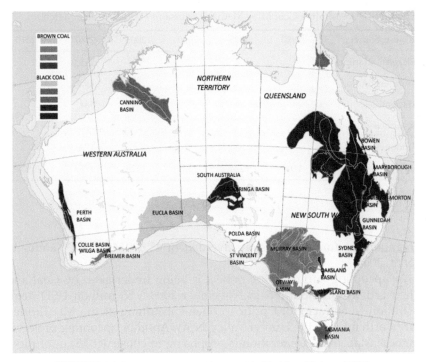

FIGURE 18.2   Australia's coal basins. *Modified from Huleatt (2009).*

owned mines in 2008 (EIA, 2009a; Australian Black Coal Statistics, 2006) (Figure 18.2).

In 2007, emissions associated with coal mining and handling and abandoned or decommissioned mines were 26.8 million metric tons of CO_2e (1877 million cubic meters of methane) (AGDCC, 2009). Australia's CMM emissions for several years are summarized below (Table 18.1).

An estimated 6.4 million metric tons of CO_2e (448 million cubic meters of methane) are recovered from coal mines annually at a variety of projects including flaring, pipeline injection, power generation, and ventilation air methane (VAM) oxidation (Methane to Markets, 2009). At this time, there are 14 CMM projects in Australia as shown below (Methane to Markets, 2010) (Table 18.2).

A first-of-its-kind CMM project has been developed in Australia by BHP Billiton, which constructed a CMM power station at the West Cliff Colliery, near Wollongong, NSW to allow the combustion of very dilute methane contained in coal mine ventilation air. The West Cliff VAM Project (WestVAMP) was commissioned in February of 2007 (Booth, 2008) and produces 6 MW of electricity (Methane to Markets, 2010). WestVAMP is estimated to reduce methane emissions by 250,000 Mt of CO_2 equivalent (CO_2e) each year. In addition, the project displaced coal-fired electricity

TABLE 18.1   Australia's Coal Mine Methane Emissions (Million Cubic Meters)

Emission Source	1990	1995	2000	2005	2007[1]	2010
Underground mines					1123	
Postunderground					49.7	
Surface mines					584	
Postsurface						
Abandoned mines					122	
Total	1108	1224	1375	1528	1757	1847[1]

[1] Projected.
USEPA (2006), UNFCCC (2009).

generation for further emissions avoided of 45,000 Mt of CO_2 annually (Booth, 2008).

An upcoming surface mine project is being undertaken by Coal and Allied Ltd who announced in November 2008 a $5 million USD demonstration to extract CBM prior to open-cut mining at the Mt. Thorley Warkworth mine in the Hunter Valley, NSW. Another upcoming project is planned; Macquarie Generation is proposing to construct 75 km of pipelines from underground coal mines to supply methane gas as a supplementary fuel to the Liddell Power Station in the Hunter Valley, NSW.

As of December 2008, the proven and probable reserves of CBM in Australia were 428 billion cubic meters (AAGA, 2009). CBM production began in Australia in 1996; and in 2008, CBM production in Australia was estimated at 3.6 billion cubic meters (AAGA, 2009), with Queensland producing 3.5 billion cubic meters (or 96% of CBM production) from the Bowen and Surat Basins, and NSW producing 140 million cubic meters from the Sydney Basin.

Some key developments have occurred recently in the CBM industry, including:

- Arrow Energy is planning to build an initial 1.5 million ton per year Liquefied natural gas (LNG) plant at Fisherman's Landing near Gladstone in Queensland with production starting in 2012.
- ConocoPhillips and Origin formed a joint venture to commercialize Origin's CBM reserves. The partners plan to develop four LNG plants at Gladstone with production from the first two 3.5 million ton per year plants, which are expected in 2014.
- In late 2009 Origin is expecting to commission a $780 million 630-MW gas-fired power station on the Darling Downs using CBM from Spring Gully, Queensland.
- In May 2008 Petronas bought a 40% stake in Santos's Gladstone LNG project. The partners plan to build a $5–$7 billion LNG project

TABLE 18.2 Australia's Coal Mine Methane Projects

Region Name	Name	Project Type	Mine Status	First Year of Project Operation	Mine Owner
New South Wales	Appin Colliery	VAM as auxiliary fuel for combustion air	Abandoned		Illawarra Coal (BHP Billiton)
New South Wales	Mandalong	Power generation	Active (underground)	2008	Centennial Coal
New South Wales	Munmorah & Endeavour	Power generation	Abandoned	2009	Centennial Coal
New South Wales	Newvale	Boiler fuel	Abandoned	2002	Centennial Coal
New South Wales	Tahmoor Colliery	Power generation	Active (underground)	1999	Centennial Coal
New South Wales	Teralba/ Bellambi Mines	Power generation	Abandoned	2003	Xstrata
New South Wales	United Colliery	Power generation	Active (underground)	2006	United Collieries Pty Ltd (Xstrata)
New South Wales	West Cliff Colliery	Power generation	Active (underground)		Illawarra Coal (BHP Billiton)
New South Wales	West Cliff Colliery	VAM as primary fuel for power generation	Active (underground)	2007	Illawarra Coal (BHP Billiton)
Queensland	Central Colliery	Flaring	Abandoned	1999	Anglo Coal
Queensland	Dawson Mine	Pipeline injection	Active (underground)	1996	
Queensland	German Creek (Grasstree)	Power generation	Active (underground)	2007	Anglo Coal (70%) and Mitsui (30%)
Queensland	Moranbah North Mine	Power generation	Active (underground)		Anglo Coal (51%) and Mitsui (49%)
Queensland	Oaky Creek	Power generation	Active (underground)	2006	Xstrata

VAM, ventilation air methane.
Methane to Markets (2010).

at Gladstone with a capacity of 3–4 million metric tons per year commencing in 2014.

- Queensland Gas plans to build an $8 billion LNG plant in Gladstone with a capacity of 3–4 million metric tons per year commencing in 2013.
- A 135-MW CBM-fired Condamine, Queensland power station was scheduled to be commissioned in 2009.
- Metgasco Ltd plans to build a 30-MW gas-fired Richmond Valley, NSW power station to be fueled by CBM over 15 years. Metgasco also plans to provide CS Energy's Swanbank Power Station in Ipswich with CBM via the Lions way pipeline.
- Eastern Star Gas plans to construct a 32-km pipeline from the Narrabri Coal Seam Gas Project to the existing Wilga Park power station in NSW, which is planned to be expanded from 12 to 40 MW.

18.2.2 Botswana

Botswana has coal reserves estimated at 212 billion metric tons, with economically viable coal reserves estimated at 3–5 billion metric tons (Abi, 2009; MacauHub, 2009). Coal reserves by coalfields are summarized below (Table 18.3) (Figure 18.3).

TABLE 18.3   Botswana's Coal Reserves in Million Metric Tons

Coalfield	Measured (Mt)	Indicated (Mt)	Inferred (Mt)	Total (Mt)
Morupule	2864	2706	12,520	18,090
Moijabana	0	0	3054	3054
Mmamabula	4325	16,394	2504	23,223
Letlhakeng	0	7213	63,140	70,353
Ncojane	0	0	4725	4725
Dukwi	0	32	1604	1636
Mmamantswe	0	0	2898	2898
Serule	0	307	9377	9684
Dutlwe	0	2070	69,670	71,740
Foley	0	0	6860	6860
Bobonong	0	0	179	179
Total	7189	28,722	176,531	212,442

Abi (2009).

FIGURE 18.3   Botswana's coalfields. *Abi (2009)*.

Botswana produced 1.036 million metric tons of coal in 2008 (EIA, 2009b) from the Morupule Colliery and supplies power to the Morupule power station. As coal is mined from relatively shallow depths and is not considered gassy, Botswana has no reported CMM emissions.

In 2001, the Botswana Department of Geologic Survey commissioned a study to investigate the CBM potential of the Central Kalahari Karoo Basin (Advanced Resources, 2003). The study estimated an in-place resource of 60 trillion cubic feet (1.7 trillion cubic meters) and an additional 136 TCF (3.8 trillion cubic meters) of gas-in-place in the associated carbonaceous shales. Kalahari Gas Corporation is actively pursuing CBM development and recently completed a five-well pilot program.

There are plans for a pilot CO_2 injection project at the site of the Kalahari Gas Corporation pilot program. The project would entail injection of a pure carbon dioxide stream brought from Sasol Secunda (South Africa) by pipeline to Botswana into the wells to enhance the pilot production of CBM (Abi, 2009).

18.2.3 Brazil

Brazil has recoverable lignite reserves estimated at 7–8 billion metric tons (EIA, 2009b; Süffert, 1997). Reserves by coalfield are summarized below (Table 18.4) (Figure 18.4).

Brazil produced 6.9 million metric tons of coal in 2008 (EIA, 2009b). Brazil has 15 coal mining companies, but the following eight companies produce most of Brazil's coal:

1. Copelmi, of Rio Grande do Sul has four surface mines and one underground mine.
2. Companhia Riograndense de Mineraçao in Rio Grande do Sul produces mainly from surface mines, but has two underground mines.
3. Carbonifera Criciuma S.A. in Santa Catarina has an existing underground mine and is building a second.
4. Industria Carbonifera Rio Deserto Ltda. of Santa Catarina has two underground room and pillar mines.
5. Carbonífera Metropolitana of Santa Catrina has two underground room and pillar mines.
6. Companhia Carbonifera Catarinense of Santa Catrina has two underground room and pillar mines.
7. Carbonífera Belluno Ltda. has three coal mines, of which two are surface mines.
8. Cumbui of Paraná has four underground mines.

No CMM recovery occurs in Brazil. Brazil's CMM emissions are summarized below (Table 18.5).

Studies have identified the Santa Terezinha coalfield, Rio Grande do Sul, as a candidate for CBM exploration based on coal distribution, and size and depth of the coal beds. A CBM test well was drilled in early 2007 near Osorio, RS, where samples from 12 coal seams and carbonaceous shales were collected. Integration of coal volume determined by

TABLE 18.4   Brazil's Coal Reserves in Million Metric Tons

Coalfield	Measured	Indicated	Inferred	Total
Candiota	209,351	978,637	232,852	1,520,840
Capané	27,912	14,020	0.26	42,192
Iruí	16,935	92,858	7708	117,914
Leão	366,945	326,502	21,708	715,854
Chico Lomã	37,518	178,289	169,138	384,945
Sul Catarinense	9603	37,671	7961	5,523,595

Süffert (1997).

FIGURE 18.4   Brazil's coal deposits. *Süffert (1997).*

TABLE 18.5   Brazil's Coal Mine Methane Emissions (Million Cubic Meters)

Emission Source	1990[1]	1994[1]	1995	2000	2005	2010
Underground mines	77	69.5				
Postunderground	8	7				
Surface mines	1	1				
Postsurface	0.1	0.1				
Total	86	78	77	92	85	78[2]

[1] *Sources: MST (2002) in Methane to Markets (2009) and USEPA (2006).*
[2] *Projected.*

3D modeling with average desorbed methane values suggest a total amount of 5 billion cubic meters of methane associated with the coal seams and carbonaceous shales for the study area investigated (Busch et al., 2008).

18.2.4 Bulgaria

Bulgaria has recoverable brown coal and anthracite reserves estimated at 2 billion metric tons (EIA, 2009b; EIB, 2009). The reserves of brown coal are 800 million metric tons with major deposits near Pernik, Bobov Dol, and Cherno More mines. The reserves of bituminous coal are slightly more than 1.2 billion tons with more than 95% of these located in the Dobroudja Coal Basin. Reserves of lignite coal are estimated to 4.5 billion metric tons and they are located in the Maritsa Iztok Coal Basin (around 70%), Sofia Coal Basin, and the Lom coalfield (EIB, 2009). Coal basins are shown below (Figure 18.5).

Bulgaria produced 28.8 million metric tons of coal in 2008 (EIA, 2009b). Bulgaria's largest coal mining complex, Maritsa East Mines, plans an output of 26 million metric tons in 2010 and up to 30 million metric tons in 2011. The mines' output in 2008 reached 24.7 million metric tons of coal, up from 23.9 million metric tons in 2007 (SeeNews, 2009).

Though surface mining dominates Bulgaria's coal production, CMM emissions result from both underground and surface mining. Bulgaria's CMM emissions are summarized below, with no current CMM recovery occurring in Bulgaria (Table 18.6).

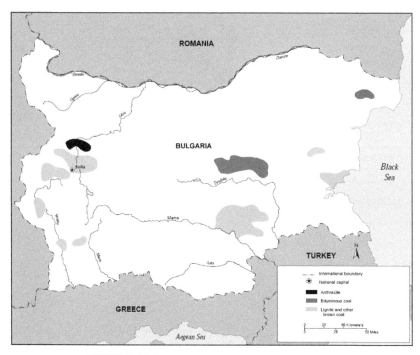

FIGURE 18.5   Bulgaria's coal deposits. *LOC (1992).*

TABLE 18.6   Bulgaria's Coal Mine Methane Emissions (Million Cubic Meters)

Emission Source	1990	1995	2000	2005	2010
Underground mines	66.64	58.55	47.10		
Postunderground	9.53	8.36	6.74		
Surface mines	35.80	23.91	20.65		
Postsurface	2.75	2.72	2.35		
Total	114.72 (111[1])	93.54 (102[1])	76.84 (84[1])	94[1]	116[1,2]

[1] USEPA (2006).
[2] Projected.
Methane to Markets (2009).

The majority of CBM exploration and quantification has taken place in the black coal deposits of the Dobroudja Basin, where the total CBM potential is 195 billion cubic meters. The average depth of the coal bearing units is over 1300 m with thicknesses ranging from 30 to 45 m (net). CBM Energy Ltd explored the potential for CBM beginning in 1997. In 2000, they were granted an exploration license for a block of land in the basin and moved forward with the evaluation process by drilling an exploration well. To date, no production has been achieved (World Coal, 2001).

18.2.5 Canada

Canada has recoverable coal reserves estimated at 6.6 billion metric tons (EIA, 2009b). The Western Canada Sedimentary Basin (WCSB) contains about 90% of Canada's mineable coal resources and Nova Scotia the other 10%. The WCSB is a vast sedimentary basin underlying 1,400,000 km[2] of Western Canada including southwestern Manitoba, southern Saskatchewan, Alberta, northeastern British Columbia, and the southwest corner of the Northwest Territories. The WCSB is outlined in the figure below (Figure 18.6).

The figure below shows coal deposits in the WCSB as well as the rest of Canada (Figure 18.7).

There were 22 coal mines operating in Canada at the end of 2007 with 17 of these mines located in British Columbia and Alberta. These are the two highest producing provinces accounting for more than 80% of Canada's coal production (NRCAN, 2008). Canada produced 68 million metric tons of coal in 2008 that includes 58 million metric tons of bituminous coal and 10 million metric tons of lignite (EIA, 2009b).

Canada's CMM emissions have declined in the recent years due primarily to a decline in underground coal production. In 1995 Canada produced 5.31 million metric tons of coal from underground mines, compared to 1.61

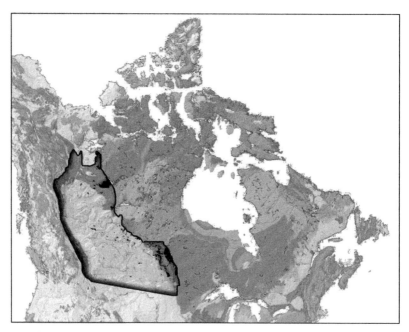

FIGURE 18.6   Outline of Western Canada Sedimentary Basin. *WCSB (2009)*.

FIGURE 18.7   Canada's coal deposits. *CAC (2006)*.

million metric tons of coal, which was mined in 2007 (UNFCCC, 2009). CMM emissions in 1995 were approximately 120 million cubic meters compared to 53.6 million cubic meters in 2007 (Table 18.7).

CBM exploration in Alberta, Canada commenced in the early 1990s, but until 2000 commercial development had not taken off. As of mid-2008, there were over 6000 producing CBM wells almost all of which were drilled since 2004 with a cumulative production rate of about 14 million cubic meters per day (Bustin et al., 2008).

The Alberta Geological Survey estimates there may be up to 14 trillion cubic meters of natural gas in Alberta's coals. It is not known how much of this gas may be economic to produce, but the majority (over 90% in 2006) of CBM development is taking place in the dry Horseshoe Canyon and Belly River coals of south-central Alberta (GOA, 2010). Estimated reserves and coal characteristics are summarized below (Table 18.8).

TABLE 18.7   Canada's Coal Mine Methane Emissions (Million Cubic Meters)

Emission Source	1990	1995	2000	2005	2007[1]	2010
Underground mines					6.65	
Postunderground						
Surface mines					46.9	
Postsurface						
Total	134	120	66.5	61.6	53.6	61.6[2]

[1] *UNFCCC (2009).*
[2] *Projected.*
USEPA (2006).

TABLE 18.8   Alberta's Coalbed Methane Deposits

Province	Region/ Coalfield	Estimated Reserves (Billion Cubic Meters)	Characteristics
Alberta	Horseshoe Canyon	2000	Thinner coal seams, produce little or no water, depth 200–800 m
Alberta	Belly River	NA	
Alberta	Mannville	6800	Thicker coal seams, substantial saline water, depth 900–1500 m
Alberta	Ardley coals of Scollard formation	1600	Both saline and nonsaline water as well as no water, depth 350–700 m
Alberta	Kootenay	NA	Small volumes of water

GOA (2010).

Alberta's coals are located in the southern part of the province as shown in the figure below (Figure 18.8).

There is no full-scale commercial production of CBM in British Columbia. As of December 2007, a total of 87 CBM wells and test holes had been drilled. British Columbia has an estimated resource potential of 2.4 trillion cubic meters of CBM from coalfields around the province (BCMEM, 2008). Estimated reserves by region or coalfield are summarized below (Table 18.9).

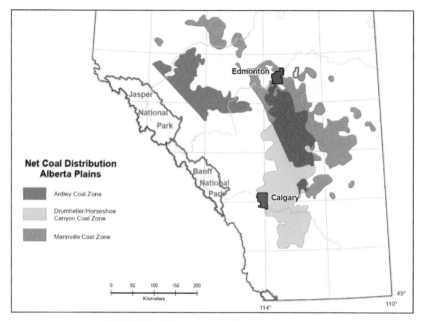

FIGURE 18.8   Coals of Alberta. *GOA (2010).*

TABLE 18.9   British Columbia's Coalbed Methane Deposits

Province	Region/Coalfield	Estimated Reserves (Billion Cubic Meters)
British Columbia	East Kootenay	400
British Columbia	Northwest British Columbia	220
British Columbia	Peace River	1690
British Columbia	Vancouver Island	8.5–45
British Columbia	Hat Creek	14

BCMEM (2004).

British Columbia has reserves of anthracite, medium- and high-volatile bituminous, and subbituminous to lignite coals. Locations of the coalfields are shown below (Figure 18.9).

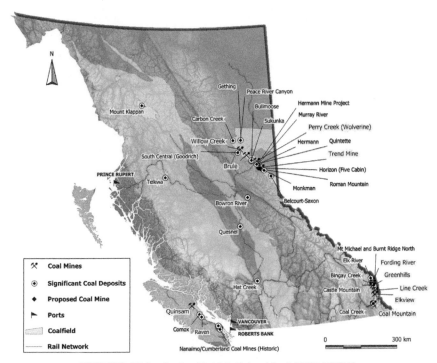

FIGURE 18.9   Coals of British Columbia. *BCMEM (2013)*.

CBM exploration has occurred in Nova Scotia in the Cumberland and Stellarton Basins. Estimated reserves are shown below.

Province	Region/Coalfield	Estimated Reserves (Billion Cubic Meters)
Nova Scotia	Cumberland and Stellarton	45

NSDOE (2009).

The map below shows CBM agreements that have been made with East Coast Energy Inc., Donkin Tenements Inc., and Stealth Ventures, Ltd (Figure 18.10).

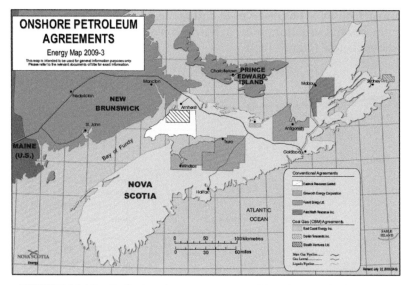

FIGURE 18.10   Coalbed methane agreements in Nova Scotia. *NSDOE (2009a).*

18.2.6 China

China has total recoverable coal reserves of 114.5 billion metric tons, of which 62.2 billion metric tons is hard coal (EIA, 2009b). There are 27 provinces in China that produce coal. Northern China, especially Shanxi Province, contains most of China's easily accessible coal and virtually all of the large state-owned mines (EIA, 2009d). The map below shows China's widespread coal reserves (Figure 18.11).

China produced 2.5 billion metric tons of hard coal in 2008 including 489 million metric tons of anthracite, 2 billion metric tons of bituminous coal, and 101 million metric tons of lignite in 2008 (EIA, 2009b). Of the estimated 27,500 mines in China, about 3000 of these mines are state owned and the majority of mines, approximately 24,500, belong to villages and towns. In 2004 underground mining accounted for 90% of China's coal production—a drop from 95% in 1996 (Methane to Markets, 2009).

China has the highest CMM emissions of any country, corresponding with being the largest coal producer in the world (USEPA, 2006; EIA, 2009d). China's CMM emissions are summarized in the table below (Table 18.10).

China has 38 CMM recovery and utilization projects at active mines and 17 in development. These projects span 26 coal mining groups and include power generation, town gas, boiler fuel, combined heat and power (CHP), and industrial use. A VAM project is operating at Zhengzhou mining group in Henan province. This project abates VAM as well as recovers energy in the form of hot water for local use. Another VAM project is in development in Anhui Province at the Huainan Mining Group's Panyi Mine.

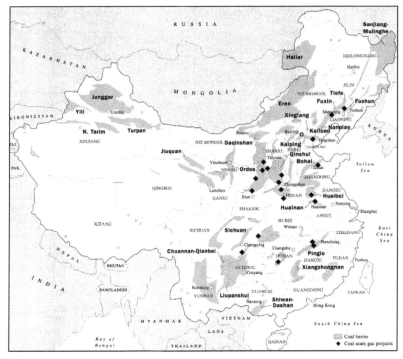

FIGURE 18.11   China's coal reserves. *Schwochow (1997).*

TABLE 18.10   China's Coal Mine Methane (CMM) Emissions (Million Cubic Meters)

	1990	1995	2000	2005	2010
Total CMM emissions	8832	10,441	8233	9500	10,767[1]

[1] *Projected.*
USEPA (2006).

China has no abandoned mine methane (AMM) recovery projects (Methane to Markets, 2010). The China Coal Information Institute established the Abandoned Mine Methane Project Advice Centre to advise and promote country's AMM use. The closing of many under-producing and unsafe mines leaves many mines abandoned, emitting a large amount of AMM (Methane to Markets, 2009).

CBM in-place resources in China are estimated to be 36 trillion cubic meters, the world's third largest (EIA, 2009d). The figure below shows the distribution of China's CBM resources (Figure 18.12).

Prior to the late 1980s, no surface CBM exploitation occurred in China. In the late 1980s CBM development activities including resource

CHINA'S COAL BASINS AND COALBED METHANE RESOURCES

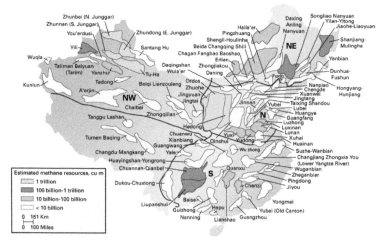

FIGURE 18.12   China's coalbed methane resources. *Liu (2007) in Methane to Markets (2009).*

evaluation, exploration, and well testing commenced. In 1996 the Chinese government formed a monopoly CBM company (CUCBM) to develop CBM with foreign companies, which enhanced exploration activities. By 2006, about 1000 CBM test and pilot wells were completed within more than 30 basins by domestic and foreign companies. Much of this activity occurred in the Qinshui basin of Shanxi Province. Since 2006, the Chinese government has started to encourage CBM development in coal mining areas and to allow coal mines to cooperate directly with international CBM players. The government is expected to revise the existing Mineral Resources Law to change the way licenses for CBM production and coal mining are granted (Guanghua, 2008; Fang, 2009).

China's CBM production in 2008 was 5.8 billion cubic meters, up from 1.4 billion cubic meters in 2006. China's 2006–2010 CBM development plan has a target of 10 billion cubic meters of CBM production annually by 2010 (Fang, 2009; Methane to Markets, 2009).

18.2.7 Colombia

Colombia's principal coal resources are found in the northwestern half of the country in the Guajira, Cesar, Magdalenas, Catatumbo, Cauca, and Bogata basins. Recoverable coal reserves are estimated at 7 billion metric tons, 6.6 billion metric tons of which is anthracite and bituminous (EIA, 2009a). With almost 79 million metric tons produced in 2008, Colombia's coal production has been steadily rising, more than doubling since 2000 (EIA, 2009a). Below is a map of Colombia's coal basins (Figure 18.13).

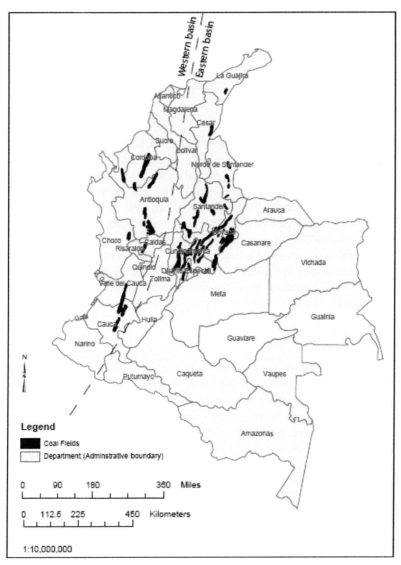

FIGURE 18.13   Colombia's coal basins. *(Tewalt et al., 2006).*

In 2010, Colombia's coal mines are projected to emit approximately 282 million cubic meters of methane (USEPA, 2006). A pilot project to measure methane emissions is underway at the Pribbenow Mine, located near La Loma in the Cesar Basin, which has estimated reserves in excess of 534 million metric tons of high-Btu, low-ash, low-sulfur coal. There have been no published results (Table 18.11).

TABLE 18.11   Colombia's Coal Mine Methane (CMM) Emissions (Million Cubic Meters)

	1990	1995	2000	2005	2010
Total CMM emissions	130	139	207	241	282[1]

[1] *Projected.*
USEPA (2006).

The current potential for CMM projects in Colombia is limited as most of the coal is mined at surface mines, although as mines target deeper seams there should be significant potential for CMM projects.

Drummond Company, Incorporated has studied a 10,000 ha area of La Loma (main coal mine in the Cesar Basin) and 30,000 ha surrounding the regions of Descanso, Guaymaral, Rincón Hondo, and Similoa. CBM reserves are estimated at 65 million cubic meters (EIA, 2009d).

A Colombian company, Andina Electrica, is conducting a CBM exploration project near the Cerrejon mine area. Several wells have been drilled and the coal had favorable gas contents. Andina plans to use the CBM for power generation and sell the power to the Cerrejon mine.

18.2.8 Czech Republic

The Czech Republic has recoverable coal reserves of 4.5 billion metric tons (EIA, 2009a). About one-fourth of the Upper Silesian Coal Basin (USCB) lies in the northeastern Czech Republic where it is known as the Ostrava–Karvina Coal Basin. Anthracite and bituminous (hard) coal production is centered on the Ostrava–Karvina district of the Czech section of the Upper Silesian coalfield, although hard coal resources occur in other areas: North Moravia; Kladno, Central Bohemia; Rosice, South Moravia; and Plzen, South Bohemia. The most important lignite (brown coal) resources are located at Plzen, South Bohemia; Northern Bohemian basin; Sokolov, West Bohemia; Hodonin, and Southern Moravia; with production centered on North Bohemia (Figure 18.14).

The Czech Republic produced 60 million metric tons of coal in 2008 (EIA, 2009b). Kladno is the oldest producing coalfield in the Czech Republic, with continuing limited output of bituminous coal; but hard coal mining in Trutnov, Rosice, and Plzen has been phased out.

Both Poland and the Czech Republic began actively draining methane, for safety reasons, in the late 1950s. All 10 of the Czech Republic's mines in the USCB have methane drainage systems (USEPA, 1995; USEPA, 1992). Methane is drained as coal extraction proceeds at the working face, from gob areas, and from development areas.

It is estimated that the Czech Republic uses about 100 million cubic meters of CMM annually for domestic purposes and for sale to European markets (USEPA, 2005). The Methane to Markets International Coal Mine

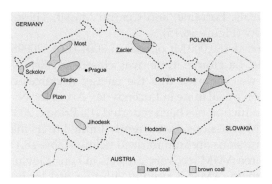

FIGURE 18.14   Czech Republic coal basin map. *Clarke and McConville (1998).*

TABLE 18.12   Czech Republic Coal Mine Methane (CMM) Emissions (Million Cubic Meters)

Emission Source	1990	1995	2000	2005	2010
Underground mines	384	297	260	239	
Postunderground	51	39	34	32	
Surface mines	90	65	53	51	
Postsurface	7.8	5.6	4.6	4.4	
Total CMM emissions	532	407	352	326	274[1,2]

[1] *USEPA (2006).*
[2] *Projected.*
UNFCCC (2009).

Methane Projects Database indicates that one project in Karviná, Brusperk, and Frenstát sells 77 million cubic meters per year of CMM to pipeline as well as 32 million cubic meters of AMM (Methane to Markets, 2010). Green Gas International BV reports to have 19 CHP projects operating in the Czech Republic for total annual emission reductions of 753,000 Mt of CO_2e (Green Gas, 2010) (Table 18.12).

The production of CBM from the Czech portion of the USCB reaches 40 million cubic meters per year. CBM reserves in the Czech portion of the USCB are estimated at a minimum of 100 billion cubic meters. During a national program to explore CBM in the Czech Republic, 17 blocks for CBM exploration were awarded in the Czech portion of the USCB. Proven reserves in these areas are reported to be 25 billion cubic meters (GSDG, 2009).

18.2.9 France

France has recoverable coal reserves estimated at 15 million metric tons (Methane to Markets, 2009) from three main coal basins: the

Nord-Pas-de-Calais, Lorraine, and Central Massif. These are delineated on the map below (Figure 18.15).

France produced 160,000 Mt of coal in 2004 (EIA, 2009b). France's last active coal mine, La Houve Mine, was closed in 2004 (Newman, 2007). In July 2006 it was reported that the energy resources company Seren planned to reopen a coal mine in Lucenay-les-Aix in the Nievre area; however, no coal production has been reported (Le Prioux, 2006).

France's CMM emissions are currently limited to its many abandoned mines. CMM emissions are summarized below (Table 18.13).

France has three AMM recovery projects in Lens-Liévin. At the Desiree & La Naville project, AMM is used as diluting fuel for boilers and in an ash dryer. At the Division project, AMM is used as fuel for a coke oven. The

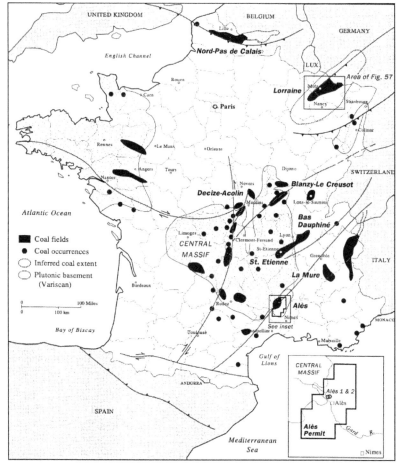

FIGURE 18.15   France's coal deposits. *Schwochow (1997).*

TABLE 18.13   France's Coal Mine Methane Emissions (Million Cubic Meters)

Emission Source	1990	1995	2000	2005	2010
Total	303	310	179	182	184[1]

[1] Projected.
USEPA (2006).

Methamine project is a pipeline injection project using AMM from three mines (Methane to Markets, 2010).

France's recoverable CBM resources are estimated at 320 billion cubic meters (Schwochow, 1997). European Gas Limited (EGL) completed exploration activities within the company's Gazonor production permit areas in the Nord-Pas-de-Calais region, with estimates of 136 billion cubic meters of CBM (EGL, 2008). Drilling at the Gazonor area began in August, 2009. EGL also reports testing activities in Lorraine (EGL, 2009).

18.2.10 Germany

Germany has recoverable coal reserves estimated at 6.7 billion metric tons and 6.6 billion metric tons of this is lignite (EIA, 2009b). Germany's most important hard coal production is from underground mines located in the Ruhr and Saar basins in western Germany, while all brown coal production is located in surface mines in basins across the country (Methane to Markets, 2009). The location of all hard coal basins is illustrated in the map below (Figure 18.16).

Germany produced 19 million metric tons of hard coal and 175 million metric tons of lignite in 2008 (EIA, 2009b). Hard coal mining is scheduled to end in 2018 as the German government decided in 2007 to gradually withdraw expensive subsidies that have kept its hard coal mines open (Whitlock, 2007).

Germany's underground CMM emissions have been on the decline with the reduction in hard coal mining. Surface mine emissions have also dropped somewhat since the early 1990s. CMM emissions are summarized below (Table 18.14).

There are 37 AMM projects in Germany, 35 of which are located in the Ruhr region. Germany has 11 CMM recovery projects at active underground mines, mostly in the Ruhr region, but also in the Saar and Ibbenburen regions. All CMM and AMM projects are either CHP or electricity generation projects (Methane to Markets, 2010).

Germany has in-place CBM resources estimated at 3 trillion cubic meters of which 2 trillion cubic meters estimated in the Ruhr basin. Recoverable reserves are unknown. Research is being done by the Technical University of Aachen, which is holding a concession area in northwest Germany in

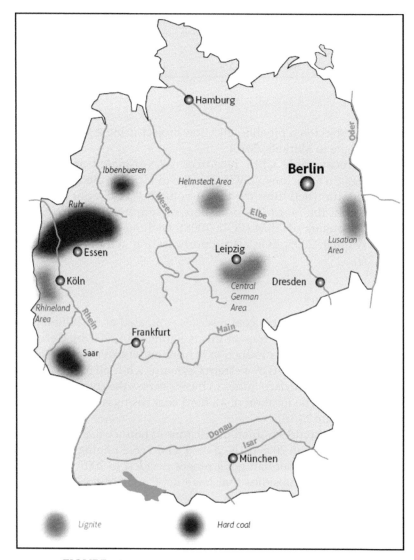

FIGURE 18.16   Germany's hard coal deposits. *Euracoal (n.d.).*

the Münsterland Basin, of which the southern part includes the heavily mined Ruhr area (Mösle et al., 2009).

The map below shows that CBM activity is developing in Germany with a number of concessions in Nordrhein-Westfalen as of 2009 (Figure 18.17).

18.2.11 India

Coal resources are found in 17 major coalfields in India. The largest coalfields in India are located in the eastern and central states of Bihar, West

TABLE 18.14   Germany's Coal Mine Methane Emissions (Million Cubic Meters)

Emission Source	1990	1995	2000	2005	2010
Underground mines	1284	877	675	395	
Postunderground	59.4	45.4	28.5	21.1	
Surface mines	5.77	3.12	2.7	2.9	
Total	1349	925	706	419	542[1,2]

[1] *USEPA (2006).*
[2] *Projected.*
UNFCCC (2009).

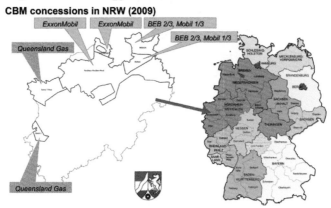

Exploration Geology and CBM

CBM concessions in NRW (2009)

FIGURE 18.17   Coalbed methane concessions in NRW 2009. *Schlueter (2009).*

Bengal, and Madhya Pradesh, but other resources are scattered throughout the country such as in the far northeast in Assam and Arunachal Pradesh (Figure 18.18).

Hard coal is found in 14 of India's states with total recoverable coal reserves of 56.5 billion metric tons with 52.2 billion metric tons of this consisting of hard coal (EIA, 2009b). Reserves in million metric tons are shown by state and coalfield in the table below (Table 18.15).

India produced 516 million metric tons of coal in 2008, of which 483 million metric tons was hard coal (EIA, 2009b). In 2000, 27% of coal production in India came from underground mines, and declined to 15% by 2009. Both underground and surface mines continue to be developed (Methane to Markets, 2009).

India's CMM emissions are projected to be more than double from 1990 levels in 2010. There are currently no commercial CMM recovery projects in India; however, a CMM demonstration project, which is being implemented at Sudamdih and Moonidih mines in the Jharia Coalfield,

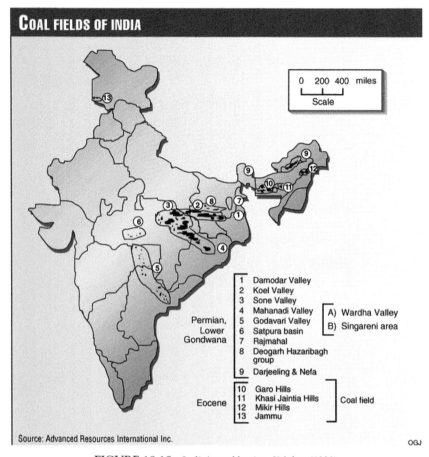

FIGURE 18.18   India's coal basins. *Kelefant (1998).*

Jharkhand has been funded by the United Nations Development Programme, the Global Environmental Facility, and the Indian Ministry of Coal. The project is reportedly providing about 400 homes of miners with electricity (UNDP, n.d.). India's CMM emissions are summarized in the table below (Table 18.16).

India holds significant prospects for commercial recovery of CBM with resources estimated around 4.6 trillion cubic meters. A total of 26 CBM blocks and 13,600 km^2 have been awarded in three rounds of CBM bidding held so far (DGH, 2009). Essar Oil plans to drill 500 wells over the next 3–4 years to produce 3 million cubic meters per day from the Raniganj block in West Bengal with a life estimated at 20 years (Press Trust of India, 2009). Other companies that have been awarded CBM blocks include Reliance Industries and Oil and Natural Gas Corporation (GEF, 2007).

TABLE 18.15   India's Coal Reserves (Million Metric Tons)

State	Coalfield	Measured	Indicated	Inferred	Total
West Bengal	Raniganj	11,538.57	7532.02	4443.91	23,514.5
West Bengal	Barjora	114.27	0.00	0.00	114.27
West Bengal	Birbhum	0.00	4071.23	611.79	4683.02
West Bengal	Darjeeling	0.00	0.00	15.00	15.00
Jharkhand	Raniganj	1538.19	466.56	31.55	2036.30
Jharkhand	Jharia	15,077.57	4352.49	0.00	19,340.06
Jharkhand	East Bokaro	3351.87	3868.10	863.32	8083.29
Jharkhand	West Bokaro	3629.03	1349.04	34.42	5012.49
Jharkhand	Ramgarh	446.27	545.15	58.05	1049.47
Jharkhand	North Karanpura	9345.70	5832.94	1864.96	17,043.60
Jharkhand	South Karanpura	2620.41	2020.82	1508.88	6150.11
Jharkhand	Auranga	213.88	2279.82	503.41	2997.11q
Jharkhand	Hutar	190.79	26.55	32.49	249.82
Jharkhand	Daltonganj	83.86	60.10	0.00	143.96
Jharkhand	Deogarh	326.24	73.60	0.00	399.84
Jharkhand	Rajmahal	2655.52	10,019.14	1411.25	14,115.91
Bihar	Rajmahal	0.00	0.00	160.00	160.00
Madhya Pradesh	Johilla	185.08	104.09	32.83	322.00
Madhya Pradesh	Umaria	177.70	3.59	0.00	181.29
Madhya Pradesh	Pench-Kanhan	1375.98	736.71	316.78	2429.47
Madhya Pradesh	Pathakhera	290.80	88.13	68.00	446.93
Madhya Pradesh	Gurgunda	0.00	47.39	0.00	47.39
Madhya Pradesh	Mohpani	7.83	0.00	0.00	7.83
Madhya Pradesh	Sohagpur	1637.97	3301.26	190.36	5129.59

Continued

TABLE 18.15   India's Coal Reserves (Million Metric Tons) — cont'd

State	Coalfield	Measured	Indicated	Inferred	Total
Madhya Pradesh	Singrauli	4365.82	6013.41	2037.28	12,416.51
Chhattisgarh	Sohagpur	94.30	10.08	0.00	104.38
Chhattisgarh	Sonhat	199.49	2463.86	1.89	2665.24
Chhattisgarh	Jhilimili	228.20	38.90	0.00	267.10
Chhattisgarh	Chirimiri	320.33	10.83	31.00	362.16
Chhattisgarh	Bishrampur	733.44	765.55	0.00	1498.99
Chhattisgarh	East of Bishrampur	0.00	41.75	0.00	41.75
Chhattisgarh	Lakhanpur	365.56	85.84	0.00	451.40
Chhattisgarh	Panchabahini	0.00	11.00	0.00	11.00
Chhattisgarh	Hasdo-Arand	1369.84	2888.25	762.69	5020.78
Chhattisgarh	Sendurgarh	152.89	126.32	0.00	279.21
Chhattisgarh	Korba	4980.58	4499.90	830.18	10,310.66
Chhattisgarh	Mand-Raigarh	2466.01	16,834.91	2552.72	21,853.64
Chhattisgarh	Tatapani-Ramkola	0.00	1414.60	202.19	1616.79
Uttar Pradesh	Singrauli	866.05	195.75	0.00	1061.80
Maharashtra	Wardha Valley	3192.73	1344.03	1466.73	6003.49
Maharashtra	Kamptee	1276.14	1079.23	505.44	2860.81
Maharashtra	Umrer-Makardhokra	308.41	0.00	0.00	308.41
Maharashtra	Nand-Bander	468.08	483.95	0.00	952.03
Maharashtra	Bokhara	10.00	0.00	20.00	30.00
Orissa	Ib-River	5703.55	9534.91	7183.33	22,421.79
Orissa	Talcher	14,240.08	21,949.14	6615.85	42,805.07
Andhra	Godavari	9193.61	6748.04	2985.27	18,926.92
Assam	Singrimari	0.00	2.79	0.00	2.79
Sikkim	Rangit Valley	0.00	58.25	42.98	101.23
Assam	Makum	315.96	11.04	0.00	327.00
Assam	Dilli-Jeypore	32.00	22.02	0.00	54.02
Assam	Mikir Hills	0.69	0.00	3.02	3.71

TABLE 18.15 India's Coal Reserves (Million Metric Tons) — cont'd

State	Coalfield	Measured	Indicated	Inferred	Total
Arunachal Pradesh	Namchick-Namphuk	31.23	40.11	12.89	84.23
Arunachal Pradesh	Miao Bum	0.00	0.00	0.00	6.00
Meghalaya	West Darangiri	65.40	0.00	59.60	125.00
Meghalaya	East Darangiri	0.00	0.00	34.19	34.19
Meghalaya	Balphakram-Pendeguru	0.00	0.00	107.03	107.03
Meghalaya	Siju	0.00	0.00	125.00	125.00
Meghalaya	Langrin	10.46	16.51	106.19	133.16
Meghalaya	Mawlong-Shella	2.17	0.00	3.83	6.00
Meghalaya	Khasi Hills	0.00	0.00	10.10	10.10
Meghalaya	Bapung	11.01	0.00	22.65	33.66
Meghalaya	Jayanti	0.00	0.00	2.34	2.34
Nagaland	Borjan	5.50	0.00	4.50	10.00
Nagaland	Jhanzi-Disai	2.00	0.00	0.08	2.08
Nagaland	Tuen Sang	1.26	0.00	2.00	3.26
Nagaland	Tiru Valley	8.76	0.00	13.18	21.94

GSI (2009).

TABLE 18.16 India's Coal Mine Methane Emissions (Million Cubic Meters)

Emission Source	1990	1995	2000	2005	2010
Total	761	956	1109	1363	1616[1]

[1] *Projected.*
USEPA (2006).

18.2.12 Indonesia

Indonesia has total recoverable coal reserves estimated at 4.3 to 6.8 billion metric tons (EIA, 2009b; MEMR, 2006), of which 1.7 billion metric tons are hard coal (EIA, 2009b). Deposits are concentrated in Sumatra in the Central and South Sumatra Basins and Kalimantan in the Kutai and Barito Basins as shown in the map below (Figure 18.19).

Reserves are listed by province in the table below (Table 18.17).

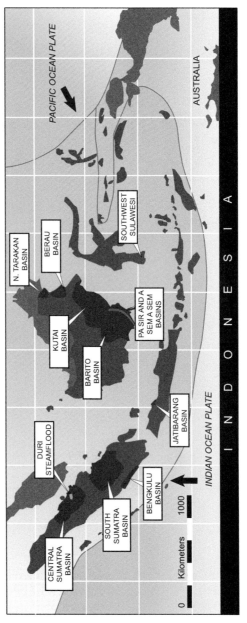

FIGURE 18.19   Coal basins of Indonesia. *CBM Asia (2009)*.

TABLE 18.17   Indonesia's Coal Reserves (Million Metric Tons)

Province	Measured	Indicated	Inferred	Hypothetic	Reserves
Java	0	0	8.66	5.47	0
Sumatra	2601.74	11,974.38	13,307.82	769.21	2770.73
Kalimantan	7716.90	6720.50	20,795.60	3035.13	3988.11
Sulawesi	23.10	33.09	146.91	0	0.06
Maluku	0	0	2.13	0	0
Papua	0	0	61.86	89.40	0
Total					6758.90

TABLE 18.18   Indonesia's Coal Mine Methane Emissions (Million Cubic Meters)

Emission Source	1990	1995	2000	2005	2010
Total	23.1	30.1	31.5	34.3	35.0[1]

[1] *Projected.*
USEPA (2006).

Indonesia produced 284 million metric tons of coal in 2008, of which 277 million metric tons was hard coal (EIA, 2009b). There were three underground active mines in Indonesia in 2004. In 1999, 99% of coal production, or roughly 131 million metric tons, came from surface mines (Methane to Markets, 2009).

There are currently no commercial CMM recovery projects in Indonesia (Methane to Markets, 2010). Indonesia's CMM emissions are summarized in the table below (Table 18.18).

Indonesia has CBM reserves estimated at 12.8 trillion cubic meters (WK CBM, 2009). The government unveiled CBM projects in 2007 with incentives that provides a relatively better production split for CBM projects than oil and gas projects, increasing CBM operators' profit sharing to 45%—much higher than the 15% and 30% that oil operators and gas operators get, respectively (MSS, 2009). CBM resources of Indonesia, provided by the Indonesia Directorate General of Oil and Gas, are summarized in the map below (DJ MIGAS, 2007) (Figure 18.20).

Ephindo and McLaren Resources Inc, as South Sumatra Energy, with Medco E&P, signed a CBM contract in May 2008, making it the first CBM contract in Indonesia (CBM World, 2009). Churchill Mining Plc and its Indonesian partner, PT Ridlatama Mining Utama, formed the "Ridlatama Consortium" and won Indonesia's first joint evaluation agreement (JEA) to explore for CBM in 2007. Companies holding coal tenements in potential CBM areas have the right to direct appointment of CBM licenses, known

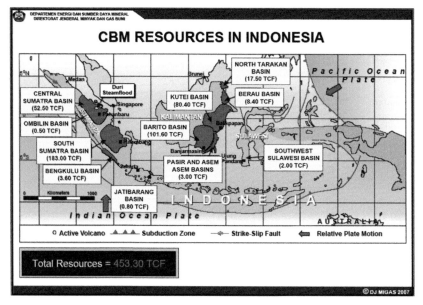

FIGURE 18.20   Indonesia coalbed methane resources map. *DJ MIGAS.*

as a JEA. Churchill and PT Ridlatama won a contract to extract gas in the Kutai Basin, East Kalimantan, Indonesia in 2008 (Churchill Mining, 2007; CBM World, 2009).

As of November, 2009 there were seven agreements were signed for working CBM acreage blocks; Sekayu CBM block, Indragiri Hulu CBM block, Barito Banjar II CBM block, Bentian Besar CBM block, Sangatta CBM block and Kutai CBM block. During the 2009 bidding round, 54 investors proposed a request for CBM development, 33 of which meet necessary requirements (WK CBM, 2009).

18.2.13 Kazakhstan

Kazakhstan has total recoverable coal reserves of 31 billion metric tons with reserves totaling 28 billion metric tons (EIA, 2009b) of hard coal. The three primary coal basins are in Northern Kazakhstan: Karaganda, Ekibastuz, and Maikyubensk. Coal rank varies from low-rank subbituminous in the Maikyubensk basin to anthracite in the southern portion of the Karaganda basin (Mining Week, 2009).

Kazakhstan produced a total of 109 million metric tons of coal in 2008, 104 million metric tons of this was hard coal (EIA, 2009b). Mining is performed in 53 mines, including 15 in the Karaganda coal basin, by 34 companies (1 joint venture, 5 foreign, and 28 local companies). The major mining companies include: Bogatyr Access Komir, Shubarkol Komir, Mittal Steel

Temirtau, the Eurasian Energy Corporation, Maykuben West, Karazhira Ltd, the Kazakhmys Corporation, and Gamma (ERK, 2008).

Kazakh coals are gassy, therefore underground mines must be degasified and ventilated for safety. In the Karaganda basin surface degasification wells are widely used (Methane to Markets, 2009). Kazakhstan has seen a significant drop in CMM emissions from underground mines as underground production has slowed down. Underground coal production in 1990 was approximately 35 million metric tons and had dropped to 11 million in 2005 (UNFCCC, 2009). Kazakhstan has one CMM recovery project which has been gradually increasing the amount of CMM recovered. This project is in the Karaganda basin at an underground mine and it uses CMM as boiler fuel (Methane to Markets, 2010). Kazakhstan's CMM emissions are summarized in the table below (Table 18.19).

Kazakhstan has CBM reserves of 1.2–1.7 trillion cubic meters (Alekseev et al., 2003). The table below shows CBM resources estimated by basin (Table 18.20).

TABLE 18.19   Kazakhstan's Coal Mine Methane (CMM) Emissions (Million Cubic Meters)

Emission Source	1990	1995	2000	2005	2010
Underground mines	1147	769	310	345	
Postunderground	35	23	8	11	
Surface mines	559	411	380	414	
Recovered	9	6	12	26	
Total CMM emissions	1732	1197	686	744	447[1,2]

[1] *USEPA (2006).*
[2] *Projected.*
UNFCCC (2009).

TABLE 18.20   Kazakhstan's Coalbed Methane (CBM) Resources by Basin

Basin or Field	Estimated CBM Resources (Billion Cubic Meters)
Karaganda Basin	550–750
Ekibastuz Basin	75–100
Zavialov field	14.6–16.8
Samarskiy field	11.0–14.2

Methane to Markets (2009).

In 2000, 12.5 million cubic meters of CBM was produced from 134 wells (Alekseev et al., 2003). In April 2003, Kazakhstan's Ministry of Energy and Mineral Resources recommended that Bogatyr Access Komyr, Ltd and Azimut Energy Services, Ltd pursue a CBM development effort in the Ekibastuz basin (Methane to Markets, 2009). Total Kazakhstan, LLP was awarded the State tender for exploring and producing CBM in the Taldykuduk area of the Karaganda Basin in 2004.

18.2.14 Mexico

Mexico's total recoverable coal reserves are 1.2 billion metric tons, with 860 million metric tons of hard coal (EIA, 2009d). There are three major coal-bearing areas in Mexico. The largest is an area that produces more than 90% of Mexican coal and comprises the Sabinas and Fuentes-Río Escondido basins of north–central Coahuila, including a small contiguous area of Nuevo León, which covers approximately 12,000 km². The Sabinas Basin is the source of coking coal of lower ash content and the Fuentes-Río Escondido basin produces mostly thermal coal. In the northwest portion of Oaxaca is another basin where seams are estimated to contain 30 million metric tons. The third area, located south of Hermosillo in Sonora also has estimated reserves of 85 million metric tons (Wallace, 2009).

Mexico produced 11.5 million metric tons of bituminous coal in 2008 (EIA, 2009b). Two major mining companies produced 82% of Mexico's coal in 2007: Minera Carbonífera Río Escondido (MICARE), formerly government owned but now a subsidiary of Altos Hornos de México (AHMSA) and Minera Monclova (MIMOSA), a 98% owned subsidiary of AHMSA and Carbonífera de San Patricio. MICARE produces primarily thermal coal from one open pit mine and two underground longwall mines while MIMOSA produces metallurgical coal from four underground mines and two open pit mines (Wallace, 2009).

MIMOSA incorporated a methane drainage system operating in advance of mining for safety reasons. In addition to this degasification program MIMOSA has planned a methane recovery project at MIMOSA's Mines 5–7 (Flores, 2007; Methane to Markets, 2009). MIMOSA also has a boiler fuel and power generation at its Esmeralda mine (Methane to Markets, 2010). Despite these recovery projects, Mexico's CMM emissions have steadily risen with increases in coal production. In 1990, Mexico produced 7.8 million metric tons compared with 11.5 million metric tons in 2008 (EIA, 2009b). Mexico's CMM emissions are summarized in the table below (Table 18.21).

There currently is no reported CBM production in Mexico though the basins located in Coahuila have a CBM resource potential of 249 billion cubic meters (Querol-Suñe, 2006). A study has been done on the resources of the Sabinas Basin concluding that 60 wells in this basin would produce

TABLE 18.21   Mexico's Coal Mine Methane Emissions (Million Cubic Meters)

Emission Source	1990	1995	2000	2005	2010
Underground mines	82	88	110		
Postunderground	5	5	7		
Surface mines	3	4	5		
Total	90	97	122	173[1]	199[1,2]

[1] *USEPA (2006).*
[2] *Projected.*
Flores (2007).

TABLE 18.22   Mongolia's Coal Reserves

Region	Coal Deposits	Reserves
Central Mongolia	Shivee Ovoo	2.7 billion metric tons brown coal
	Tugrugnuur and Tsaidannuur	2 billion metric tons brown coal
East Mongolia	Adduun Chuluun	100 million metric tons brown coal
	Tugalgatai	3 billion metric tons of subbituminous coal
West Mongolia	Hushuut	300 million metric tons of bituminous and metallurgical coal
South Gobi	Tavan Tolgoi	6.4 billion metric tons of bituminous coal
	Baruun Naran	155 million metric tons of thermal and metallurgical coal
	Naryin Suhait	250 million metric tons of bituminous coal
	Ovoot Tolgoi	150 million metric tons

Badgaa (n.d.).

approximately 510,000 cubic meters of gas per day for nearly 20 years (Eguiluz de Antuñano and Amezcua, 2003).

18.2.15 Mongolia

Mongolia has proven coal reserves of 12.2 billion metric tons, which includes 2 billion metric tons of coking or metallurgical coal and 10.1 billion metric tons of thermal coal (Badgaa, n.d.). There are 200 coal deposits within 15 coal basins in Mongolia (Methane to Markets, 2009). The table below summarizes the reserves of coal in the major coal deposits of Mongolia (Table 18.22).

There are numerous additional basins throughout the country. Mongolia's coal basins are shown in the map below (Figure 18.21).

Mongolia produced 10 million metric tons of coal in 2008, which includes 1 million metric tons of hard coal and 9 million metric tons of lignite (EIA, 2009b) from over 30 surface mines. Mongolia's only underground mine, the Nailakh mine, was closed in 1993, but it is reported that the mine may reopen (Methane to Markets, 2009). The table below summarizes some of Mongolia's major coal mines (Table 18.23).

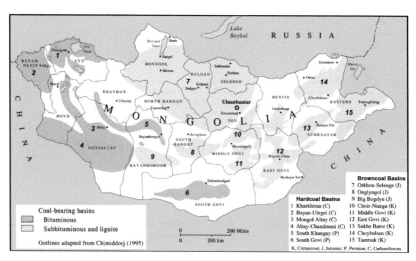

FIGURE 18.21   Mongolia's coal basins. *Schwochow (1997)*.

TABLE 18.23   Mongolia's Major Coal Mines

Mine	Province	Region	Annual Output (Mt)
Baganuur	Tov	Central	3,000,000
Shivee Ovoo	Govisümber	Central	2,000,000
Sharyn Gol	Selenge	Northern	1,000,000
Changana Tal	Khentii	Eastern	20,000
Eldev	Dornogovi	Southeastern	500,000
Khotgor	Uvs	Western	–
Naryn Sukhait	Ömnögovi	Southwest	–
Tavan Tolgoi	Ömnögovi	Southern	787,100

BusinessMongolia.com (2009), Red Hill (2010), Mongolyn Alt Corporation (2010), Tavan Tolgoi (2010), Altan San Securities (2010).

TABLE 18.24   Mongolia's Coal Mine Methane Emissions (Million Cubic Meters)

Emission Source	1990	1995	2000	2005	2010
Total	14	7	5	4	3[1]

[1] *Projected.*
USEPA (2006).

Mongolia has relatively low CMM emissions as all mining is from surface mines. No CMM recovery projects have been developed in Mongolia; however, there is interest in CMM recovery and utilization at the Nailakh mine, which may facilitate reopening of the mine (Methane to Markets, 2009). Studies of the Nailakh mine's feasibility as a CMM project are reportedly underway (Badarch, 2009). Mongolia's CMM emissions are summarized below (Table 18.24).

As no natural gas infrastructure exists in Mongolia, CBM activity is in its infancy. A Canadian firm, Storm Cat Energy Corp., was granted a Production Sharing Contract with the Petroleum Authority of Mongolia and completed exploration activities in the Noyon Uul region of the South Gobi Basin in 2004. Results showed potential CBM resources in the region of 17 billion cubic meters to 34 billion cubic meters (Storm Cat, 2005). Sproule, a Canadian consulting company, also reports to have evaluated CBM resources in Mongolia (Sproule, 2010).

18.2.16 Poland

Poland has total recoverable coal reserves of 7.5 billion metric tons of which 6 million metric tons is hard coal (EIA, 2009b). Hard coal deposits occur in three basins in Poland, the most important being the USCB. The others are the Lublin Hard Coal Basin and the Lower Silesian Coal Basin. Poland's hard coal reserves are shown in the map below (Figure 18.22).

Lignite reserves occur mainly in the western, southern, and central parts of the country in the Adamów, Belchatów, Konin, and Turów basins (Volkmer, 2008). Poland's recoverable lignite reserves are 1.5 billion metric tons (EIA, 2009b) and are shown in the map below (Figure 18.23).

Poland produced 143 million metric tons of coal in 2008 (EIA, 2009b). At the end of 2009 Poland had 31 underground hard coal mines and 12 surface lignite mines (SMA, 2009).

Poland began actively draining methane, for safety reasons, in the late 1950s. Methane is drained as coal extraction proceeds at the working face, from gob areas, and from development areas. Poland's CMM emissions are attributable to its numerous underground coal mines. There are 21 CMM recovery projects at underground mines in Poland, all located in the USCB, and consists of eight boiler fuel projects, five CHP projects, four

FIGURE 18.22   Poland's hard coal basins. *Schwochow (1997).*

FIGURE 18.23   Poland's lignite deposits. *Volkmer (2008).*

TABLE 18.25 Poland's Coal Mine Methane Emissions (Million Cubic Meters)

Emission Source	1990	1995	2000	2005	2010
Underground mines	961	860	745	622	
Postunderground	68	61	51	46	
Surface mines	1	1	1	1	
Total	1030	922	797	670	754[1,2]

[1] USEPA (2006).
[2] Projected.
UNFCCC (2009).

coal drying projects, three industrial use projects, and one power generation project (Methane to Markets, 2010). Poland's CMM emissions are summarized in the table below (Table 18.25).

The Polish Institute of Geology estimates economically recoverable CBM reserves of 3 billion cubic meters. Estimated gas-in-place methane resources in the USCB are estimated to be 350 billion cubic meters. Resources are much lower in the Lower Silesian Coal Basin where the perspective resources are estimated at 5 billion cubic meters (PGI, 2005). McCallan Oil & Gas (UK) Ltd signed an Initial Joint Operation Agreement to earn exclusive rights to develop a separate CBM concession area in the USCB in Poland with the potential to produce several billion cubic meters of methane gas. Composite Energy Limited's Polish subsidiary, Composite Energy sp.z o.o (Poland) was awarded two concessions to explore for CBM in Poland in Chelm and Werbkowice–Tyszowce, located in the Lublin Coal Basin in Southeast Poland (Composite, 2008).

18.2.17 Romania

Romania has total recoverable coal reserves of 422 million metric tons of which 12 million metric tons is hard coal (EIA, 2009b). Hard coal is mined from the Jiu Valley and lignite is found in the Oltenia and Ploiesti basins as illustrated by the map below (Figure 18.24).

Romania produced a total of 35 million metric tons of predominately lignite coal in 2008. Approximately 90% of lignite mined in 2007 was extracted from the surface mine sites of Rovinari, Rosia, Pesteana, Pinoasa, Motru, Berbesti, and Mehedinti (Euracoal, n.d.) and is carried out by two companies: the National Lignite Society of Oltenia—Târgu Jiu, and the National Coal Society—Ploiesti.

Hard coal production from the Jiu Valley totaled 12,000 Mt (EIA, 2009b; Euracoal, n.d.). In the Jiu Valley, coal mining is carried out by the National Hard Coal Company—Petrosani, which operates seven mines: Lonea, Petrila, Livezeni, Vulcan, Paroseni, Uricani, and Lupeni.

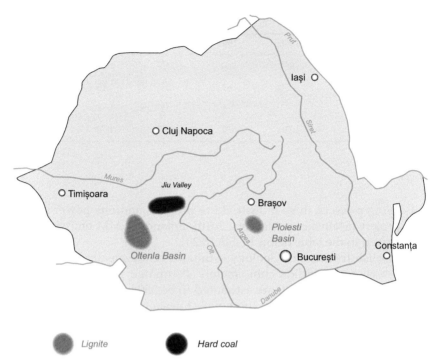

FIGURE 18.24   Romania's coal basins. *Euracoal (n.d.).*

TABLE 18.26   Romania's Coal Mine Methane Emissions (Million Cubic Meters)

Emission Source	1990	1995	2000	2005	2010
Underground mines	193	207	139	127	
Postunderground	27	29	19	18	
Surface mines	34	36	26	27	
Postsurface	3	3	2	2	
Total	257	275	186	174	193[1,2]

[1] *USEPA (2006).*
[2] *Projected.*
UNFCCC (2009).

Methane drainage occurs at all underground mines in Romania. There is one CMM recovery project at the Lupeni mine where 39% of drained methane is used to fuel two boilers inside the mine yard (Lupu and Ghicioi, 2006). Though underground mining contributes a small percentage of total coal production, methane emissions from underground mines are a large part of total CMM emissions due to the gassiness of the coal in the Jiu Valley. CMM emissions are summarized in the table below (Table 18.26).

Galaxy Energy Corporation, through wholly owned subsidiary Pannonian International, Ltd and Falcon Oil & Gas, has a 30-year concession from the Romanian government on 21,500 acres in the Jiu Valley (Falcon, 2005; Galaxy, n.d.). Falcon drilled the Lupeni South-1 well in 2005 and Pannonian reported potential CBM resources of 14–21 billion cubic meters (Falcon, 2005; Fails, 2005).

18.2.18 Russia

Russia has total recoverable coal reserves of 157 billion metric tons, 49 billion metric tons of which is hard coal (EIA, 2009b). Coal is located in the Kuzbass, Kansko-Achinsky, Pechora, Irkutsk, and South-Yakutia basins as well as some smaller basins as shown in the map below (Figure 18.25).

Russia produced a total of 323 million metric tons of coal in 2008 with most of this being hard coal and accounting for 247 million metric tons of production (EIA, 2009b) from 98 underground mines and 148 surface mines (OECD/IEA, 2009).

CMM in Russia is primarily located in three coal basins: Kuznetsk (Kuzbass), Pechora, and Donetsk. In 2008, about 317 million cubic meters was recovered by degasification or methane drainage systems at underground mines (OECD/IEA, 2009). The amount of methane utilized by mines is approximately 40 million cubic meters per year, although values for annual recovery are estimated to be as high as 82 million cubic meters (Ruban et al., 2005 in OECD/IEA, 2009; UNFCCC, 2009). CMM is utilized for five boiler fuel projects (four in the Kuzbass, one in the Pechora Basin) and two power generation projects (Methane

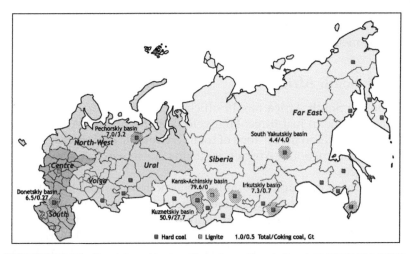

FIGURE 18.25   Russia's coal basins. *Coal mine methane in Russia © OECD/IEA (2009).*

TABLE 18.27   Russia's Coal Mine Methane Emissions (Million Cubic Meters)

Emission Source	1990	1995	2000	2005	2010
Underground mines	3325	2091	1688	1886	
Postunderground	1	0	0	0	
Surface mines	1381	958	1059	1230	
Postsurface	2	2	2	2	
Recovered	37	30	31	82	
Total	3325	2091	1688	1886	1927[1,2]

[1] USEPA (2006).
[2] Projected.
UNFCCC (2009).

to Markets, 2010). Russia's CMM emissions are summarized in the table below (Table 18.27).

Russia's CBM reserves are estimated at 49 trillion cubic meters. The Kuzbass is the largest CBM basin in Russia with estimated CBM reserves at 13 trillion cubic meters of and the Pechora Basin, Russia's second largest CBM basin, with CBM reserves estimated at nearly two trillion cubic meters of gas (Gazprom, 2008).

In 2003 Gazprom began evaluating opportunities for CBM production in the Kuzbass in the Erunakovsky region of the Kemerovo Oblast in the Taldinskaya area of the Kuzbass; four test wells were drilled and gas was produced in 2004. Currently these wells are a pilot operation. In June 2007, Gazprom purchased a control stake in Geologopromyslovaya Company Kuznetsk (GPK Kuznetsk), which holds license for CBM prospecting, exploration, and production within the Yuzhno-Kuzbasskaya Group of coalfields. These coalfields contain methane reserves estimated at 6.1 trillion cubic meters (Gazprom, 2008).

18.2.19 South Africa

Although South Africa has 19 official coalfields, 70% of recoverable coal reserves lie in just three: Highveld, Waterberg, and Witbank (EIA, 2008) with 48 billion metric tons of total recoverable coal reserves of hard coal (EIA, 2009b). South Africa's coalfields are shown in the map below (Figure 18.26).

South Africa produced 236 million metric tons of coal in 2008 (EIA, 2009b) from 60 operating coal mines. About 46.5% of South African coal mining is done underground and about 53.5% is produced from surface mines. Five coal mining companies account for 90% of South Africa's coal production: Anglo Coal, BHP Billiton, Sasol Mining, Exxaro Coal, Kumba

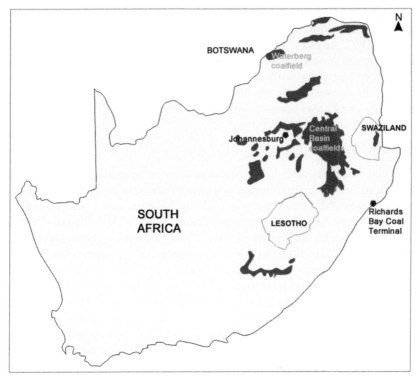

FIGURE 18.26   South Africa's coal fields. *Welz (2013).*

TABLE 18.28   South Africa's Coal Mine Methane Emissions (Million Cubic Meters)

Emission Source	1990	1995	2000	2005	2010
Total	471	466	496	518	505[1]

[1] *Projected.*
USEPA (2006).

Coal, and Xstrata Coal. South Africa's eight largest mines account for 61% of coal production (GCIS, 2009).

South Africa has no CMM recovery projects. Some South African mines drain methane prior to mining and utilization of this methane has reportedly been investigated (UNFCCC, 2000 in Methane to Markets, 2009). As mining in South Africa moves underground, CMM emissions are expected to rise. Africa's CMM emissions are summarized in the table below (Table 18.28).

Estimates of CBM potential in South Africa vary widely from as low as 140 billion cubic meters up to 1.1 trillion cubic meters (Methane to

Markets, 2009; Reuters, 2009). In 2008 it was reported that there were 23 rights for CBM exploration in South Africa with many applicants (Van der Merwe, 2008). Anglo Coal conducted a CBM exploration program in the Waterberg Basin and a five-well pilot production project was completed. Anglo is progressing with a 39-well project with plans for a feasibility study in the future (Creamer, 2009). Badimo Gas has applied for exploration rights in 14 areas across South Africa (Badimo, 2007).

18.2.20 Turkey

Turkey has mineable hard coal reserves of approximately 1 billion metric tons and mineable lignite reserves of about 3.2 billion metric tons. Turkey's hard coal deposits are located in the Zonguldak Basin in northwestern Turkey. Lignite deposits are found in the Afsin-Elbistan, Soma, Bursa, Çan, and Mugla basins. About 40% of Turkey's lignite reserves are located in the Afsin-Elbistan basin (Euracoal, n.d.). Turkey's coal basins are shown in the map below (Figure 18.27).

Turkey's annual coal production has been on the rise reaching nearly 76 million metric tons in 2008 with mostly lignite, over 72 million metric tons, and only 3.3 million metric tons of hard coal produced in 2008 (EIA, 2009b). The state-owned Turkish Hard Coal Enterprises (TTK) has a monopoly on the production, processing, and distribution of hard coal and operates five underground mines in the gassy Zonguldak coal basin.

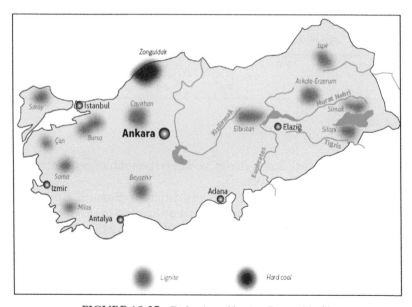

FIGURE 18.27   Turkey's coal basins. *Euracoal (n.d.).*

TABLE 18.29   Turkey's Coal Mine Methane Emissions (Million Cubic Meters)

Emission Source	1990	1995	2000	2005	2010
Underground mines	47	39	41	37	
Surface mines	53	62	72	66	
Total	100	101	113	103	137[1,2]

[1] *USEPA (2006).*
[2] *Projected.*
UNFCCC (2009).

Lignite is produced from 30 surface mines and nine underground mines operated by Turkish Coal Enterprises (TKI). Additionally, some private sector operations as well as Turkey's Electricity Generating Authority provide lignite to power plants (Euracoal, n.d.).

Turkey has no CMM recovery projects. Emissions from Turkey's underground and surface mines are summarized in the table below (Table 18.29).

The CBM in-place resources in two districts of the Zonguldak hard coal region are presently estimated to be at least 3 trillion cubic meters (Balat and Ayar, 2004; Methane to Markets, 2009). Another area of CBM investigation is the Sorgun and Suluova basins. Mines in these areas have also experienced explosions indicating a need to evaluate CMM recovery potential (Schwochow, 1997).

18.2.21 Ukraine

Ukraine has total recoverable coal reserves of 34 billion metric tons with hard coal reserves of 15 billion metric tons (EIA, 2009b). Ukraine's most significant coal reserves are located in the Donetsk coal basin with about 46% of the coal reserves. The Luhansk region holds approximately 34% of reserves and the Dnipropetrovsk region holds 15%. The remaining 5% is located in the regions of Lviv, Volyn, and Kirovograd. Ukraine's main hard coal reserves are shown in the map below (Figure 18.28).

Ukraine has 160 operating mines. Of these, the Ukrainian Ministry of the Coal Industry supervises 139 and the remaining 21 are privately operated (Euracoal, n.d.). Most of Ukraine's coal is mined in the Donetsk Basin (EIA, 2007a). Ukraine produced a total of 60 million metric tons of coal in 2008, over 99% of which was hard coal (EIA, 2009b).

Ukraine has eight operating CMM projects at active underground mines. Seven of these are in the Donetsk Basin and comprise four boiler fuel projects, one power generation project, one industrial use project, and one CHP project. The remaining project is a CHP project in the Lugansk Basin. There are an additional three projects in development in the

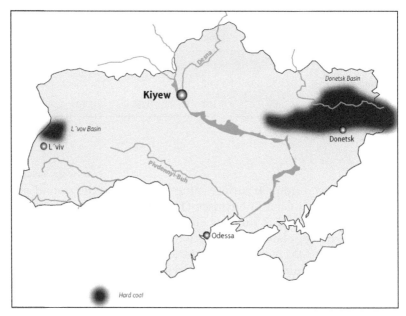

FIGURE 18.28   Ukraine's hard coal deposits. *Euracoal (n.d.).*

TABLE 18.30   Ukraine's Coal Mine Methane Emissions (Million Cubic Meters)

Emission Source	1990	1995	2000	2005	2010
Underground mines	3557	1945	2039	1839	
Postunderground	306	160	156	155	
Surface mines	13	3	1	0	
Postsurface	2	0	0	0	
Recovered	145	89	73	147	
Total	3733	2019	2123	1847	1714[1,2]

[1] *USEPA (2006).*
[2] *Projected.*
UNFCCC (2009).

Donetsk Basin. A CHP project is planned at the Krasnoarmeiskaya Zapadnaya mine. A vehicle fuel project and a power generation project are both planned at the Zasyadko mine (Methane to Markets, 2010). Mine closures and a drop in coal production have led to a decrease in Ukraine's CMM emissions (summarized below) (Table 18.30).

Total in situ CBM reserves in the Donetsk Basin are estimated to be approximately 12 trillion cubic meters that are 500–1800 m deep. Euro-Gas, Inc. has entered into a Memorandum of Understanding with OJSC

ZNVKIF New Technologies (ZNT) of Kiev, Ukraine to form a joint venture that will explore and develop large CBM reserves in East and West Ukraine (EuroGas, 2009).

18.2.22 United Kingdom

The UK has total recoverable hard coal reserves of 155 million metric tons (EIA, 2009b; BP, 2009). The UK's coal mines are located in central and northern England, south Wales and central and southern Scotland. Southern Scotland has the highest concentration of surface mines (Euracoal, n.d.). The UK's coalfields are illustrated in the map below (Figure 18.29).

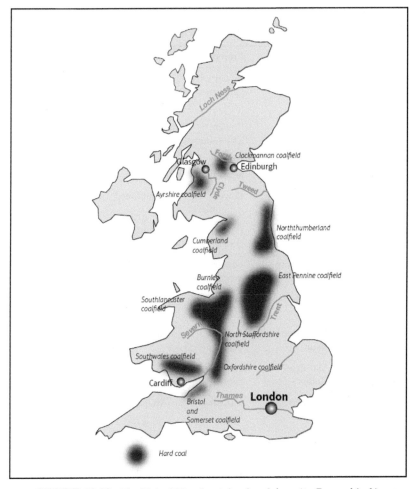

FIGURE 18.29   The United Kingdom's hard coal deposits. *Euracoal (n.d.).*

In 2008, the UK produced almost 18 million metric tons of coal (EIA, 2009b). There are six large underground mines operating in the UK. Four of these are owned by UK Coal Plc (Daw Mill, Thoresby, Welbeck, and Kellingley). Maltby is owned by Hargreaves Plc. Hatfield Colliery is owned by Powerfuel Plc and recommenced production in 2008. There are also 10 smaller underground mines in the UK (Euracoal, n.d.). In 2006 there were 35 surface mines in operation (Berr, 2006 in Methane to Markets, 2009).

Large-scale drainage and utilization of CMM began in Britain in the 1950s at the Point of Ayr Colliery in North Wales (Young et al., 1994). There are currently 29 CMM projects operating or in development at active and abandoned mines in the UK. Projects are summarized in the table below (Table 18.31).

TABLE 18.31   Coal Mine Methane Projects in the UK

Region	Mine	Project Type	Mine Status	Project Status
Nottinghamshire	Annesley Bentinck Colliery	Power generation	Abandoned	Operating
South Yorkshire	Bentley	Power generation	Abandoned	In development
Nottinghamshire	Bevercotes	Power generation	Abandoned	In development
Yorkshire	Brodsworth	Power generation	Abandoned	In development
Yorkshire	Frickley	Power generation	Abandoned	In development
Yorkshire	Grimethorpe	Power generation	Abandoned	In development
South Yorkshire	Hickleton Colliery	Power generation	Abandoned	Operating
Yorkshire	Houghton Main	Power generation	Abandoned	In development
Derbyshire	Markham Colliery	Heating or cooling	Abandoned	Unknown
Yorkshire	Monk Bretton	Industrial use	Abandoned	Operating
Derbyshire	Shirebrook Colliery	Power generation	Abandoned	Operating
Staffordshire	Silverdale	Power generation	Abandoned	Operating

TABLE 18.31 Coal Mine Methane Projects in the UK — cont'd

Region	Mine	Project Type	Mine Status	Project Status
Staffordshire	Silverdale	Pipeline injection	Abandoned	Operating
Nottinghamshire	Steetley Colliery	Power generation	Abandoned	Operating
Nottinghamshire	Warsop	Power generation	Abandoned	In development
Yorkshire	Wheldale	Power generation	Abandoned	Operating
Derbyshire	Whitwell	Power generation	Abandoned	In development
North Yorkshire	Harworth Colliery	Power generation	Active (underground)	Operating
North Yorkshire	Harworth Colliery	Boiler fuel	Active (underground)	Operating
North Yorkshire	Kellingley Colliery	Power generation	Active (underground)	Operating
North Yorkshire	Kellingley Colliery	Boiler fuel	Active (underground)	Unknown
South Yorkshire	Maltby Colliery	Power generation	Active (underground)	Operating
South Yorkshire	Maltby Colliery	Boiler fuel	Active (underground)	Operating
South Yorkshire	Maltby Colliery	Flaring	Active (underground)	Operating
Nottinghamshire	Thoresby Colliery	Power generation	Active (underground)	Operating
Nottinghamshire	Thoresby Colliery	Flaring	Active (underground)	Operating
Nottinghamshire	Welbeck Colliery	Power generation	Active (underground)	Operating
Nottinghamshire	Welbeck Colliery	Flaring	Active (underground)	Operating
Nottinghamshire	Welbeck Colliery	Boiler fuel	Active (underground)	Operating

Methane to Markets (2010).

TABLE 18.32   The UK's Coal Mine Methane Emissions (Million Cubic Meters)

Emission Source	1990	1995	2000	2005	2010
Underground mines	1148	814	453	264	
Postunderground	122	58	29	16	
Surface mines	9	8	7	5	
Recovered	120	58	47	72	
Total	1159	822	442	213	462[1,2]

[1] *USEPA (2006).*
[2] *Projected.*
UNFCCC (2009).

CMM emissions in the UK have dropped due to increases in recovery as well as a decline in coal mining. CMM emissions are summarized in the table below (Table 18.32).

UK coals exhibit low permeability, limiting the potential for CBM development; however, several projects are underway (BGS, 2006). In 2008 Eden Energy completed the first phase of exploration drilling in its South Wales Coal Seam Methane project. A resource estimate based on wellbore data from three holes drilled and available regional data from 33 offset wells previously drilled by British Coal estimated recoverable CBM resources of between 11 and 19 billion cubic meters of prospective resources (Eden Energy, 2008). Two CBM projects were approved by the UK Department of Energy and Climate Change in 2009. The Doe Green and Potteries projects will be operated by Nexen (DECC, 2010). Island Gas, along with Nexen, is selling electricity generated from CBM at its pilot Doe Green site in Cheshire.

18.2.23 Vietnam

Vietnam has total recoverable hard coal reserves of 150 million metric tons, the majority of which is anthracite (EIA, 2009b; EIA, 2007b). Vietnam's most significant coal reserves are found in the northern part of the country in the Quang Yen anthracite region near the Red River Delta. Reserves in this area are estimated to be 190–272 billion metric tons (Methane to Markets, 2009).

Vietnam produced almost 40 million metric tons of coal in 2008. Coal production has steadily risen since the early 1990s (EIA, 2009b). With increasing coal production CMM emissions have also increased. No CMM recovery projects are currently operating or in development in Vietnam. CMM emissions are summarized in the table below (Table 18.33).

TABLE 18.33   Vietnam's Coal Mine Methane Emissions (Million Cubic Meters)

Emission Source	1990	1995	2000	2005	2010
Total	32	58	70	83	99[1]

[1] *Projected.*
USEPA (2006).

Keeper Resources has been working on the first CBM extraction projects in Vietnam. The negotiated CBM concession area covers approximately 3600 km² of the Red River Basin to the southeast of Hanoi. Three years of negotiations were concluded with PetroVietnam and PetroVietnam Exploration Production Corporation (PVEP) for the CBM Production Sharing Contract (PSC) in early 2010. The project advanced with the signing of drill site construction and preparation contracts after acquiring land access approvals (Dragon Capital, 2008; Dragon Capital, 2010).

A scoping study of the CBM potential in the Red River Basin commissioned by Keeper estimated prospective gas resources to be 55 billion cubic meters in the study area, which represents 28% of Keeper's concession (Keeper, 2005).

In addition, Arrow Energy has signed a PSC with PVEP in a CBM concession of 2743 km² in the Red River Basin. The PSC requires Arrow to drill eight wells on the block. Exploration drilling began in January 2009 (Arrow, 2009).

References

AAGA, December 2009. Australia's Identified Mineral Resources 2009. Australian Government Geoscience Australia. http://www.australianminesatlas.gov.au/aimr/index.jsp.

Abi, 2009. The Need for Clean Coal in Africa: The Case of Botswana. Kgomotso Abi, Ministry of Minerals, Energy, and Water Resources, Botswana. Presented April 1, 2009 at The World Bank Group Energy Week 2009, Energy Parallel Session 7: Advanced Coal-based Power Generation Technologies and Carbon Capture and Storage Options. http://siteresources. worldbank.org/INTENERGY/Resources/335544-1232567547944/parallel_8.htm.

Advanced Resources, 2003. Results of the Central Kalahari Karoo Basin Coalbed Methane Feasibility Study. Department of Geologic Survey, Botswana, Lobatse.

AGDCC, May 2009. Australia's National Greenhouse Accounts National Greenhouse Gas Inventory. Australian Government, Department of Climate Change. http://www.climat echange.gov.au/en/climate-change/emissions.aspx.

Alekseev, E.G., Mustaffin, R.K., Umarhajleva, N.S., 2003. Coal Methane: Potential Energy Prospects for Kazakhstan. Presented to UNECE Ad Hoc Group of Experts on Coal in Sustainable Development, Almaty, November 2003. http://www.unece.org/ie/se/pp/ coal/mustafin.pdf.

AltanSanSecurities, 2010. Top 30 Index Companies. AltanSanSecurities. http://www.mongolia-investment.com/top-30-index/companies/. (accessed February 2010).

Arrow, 2009. Vietnam – Arrow Energy. http://www.arrowenergy.com.au/page/Projects/ Vietnam/.

Australian Black Coal Statistics 2006. Coal Services Pty Ltd and Queensland Department of Mines and Energy.

Badarch, 2009. Coal Mine Methane Development in Mongolia. Mongolian Nature and Environment Consortium. Presented at Methane to Markets Partnership Meeting, Mexico, January 27–29, 2009. http://www.methanetomarkets.org/m2m2009/documents/events _coal_20090127_subcom_mongolia.pdf.

Badgaa, n.d. Current Status and Prospects for Energy Resources and Infrastructure Development of South Gobi in Mongolia. Presented by Mr. Ganbaatar Badgaa, Department of Fuel Policy and Regulation, Ministry of Fuel and Energy Mongolia. http://www.keei.re. kr/keei/download/seminar/080703/s1-4.pdf.

Badimo, 2007. Badimo Gas Operations. http://www.badimogas.co.za/operations.html.

BCMEM, June 2004. Petroleum Geology Paper 2004-1, Coalbed Gas Potential in British Columbia. Ministry of Energy and Mines, British Columbia. http://www.em.gov.bc.ca/ DL/Oilgas/CBM/CBGPotential_2004-1.pdf.

BCMEM, 2008. Q&As Coalbed Gas. Ministry of Energy and Mines, British Columbia. http://www.em.gov.bc.ca/subwebs/coalbedgas/FAQs/Q&ACBG.htm#How_much_ coalbed_gas_development_is_there_in_B.C.

BCMEM, 2013. Coal. Ministry of Energy and Mines, British Columbia. http://www.empr.g ov.bc.ca/mining/geoscience/coal/Pages/default.aspx.

BERR, 2006. Coal Industry in the UK. Department of Business Enterprise & Regulatory Reform, United Kingdom. http://www.berr.gov.uk/energy/sources/coal/industry/ page13125.html.

BGS, October 2006. Coal and Coalbed Methane Mineral Planning Factsheet. British Geological Survey Natural Environment Research Council. http://www.bgs.ac.uk/downloads/ start.cfm?id=1354.

Booth, 2008. WestVAMP BHPB Illawarra Coal. Presented at the U.S. Coal Mine Methane Conference, October 2008 Patrick Booth, BHP Billiton. http://epa.gov/cmop/conf/ cmm_conference_oct08.htm.

BP, 2007. Conjuring with coal. In: Knott, T. (Ed.), Frontiers Issue 18: April 2007. http://www. bp.com/liveassets/bp_internet/globalbp/globalbp_uk_english/reports_and_publicati ons/frontiers/STAGING/local_assets/pdf/bpf18_28-32_coalbedmethane.pdf.

BP, June 2009. Statistical Review of World Energy 2009. www.bp.com.

Busch, A., Casagrande, J., Holz, M., Kalkreuth, W., Kern, M., Krooss, B., Levandowski, J., Oliveira, T., 2008. Coalbed Methane Potential of Paraná Basin Coals, Brazil – Results from Test Well CBM001-ST-RS. Presented at the International Geological Congress, Oslo, August 6–14, 2008. http://www.cprm.gov.br/33IGC/1350825.html.

BusinessMongolia.com, November 19, 2009. Coal Industry Update (2): Baganuur Coal Mine Has Debt of 89.9 Billion Tugrik. BusinessMongolia.com. http://www.business-mongolia.com/mongolia-business/mongolia-economy/coal-industry-update-2-bagan-uur-coal-mine-has-debt-of-%E2%82%AE89-9-billion/.

Bustin, M., Hancock, B., Solinger, R., Taylor, M., 2008. Coalbed Methane Development in Canada – Challenges and Opportunities. Presented at the International Geological Congress, Oslo, August 6–14, 2008. http://www.cprm.gov.br/33IGC/1324175.html.

CAC, 2006. Coal Related Maps. The Coal Association of Canada. http://www.coal.ca/conte nt/index.php?option=com_content&task=view&id=32&Itemid=55.

CBM Asia, 2009. CBM Asia Development Corp. http://www.cbmasia.ca/s/Home.asp.

Coalbed Methane World, January 20, 2009. CBM Contracts Offered in Indonesia. http://www.futureenergyevents.com/coalbedmethane/2009/01/20/coalbed-methane-cbm-contracts-offered-in-indonesia/.

Churchill Mining, September 5, 2007. Churchill Awarded Indonesia's First Coal Bed Methane JEA License. Churchill Mining PLC. http://www.churchillmining.com/library/file /05_09_07.pdf.

Clarke, L.B., McConville, A., December 1998. Coal in the Czech Republic. IEA Coal Research, London (Clean Coal Centre).

Composite, September 2008. Composite Energy Awarded Coalbed Methane Exploration Concessions in Poland. Composite Energy Ltd. http://www.composite-energy.co.uk/news.html.

Creamer, July 15, 2009. Anglo Coal Intensifying Search for Coal-Bed Methane in Waterberg. Martin Creamer, MiningWeekly.com. http://www.miningweekly.com/article/anglo-coal-intensifying-search-for-coal-bed-methane-in-waterberg-magara-2009-07-15.

DECC, 2010. Approved Fields, Field Development, Upstream, Oil and Gas. UK Department of Energy and Climate Change. https://www.og.decc.gov.uk/index.htm. (accessed March 2010).

DGH, 2009. CBM Acreages. Directorate General of Hydrocarbons, Ministry of Petroleum and Natural Gas, Government of India. http://www.dghindia.org/EandPAcreages.aspx.

DJ MIGAS, 2007. Directorate General of Oil and Gas. Department of Energy and Mineral Resources. http://www.migas.esdm.go.id.

Dragon Capital, 2008. Vietnam Resource Investments. Dragon Capital Markets Limited. Monthly Report March 2008 http://www.dragoncapital.com/UserFiles/File/monthly%20report/2008/Mar/MR_200803VRI.pdf?PHPSESSID=e1fc9daa1d3bfa3ae589ab41f43ad982.

Dragon Capital, January 2010. Vietnam January 2010 Update. Dragon Capital Markets Limited. http://www.dragoncapital.com/UserFiles/File/monthly%20report/2010/Jan/MR_201001.pdf.

Eden Energy, September 1, 2008. South Wales (UK) Coal Seam Methane Project (Eden – 50%) Promising Estimate of Recoverable Resources in PEDL 100. http://www.edenenergy.com.au/pdfs/ASX_Announcement%2020080901%20-SW%20CSM%20promising%20estimates.pdf.

European Gas Limited, October 7, 2008. News Release – Initial Coal Bed Methane Contingent Resources Gazonor Project, France.

European Gas Limited, August 19, 2009. News Release – Commencement Of Production Test Program and Drilling Activities, France.

Eguiluz de Antuñano, S., Amezcua Torres, N., 2003. Coalbed methane resources of the Sabinas Basin, Coahuila, México. In: Bartolini, C., Buffler, R.T., Blickwede, J. (Eds.), The Circum-Gulf of Mexico and the Caribbean: Hydrocarbon Habitats, Basin Formation, and Plate Tectonics. AAPG Memoir 79, pp. 395–402.

EIA, August 2007a. EIA Country Analysis Brief: Ukraine – Coal. U.S. Energy Information Administration, Washington, DC. http://www.eia.doe.gov/emeu/cabs/Ukraine/Coal.html.

EIA, July 2007b. EIA Country Analysis Brief: Vietnam – Coal. U.S. Energy Information Administration, Washington, DC. http://www.eia.doe.gov/emeu/cabs/Vietnam/Coal.html.

EIA, October 2008. EIA Country Analysis Brief: South Africa – Coal. U.S. Energy Information Administration, Washington, DC. http://www.eia.doe.gov/cabs/South_Africa/Coal.html.

EIA, September 2009a. EIA Country Analysis Brief: Australia – Coal. U.S. Energy Information Administration, Washington, DC. http://www.eia.doe.gov/emeu/cabs/Australia/Coal.html.

EIA, 2009b. International Energy Statistics. U.S. Energy Information Adminstration, Independent Statistics and Analysis. http://tonto.eia.doe.gov/cfapps/ipdbproject/IEDIndex3.cfm.

EIA, 2009c. EIA Country Analysis Brief: China – Coal. U.S. Energy Information Administration, Washington, DC. July 2009. http://www.eia.doe.gov/emeu/cabs/China/Coal.html.

EIA, 2009d. EIA Country Analysis Brief: Columbia. U.S. Energy Information Administration, Washington, DC. February 2009. http://www.eia.doe.gov/emeu/cabs/Australia/Coal. html.

EIB, December 2009. Energy in Bulgaria. Wikipedia. http://en.wikipedia.org/wiki/Energy _in_Bulgaria.

Embassy of the Republic of Kazakhstan, 2008. Energy Sector, Discover Kazakhstan. http:// www.kazakhembus.com/index.php?page=energy-sector.

Euracoal, n.d. Country Profiles. Euracoal European Association for Coal and Lignite. http:// www.euracoal.be/pages/layout1sp.php?idpage=77.

EuroGas, December 2009. CBM & Natural Gas Joint Venture Ukraine. David Rotman. EuroGas, Inc. http://www.eurogasinc.com/joint-venture-ukraine_85.html.

Fails, T., 2005. Pannonian International – Germany & Romania. North American Prospect Expo Promotion Forum 2005. http://energy.ihs.com/NR/rdonlyres/ 43420221-B228-4F15-A5AB-C739C8F6F9C3/0/pannonian_fails.pdf.

Falcon, September 12, 2005. Falcon Oil & Gas Ltd (TSXV: FO) Announced Today the Pre-liminary Results of Its First Romanian Coalbed Methane Well. Falcon Oil & Gas. http:// www.falconoilandgas.com/releases/9-12-05.htm.

Fang, Y. (Ed.), August 24, 2009. China's CBM Producers Might See Favorable Policies. China Daily. http://news.xinhuanet.com/english/2009-08/24/content_11934595.htm.

Flores, April 3, 2007. Recovery and Use of Methane Associated to Mexican Coal Mines. Ramón Carlos Torres Flores, General Director for Energy and Mining, Secretariat of Envi-ronment and Natural Resources, Mexico. http://www.methanetomarkets.org/events/ 2007/coal/docs/2apr07-mexico.pdf.

Galaxy, n.d. Galaxy Energy Corporation Investor Spec Sheet. http://www.investorspecsheet. com/alerts/galaxy012004.htm.

Gazprom, March 21, 2008. Prospects for Coalbed Methane Reserves Development of Russia. Gazprom. http://old.gazprom.ru/eng/articles/article24760.shtml.

GCIS, 2009. South Africa Yearbook 2008/2009. South Africa Government Communication and Information System. http://www.gcis.gov.za/resource_centre/sa_info/yearbook/ 2008-09.htm.

GEF, 2007. India: Coal Bed Methane Recovery and Commercial Utilization. Global Environment Facility, Washington, DC. http://www.gefonline.org/projectDetailsSQL. cfm?projID=325.

GOA, 2010. About Coalbed Methane. Government of Alberta Energy. Reviewed/revised January 19, 2010. http://www.energy.gov.ab.ca/NaturalGas/754.asp.

Green Gas, 2010. Global Operations: Czech Republic. Green Gas International BV. http:// www.greengas.net/output/page199.asp.

GSDG, January 2009. Assessing European Capacity for Geological Storage of Carbon Dioxide. Geological Survey of Denmark and Greenland. http://www.geology.cz/geocapacity/ publications/D16%20WP2%20Report%20storage%20capacity-red.pdf.

GSI, January 4, 2009. Inventory of Geological Resource of Indian Coal. Geological Survey of India. http://www.portal.gsi.gov.in/gsiDoc/pub/National_Inventory_Coa l_09.pdf.

Guanghua, L., 2008. CBM Development in China. Presented at the International Geo-logical Congress, Oslo, August 6–14, 2008. http://www.cprm.gov.br/33IGC/13154 13.html.

Huleatt, M.B., December 2009. Australian Coal Resources Map. Australian Government Geo-science Australia. Catalogue Number 69736. https://www.ga.gov.au/products/servlet/ controller?event=GEOCAT_DETAILS&catno=69736.

Kelefant, May 25, 1998. Coalbed Methane Could Cut India's Energy Deficit. John Kele-fant and Mark Stern. Oil & Gas Journal. http://www.ogj.com/articles/print/vol ume-96/issue-21/in-this-issue/general-interest/coalbed-methane-could-cut-india39s-energy-deficit.html.

Keeper, 2005. Vietnam CBM Review. Keeper Resources. http://www.keeperresources.com/Vietnam%20CBM%20Review.pdf.

Le Prioux, C., August 20, 2006. France's Coal Mining Industry to Get Second Wind with New Power Project. AFP. http://www.terradaily.com/reports/France_Coal_Mining_Industry_To_Get_Second_Wind_With_New_Power_Project_999.html.

Liu, W., 2007. Case Study on CMM/CBM Projects in China. China Coal Information Institute. Presented at CMM Development in the Asia–Pacific Region: Perspectives and Potential, Brisbane, Australia, October 4–5, 2006.

LOC, June 1992. Library of Congress Country Studies – Bulgaria. Chapter 3: The Economy, Resource Base, Coal and Minerals, Figure 18.10. Energy and Mineral Resources, Library of Congress. http://lcweb2.loc.gov/frd/cs/bulgaria/bg03_01a.pdf.

Lupu, C., Ghicioi, E., 2006. Methane Emissions from Hard Coal Mines of Jiu Valley – Romania – Use Possibilities. INSEMEX, Romania. Second Session of the Ad Hoc Group of Experts on Coal Mine Methane Cooperation, Geneva, January 31–February 1, 2006. http://www.unece.org/energy/se/pp/coal/cmmjanfeb06/cmm06coop.html.

MacauHub, February 16, 2009. Mozambique: Botswana Analyses Mozambican Ports for Coal Exports. MacauHub Economic Information Service. http://www.macauhub.com.mo/en/news.php?ID=6896.

MEMR, 2006. Indonesia Lower Rank Coal Resources Development Policy. Geological Agency, Ministry of Energy and Mineral Resources. http://www.dim.esdm.go.id/makalah/BD-kabangeologi.pdf.

Methane to Markets, January 2009. Global Overview of CMM Opportunities. Methane to Markets, USEPA Coalbed Methane Outreach Program . http://www.methanetomarkets.org/m2m2009/tools-resources/coal_overview.aspx.

Methane to Markets, 2010. Methane to Markets International Coal Mine Methane Projects Database. http://www2.ergweb.com/cmm/index.aspx.

Mining Week, 2009. Mining Week Kazakhstan 6th International Exhibition for Mining & Exploration, Mineral & Coal Processing and Metallurgical Technologies. Brochure. http://www.tntexpo.com/docs/1428_Brochure_MiningWeek_2010_en.pdf.

Mongolyn Alt Corporation, 2010. Mongolyn Alt Corporation "Mak". http://www.mak.mn/. (accessed February 2010).

Mösle, B., Kukla, P., Stollhofen, H., Preuße, A., April 2009. Coal bed methane production in the Münsterland Basin, Germany – past and future. Geophysical Research Abstracts 11. EGU2009–4267, European Geosciences Union General Assembly. http://meetingorganizer.copernicus.org/EGU2009/EGU2009-4267.pdf.

MSS, February 4, 2009. Government Plans to Provide Incentives for Coalbed Methane Project Development. Minister of the State Secretary, Government of Indonesia. http://www.indonesia.go.id/en/index.php?option=com_content&task=view&id=7785&Itemid=701.

MST, 2002. First Brazilian Inventory of Anthropogenic Greenhouse Gas Emissions. Ministry of Science and Technology.

Mustafa, B., Ayar, G., 2004. Turkey's coal reserves, potential trends and pollution problems of Turkey. Energy Exploration & Exploitation 22 (1).

Newman, H.R., December 2007. The Mineral Industry of France. 2005 Minerals Yearbook. U.S. Geological Survey. p. 3. http://minerals.usgs.gov/minerals/pubs/country/2005/frmyb05.pdf.

NRCAN, December 2008. Energy Sources – Coal. Natural Resources Canada. http://www.nrcan.gc.ca/eneene/sources/coacha-eng.php.

NSDOE, January 2009. Nova Scotia Prospect Profile Onshore 2009. Nova Scotia Department of Energy. http://www.gov.ns.ca/energy/resources/RA/onshore/NS-Prospect-Profile-Onshore-Jan-2009.pdf.

NSDOE, June 2009a. Onshore Petroleum Agreements Map. Nova Scotia Department of Energy. http://www.gov.ns.ca/energy/oil-gas/explore-invest/maps.asp.

OECD/IEA, 2008. Workshop Report New Trends in Coalmine Methane Recovery and Utilization. Szczyrk, Poland. February 27–29, 2008. Report written by Dr. Ming Yang, International Energy Agency, April 7, 2008, Paris, France. http://www.iea.org/papers/2008/methane_recovery.pdf.

OECD/IEA, December 2009. Coal Mine Methane in Russia Capturing the Safety and Environmental Benefits. International Energy Agency, Paris, France. http://www.iea.org/papers/2009/cmm_russia.pdf.

PGI, December 2005. Mineral Resources of Poland – Coalbed Methane. Poland Geological Institute, Department of Economic Geology. http://www.pgi.gov.pl/mineral_resources/cbm.htm.

Press Trust of India, September 9, 2009. Essar Oil's Raniganj Block Reserves Raised after Mapping. Business Standard. http://www.business-standard.com/india/news/essar-oil%5Cs-raniganj-cbm-block-reserves-raised-after-mapping/73010/on.

Querol-Suñe, F., 2006. Status of CBM Development in Mexico. Director of Mining Promotion Mexico. Presented at the Methane to Markets Coal Subcommittee Meeting, Tuscaloosa, Alabama, May 22–23, 2006. http://www.methanetomarkets.org/m2m2009/documents/events_coal_20060525_mexico_update.pdf.

Red Hill, 2010. Chandgana Tal & Khavtgai. http://www.redhillenergy.com/projects/coal/chandgana_tal/. (accessed February 2010).

Reuters, June 5, 2009. SA Slow in Tapping Coalbed Methane Potential. http://www.miningmx.com/news/energy/SA-slow-in-tapping-coal-bed-methane-potential.htm.

Ruban, A.D., Zabourdyaev, V.S., Zabourdyaev, G.S., 2005. Оценка и разработка ресурсов и объемов извлечения метана при подземной разработке угольных месторождений России (Evaluation and Recovery of Methane Resources in Underground Mining of Coal Fields in Russia). IPKON RAN, Moscow.

Schlueter, 2009. Hazard – Emission – Source of Energy Services for Greenhouse Gas Mitigation Projects. Dipl.-Geol. Ralph Schlueter (DMT GmbH & Co. KG, Essen). Presented at the UNECE – Ad Hoc Group of Experts on Coal Mine Methane, Geneva, October 12–13, 2009   http://www.unece.org/energy/se/pp/coal/cmm/5cmm_oct09/19_schlueter_dmt_oct09.pdf.

Schwochow, S. (Ed.), 1997. The International Coal Seam Gas Report. Cairn Point Publishing.

SeeNews, December 23, 2009. Bulgaria's Maritsa East Coal Mines Plans 26 Mln T Output in 2010, 30 Mln T in 2011. SeeNews – The Corporate Wire. http://www.seenews.com/.

SMA, December 31, 2009. Supervised Plants. Polish Mining Industry, Poland State Mining Authority. http://www.wug.gov.pl/index.php?english/supervised_plants.

Sproule, 2010. Sproule Projects Asia. http://www.sproule.com/Asia#title. (accessed February 2010).

Storm Cat, July 13, 2005. Storm Cat Update on Mongolia CBNG Projects. Storm Cat Energy Corporation. http://www.stormcatenergy.com/news/news_071305.html.

Süffert, T., November 1997. Coal in the States of the Rio Grande Do Sul and Santa Catarina (Portuguese). Federative Republic of Brazil, Ministry of Mines and Energy, Research Company of Mineral Resources, Directorate of Geology and Mineral Resources, Department of Mineral Resources. http://www.cprm.gov.br/opor/pdf/carvaorssc.pdf.

Tavan Tolgoi, 2010. Tavan Tolgoi Incorporated Company. http://www.tavantolgoi.mn/?lang=english#153622MetawiseCMS. (accessed February 2010).

Tewalt, S.J., Finkelman, R.B., Torres, I.E., Simoni, F., 2006. World Coal Quality Inventory: Colombia. United States Geological Survey. USGS Open File Report 2006-141. http://pubs.usgs.gov.

UNDP, n.d. Coalbed Methane Recovery: An Innovative Stride for Countering Climate Change. United Nations Development Programme, India. http://www.undp.org.in/index.php?option=com_content&view=article&id=354&Itemid=584.

UNFCCC, October 2000. South Africa "Initial National Communication to UN Framework Convention on Climate Change," p. 77.

UNFCCC, 2009. National Inventory Submissions 2009. http://unfccc.int/national_reports/annex_i_ghg_inventories/national_inventories_submissions/items/4771.php.

USEPA, 1992. Assessment of the Potential for Economic Development and Utilization of Coalbed Methane in Czechoslovakia. U.S. Environmental Protection Agency. EPA 430-R-92–1008, 76 p.

USEPA, 1995. Reducing Methane Emissions from Coal Mines in Poland – A Handbook for Expanding Coalbed Methane Recovery and Use in the Upper Silesian Coal Basin. U.S. Environmental Protection Agency. EPA 430-R-95–003, 122 p.

USEPA, 2005. Coalbed Methane Outreach Program – Czech Republic. U.S. Environmental Protection Agency.

USEPA, June 2006. Global Anthropogenic Non-CO_2 Greenhouse Gas Emissions: 1990–2020, Appendix B-2: Methane Emissions from Fugitives from Coal Mining Activities. USEPA, Office of Atmospheric Programs, Climate Change Division. http://www.epa.gov/climatechange/economics/downloads/GlobalAnthroEmissionsReport.pdf.

van der Merwe, C., July 4, 2008. Coal-bed Methane Exploration Booms as SA's Energy Demand Grows. MiningWeekly.com. http://www.miningweekly.com/article/coalbed-methane-exploration-booms-as-sas-energy-demand-grows-2008-07-04-1.

Volkmer, G., May 5, 2008. Coal Deposits of Poland. TU Bergakademie Freiberg. http://www.geo.tu-freiberg.de/oberseminar/oberseminar_07_08.html.

Vorster, C.J., 2003. Coal Fields of the Republic of South Africa. South Africa Council for Geoscience.

Wallace, R.-B., 2009. Coal in Mexico. Facultad de Economía, Universidad Nacional Autónoma de México. http://www.economia.unam.mx/publicaciones/econinforma/pdfs/359/brucelish.pdf.

WCI, n.d. Coalbed Methane – World Coal Institute. http://www.worldcoal.org/coal/coal-seam-methane/coal-bed-methane/.

WCSB, December 2009. Geological Map of Canada with Western Canadian Sedimentary Basin Highlighted. Wikipedia. http://en.wikipedia.org/wiki/Western_Canadian_Sedimentary_Basin.

Welz, December 12, 2013. In South Africa, Renewables Vie With the Political Power of Coal. Environment 360. http://e360.yale.edu/feature/in_south_africa_renewables_vie_with_the_political_power_of_coal/2719/.

Whitlock, C., July 20, 2007. German Hard-Coal Production to Cease by 2018. Washington Post Foreign Service. www.washingtonpost.com.

WK CBM, 2009. About WK CBM. WK CBM, Directorate General of Oil and Gas, Directorate of Upstream Business Development. http://www.wkmigas.com/organisation/about-wk-cbm/.

World Coal, September 2001. Coalbed Methane Opportunities in Bulgaria. Raven Ridge Resources. Published in "World Coal". Palladin Publications Ltd.

Young, S.R., Baily, H.E., Holloway, S., Glover, B.W., 1994. The History of Mine Gas Utilisation and Status of Coalbed Methane Development in the U.K. In: The Silesian International Conference on Coalbed Methane Utilization Proceedings, Katowice, Poland. Polish Foundation for Energy Efficiency, Katowice, Poland.

Index

Note: Page numbers followed by "f" indicate figures; "t" tables; "b" boxes.

Printed and bound by CPI Group (UK) Ltd, Croydon, CR0 4YY

08/05/2025

01864914-0001